MATHEMATICS IN INDUSTRY   2

Editors
Hans-Georg Bock
Frank de Hoog
Avner Friedman
William Langford
Helmut Neunzert
William R. Pulleyblank
Torgeir Rusten
Anna-Karin Tornberg

THE EUROPEAN CONSORTIUM
FOR MATHEMATICS IN INDUSTRY

SUBSERIES

Managing Editor
Vincenzo Capasso

Editors
Robert Mattheij
Helmut Neunzert
Otmar Scherzer

Springer

Berlin
Heidelberg
New York
Hong Kong
London
Milan
Paris
Tokyo

Vincenzo Capasso

Editor

Mathematical Modelling for Polymer Processing

Polymerization, Crystallization, Manufacturing

With 90 Figures, 7 in Color

 Springer

Chemistry Library

Editor

Vincenzo Capasso

MIRIAM – Milan Research Centre for Industrial
and Applied Mathematics
and
Department of Mathematics
University of Milano
Via C. Saldini 50
20133 Milano, Italy
e-mail: capasso@mat.unimi.it

Cataloging-in-Publication Data applied for

Bibliographic information published by Die Deutsche Bibliothek

Die Deutsche Bibliothek lists this publication in the Deutsche Nationalbibliografie;
detailed bibliographic data is available in the Internet at <http://dnb.ddb.de>.

Cover figure: A random tessellation generated by an experiment of polymer crystallization
carried out at BASELL – *G. Natta* Research Centre in Ferrara, Italy.

Mathematics Subject Classification (2000): 34-XX, 35-XX, 35Kxx, 35Rxx, 60-XX,
60Kxx, 60D05, 62Gxx, 70-XX, 74Exx, 80-XX, 80Axx, 92Exx

ISBN 3-540-43412-7 Springer-Verlag Berlin Heidelberg New York

Springer-Verlag Berlin Heidelberg New York
a member of BertelsmannSpringer Science+Business Media GmbH

http://www.springer.de

© Springer-Verlag Berlin Heidelberg 2003
Printed in Germany

Typeset by the authors using a Springer TEX macro-package
Cover design: *design & production* GmbH, Heidelberg
SPIN: 10757829   46/3142LK - 5 4 3 2 1 0   Printed on acid-free paper

ac 2·28·03

Preface

Polymers are substances made of macromolecules formed by thousands of atoms organized in one (homopolymers) or more (copolymers) groups that repeat themselves to form linear or branched chains, or lattice structures. The concept of polymer traces back to the years 1920's and is one of the most significant ideas of last century. It has given great impulse to industry but also to fundamental research, including life sciences. Macromolecules are made of small molecules known as monomers. The process that brings monomers into polymers is known as polymerization. A fundamental contribution to the industrial production of polymers, particularly polypropylene and polyethylene, is due to the Nobel prize winners Giulio Natta and Karl Ziegler. The ideas of Ziegler and Natta date back to 1954, and the process has been improved continuously over the years, particularly concerning the design and shaping of the catalysts.

Chapter 1 (due to A. Fasano) is devoted to a review of some results concerning the modelling of the Ziegler- Natta polymerization. The specific example is the production of polypropilene. The process is extremely complex and all studies with relevant mathematical contents are fairly recent, and several problems are still open.

From a fundamental point of view the organizational behaviour of macromolecules in solidification, crystallization in particular, is a highly open ended issue. Even the crystallization behaviour of quiescent melts, to ignore the effect of flows and stresses, which is usually taken as an indicator for the crystallization behaviour under more general conditions, is only known in fragments. It is well known that solidification starts at certain spots (nuclei) within the material, thus forming the starting points of a subsequent growth process of crystalline aggregates, which under quiescent solidification conditions and low temperature gradients often have a spherical shape (spherulites) and under flow conditions may form more complex structures, like shish-kebabs.

The concept of nucleation as a first step in phase transitions was originally developed in the 20's and the 30's by various authors. Attempts to apply the classical, one dimensional theory in its original form appear inadequate for description of polymer crystallization.

In Chapter 2 (due to J.C. Neu and L.L. Bonilla) a classical kinetic theory of nucleation is explained, in a context simpler than polymer crystallization. A convenient framework for describing nucleation and coarsening is the classical Becker-Döring kinetic theory. The growth of the radius of a supercritical cluster is obtained in terms of macroscopic parameters that can be calculated from experimental data.

Chapter 3 (due to A. Ziabicki) is dedicated to an extension of the classical, one dimensional nucleation theory. Modifications of the classical theory are presented based on a more general model of nucleation due to the author, including effects of cluster shape, orientation, position and internal structure in a multidimensional setting. Motivation of the new model is associated with orientation and stress controlled crystallization of polymers.

Chapter 4 (due to M.Rubi and collaborators) is devoted to the presentation of a kinetic theory for nucleation and growth in non-uniform temperature, stress and flow fields, in order to provide an accurate framework for the analysis of nucleation and growth rates which are the essential parameters for the Kolmogorov-Avrami-Evans theory of crystallization under realistic conditions.

Due to the release of the latent heat and the low conductivities of polymers, the actual crystallization temperature cannot be directly imposed upon the material, but it is the result of a heat transfer problem which has to be solved in order to get physically significant information on crystallization kinetic parameters. A further problem arises in presence of steep temperature gradients during the transformation, which may give rise to non-spherical crystalline aggregates, which usually cannot be easily observed directly but which, from theoretical considerations, may speed up the overall crystallization rate considerably. Recently a discussion has been raised about the relevance of the obstacle posed by the maximum level of crystallinity attainable by the material; this parameter strongly depends upon the material itself.

It is generally accepted that the morphology of solidified parts determines their mechanical and physical properties (see e.g. the well known Hall-Petch relation) and thus also the lifetime of the products. An important feature of polymer crystallization is crystalline morphology, i.e. the way in which long macromolecules are built into crystals. An accurate modelling of the evolution of the morphology of the solidified material may have a great impact on industrial processing of polymeric material, since it opens the possibility of controlling specific changes in the processing conditions so to obtain desired morphologies.

Chapter 5 (due to V. Capasso and collaborators) is dedicated to gaining insight in the generation of morphologies during the crystallization process in non-uniform temperature fields by taking into account the coupling of the heat transfer problem with the crystallization process (nucleation and growth). A detailed analysis of the mathematical modelling of the evolution of the stochastic geometries coupled with a non-uniform temperature field is presented, as an extension of the classical semiempirical Kolmogorov-Avrami-Evans theory which has been used rather successfully under isothermal conditions. A multiple scale approach is presented leading either to an hybrid model in which temperature evolves as a deterministic entity coupled with the stochastic crystallization process, or to a fully deterministic system, depending upon the level of accuracy required for industrial applications. Numerical simulations based on the above approach are presented. This chapter includes two already mentioned relevant problems for real applications. Namely inverse problems for the identification of the functional dependence of the kinetic parameters of the crystallization process upon temperature, and starting problems of optimal control of the morphological parameters of the final crystallization geometries at different Hausdorff dimensions, such as the crystallinity degree and the density of interfaces among crystals.

In Chapter 6 (due to Capasso and collaborators) an alternative approach for modelling the crystallization process coupled with an unederlying temperature field is presented based on the birth-and-growth of pointwise crystals, thus emphasizing the double scale of crystals and temperature evolution. By means of suitable laws of large numbers for "moderately" interacting particles, an asymptotic continuum model is obtained. This second approach has the advantage of allowing secondary crystallization to be taken into account, but has the disadvantage of loosing the geometrical dimension of the crystallization process; it can be anyway recovered by suitable smoothing and segmentation techniques as shown in the numerical simulations of the process.

There are various ways in which polymers are processed, depending upon the nature of the polymer, and the application for which it is intended. Thousands of tons of thermoplastic polymers are produced per year, and manufactured into plastic products as fibers, sheets, tubes, compact discs and many other products. The mathematical modelling and numerical simulation of the underlying industrial processes is a main activity within the research on polymers..

Chapter 7 (due to F.van de Ven and collaborators) deals with some aspects of the modelling of industrial processes for the manufacture of polymeric (plastic) products. The extrusion of polymer melts through a capillary die is considered, especially with regard to extrudate distortions. Two models are presented to explain so called spurt oscillations (or instabilities). Moreover injection moulding between two parallel plates is considered; the mechanical and thermal features of this process are modelled, and the onset to the instability ("wobbling") of the free flow front is discussed .

All the research which have lead to the results presented here has been carried out in strict collaboration with major polymer industries; among others we may quote MONTELL (now BASELL) based in Italy ; DOW Chemicals, Geleen and Axxicon Moulds, DSM research, all based in the Nertherlands; PCD based in Linz.

The present volume is the result of a long term cooperation among different research teams in Europe in the framework of the ECMI (European Consortium for Mathematics in Industry) Special Interest Group on "Polymers", supported by the EU within the HCM and the TMR programmes.

Specific acknowledgements are reported at the end of each chapter. But I will not forget to mention the relevant contribution given by Dr. Gerhard Eder, from Linz, to the understanding of the chemical process. Unfortunately due to serious illness problems he could not contribute explicitly with a chapter of his own in this volume.

Last but not the least I wish to thank Dr. R.Caselli and Dr. A.Micheletti, together with the editorial staff of Springer-Verlag, for the excellent work leading to the final editing of the volume.

Milan, July 2002   *Vincenzo Capasso*

Table of Contents

Crystallization

Manufacturing

List of Contributors

Luis L. Bonilla
Escuela Politécnica Superior
Universidad Carlos III de Madrid
Avda. Universidad 30
28911 Leganés, Spain
Email:
bonilla@ing.uc3m.es

Martin Burger
Industrial Mathematics Institute
Johannes Kepler Universität
Altenbergerstrasse 69
A4040 Linz, Austria
Email:
burger@indmath.uni-linz.ac.at

Vincenzo Capasso
MIRIAM - Milan Research Centre for
Industrial and Applied Mathematics
and Department of Mathematics
University of Milano
Via Saldini 50
I-20133 Milano,Italy
Email:
Vincenzo.Capasso@mat.unimi.it

Antonio Fasano
Department of Mathematics "U. Dini"
University of Florence
viale Morgagni 67a

50134 Florence, Italy
Email:
fasano@math.unifi.it

Alessandra Micheletti
MIRIAM - Milan Research Centre for
Industrial and Applied Mathematics
and Department of Mathematics
University of Milano
Via Saldini 50
I-20133 Milano, Italy
Email:
micheletti@mat.unimi.it

Daniela Morale
MIRIAM - Milan Research Centre for
Industrial and Applied Mathematics
University of Milano
Via Saldini 50
I-20133 Milano,Italy
and
Department of Mathematics
University of Torino
via Carlo Alberto 10, 10123 Torino,
Italy
Email:
morale@mat.unimi.it

John C. Neu
Department of Mathematics

University of California at Berkeley
Berkeley, CA 94720, USA
Email:
neu@math.berkeley.edu

David Reguera Lopez
Departament de Física Fondamental
Universitat de Barcelona
Diagonal 647
08028 Barcelona, Spain
Email:
davidr@ffn.ub.es

Miguel Rubí Capaceti
Departament de Física Fondamental
Universitat de Barcelona
Diagonal 647
08028 Barcelona, Spain
Email:
mrubi@ffn.ub.es

Claudia Salani
MIRIAM - Milan Research Centre for
Industrial and Applied Mathematics
and Department of Mathematics
University of Milano
Via Saldini 50
I-20133 Milano, Italy
Email:
salani@mat.unimi.it

Alfons A.F. van de Ven
Eindhoven University of Technology
P.O.Box 513
5600 MB Eindhoven, The Netherlands
Email:
a.a.f.v.d.ven@tue.nl

Andrzej Ziabicki
Polish Academy of Sciences
Institute of Fundamental Technological Research
21 Swietokrzyska St.
00-049 Warsaw, Poland
Email:
aziab@ippt.gov.pl

Polymerization

Polycondensation

1 Mathematical Models for Polymerization Processes of Ziegler-Natta Type*

Antonio Fasano

Department of Mathematics U. Dini, Florence, Italy

1.1 Introduction

Polymer science is an extraordinary source of very interesting mathematical problems, as it is largely testified by the recent literature. Research is extremely active in various directions: understanding the mechanical and thermodynamical behaviour of known polymers, improving production processes, designing new molecules and new materials with specific properties. A strong and continuous stimulus comes from technological applications in various fields, in which the demand of better performing materials is more and more intense.

Polymers behave in a very peculiar way when we analyse their mechanical and thermodynamical properties.

For instance, the dynamics of polymer melts and solutions presents a strong coupling between the macroscopic rheological parameters and the evolution of the microscopic structure (see e.g. the survey paper [14]).

Another intensively studied field is the solidification of polymer melts. This study has various branches: nucleation, crystal growth, inner structure of crystalline aggregates. A wide literature exists, characterised by a surprising variety of crystallization kinetic laws, usually based on heuristic arguments. For a review and for bibliographical information we refer to the survey papers [8], [9] and to the book [18]. The most recent developments on this subject are presented in [21] and new contributions for the understanding of polypropylene solidification are discussed in [10]. In our paper we are not going to deal with polymer crystallization, since this subject is treated in other chapters of this same book. Instead we will concentrate on diffusion driven polymerization processes.

At my knowledge the Ziegler-Natta process for the production of polypropylene or polyethilene has received the attention of applied mathematicians only in recent years. The ideas of Ziegler and Natta date back to 1954 and the process has been improved continuously over the years, particularly concerning the design and shaping of the catalysts. For a review of the experimental and theoretical work on the various aspects of this fine technology we refer

---

* Work partially supported by ASI and the MURST National Project "Mathematical Analysis of Phase Transitions ... ".

to [18], [19], [7] and to the proceedings [15], which contains also an hystorical overview [11].

The last generation of catalysts is made of particles having a crystalline structure and diameters of the order of $10^{-2}\mu$, that can be considered spherical and are tightly packed together in spherical aggregates (radius $\simeq 70\mu$). Thus the number of catalytic sites in one aggregate is of the order of 10^{10}. They are the active centers of polymerization, making the monomer molecules arriving at their surface enter a polymeric chain. If the process develops correctly a polymer shell grows around each catalytic particle and the whole system preserves its spherical symmetry.

Obtaining spherical shape of the catalyst is by no means trivial and must be considered a remarkable technological achievement, now referred to as "the replication phenomenon" (see [23], [16] for a discussion). The final product coming out of one aggregate is a sphere with a diameter of about 3mm (we are reporting average figures). Getting the final product in this shape and size is an important advantage since it can be commercialized in this form without further costly transformations.

A basic reference about the various technological details is the already quoted Proceedings volume [15], that also describes mathematical models based on the so-called multigrain scheme (see in particular [17]). In the present paper we consider a different approach first introduced in [1], which eliminates some geometrical difficulties of the multigrain model taking advantage of large difference in size between the aggregate and its components. The basic idea of [1] is that the system can be treated as a two-scale body. At the large scale we look at the agglomerate as a continuum characterized by a density of catalytic sites, while at the small scale we study the growth of each individual sphere. A general sketch can be found in the survey paper [4]. Here we review the various aspects of the model with more details and we perform an accurate dimensional analysis with the aim of understanding the relative importance of the different phenomena involved. In this way, not only we are able to formulate the model using non-dimensional variables, but we shall see that many important simplifications can be safely introduced.

Before illustrating the mathematical problem, we describe the details of the process in the next section. Sections 3, 4 and 5 are devoted respectively to the equations governing the evolution of the large scale quantities, the equations regulating the small scale processes, including polymerization, and the way the two groups of equations are coupled. Rescaling is performed in Section 6, and Section 7 summarizes the final version of the model with a sketch of existence and uniqueness proof.

Section 8 deals with a preliminary (but fundamental) stage of the process, the so-called fragmentation of the aggregate, during its first exposure to the monomer. The paper is concluded by a brief discussion on open problems.

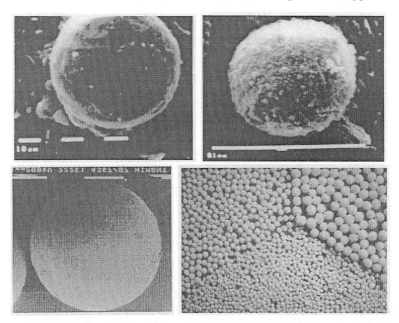

Fig. 1.1. top left: catalyst agglomerate before reaction; top right: after prepolymerization; bottom left: at the end of the process; bottom right: final products (Courtesy of Montell)

1.2 Description of the Process

In the reactor the agglomerate of catalytic particles is kept in contact with a monomer at a given concentration and temperature. If the process is performed at a sufficiently high pressure the bulk monomer in the reactor is liquid, but there are also reactors working with gaseous monomers. Here we are going to deal with the latter case, and we believe that a complete description of the former case is not yet available. The polymerization reaction takes place at the surface of the catalytic particles, so that the monomer has to penetrate the agglomerate (by diffusion) and then (still by diffusion) the growing polymeric shells, in order to feed the reaction at the inner surface of the shell. However, an extremely important feature for guaranteeing a regular growth of the system (thus preserving the spherical shape) is that the first stage of the process (pre-polymerization) is performed under controlled conditions of sufficiently mild polymerization, producing the so-called *fragmentation* of the aggregate. Fragmentation consists in a quick polymerization reaction (that we consider instantaneous) with absorption of a finite amount of monomer and the creation of a thin layer of polymer around the catalytic particles at their first exposure to the monomer. In this way a *fragmentation front* proceeds towards the center of the agglomerate. After this stage the agglomerate is transformed into a system of a large number of spherules making a porous medium through which the monomer diffuses.

This stage of the process is extremely delicate and conditions of low activity must be mantained to prevent ruptures of the catalyst.

In its simplest form fragmentation has been incorporated in mathematical models only very recently [20]. Indeed the polymerization process after fragmentation is already so complicated that it is convenient to study it separately, supposing that fragmentation has left the system with a uniform structure. This is of course not completely true, but it is not difficult to introduce the necessary changes, once the real outcome of fragmentation has been calculated.

For this reason in our exposition we follow the hystorical development of the theory, postponing the discussion about fragmentation to the last chapter.

Thus, in this section we deal with an ideal situation in which at time $t = 0$ the catalyst particle have identical radii with their surface active for the polymerization reaction (i.e. we neglect the initial thickness of the thin polymer coating around them). This is however a simplifying assumption that can be easily removed, as we shall see later.

The starting point of our analysis is the so-called *multigrain model*, which is basically a two scale approach, first proposed in [13]. Differently from [13], where it is clearly difficult to keep track of the position of the growing particles within the swelling agglomerate, we take advantage of the substantial length scale difference to describe the system as a continuum from the macroscopic point of view. This is done by introducing the *density of catalytic sites* (i.e. number of particles per unit volume), depending on time t and on the radial macroscopic coordinate r.

Other macroscopic quantities are the *interstitial monomer density* $M(r, t)$, the *radial velocity field* $v(r, t)$, the *temperature* $T(r, t)$.

The importance of the latter quantity derives from the fact that polymerization is strongly exothermic and consequently heat transport is one of the leading phenomena.

The expansion velocity $v(r, t)$ is the result of the growth of the particles that at time t are contained in a sphere of radius r. As we shall see, the assumption that the system keeps its spherical symmetry at each time is crucial for the determination of $v(r, t)$. The *outer radius* $R(t)$ of the whole system is also unknown.

At the microscopic scale we have to describe the evolution of each spherical element, consisting of the catalytic core (radius y_0) sorrounded by a polymer shell, considered as a porous material through which the monomer can diffuse.

Therefore, denoting by y the radial coordinate within each particle, we must find the *pore concentration of the monomer* $m(y, t)$ (the polymer is also schematized as a porous material), the *radial expansion velocity* $v_p(y, t)$, the *particle radius* $s(t)$, and finally the *temperature* $T(t)$.

It is important to remark that all microscopic quantities are also labelled by the macroscopic radial coordinate r, which will be frequently omitted to simplify notations.

Of course there is a strong coupling between the two scales because of the following reasons

- the macroscopic velocity field is determined by the growth of polymer layers around the catalitic particles,
- the monomer concentration at the boundary of the small particles is determined by the local value of the interstitial concentration in the agglomerate,
- the monomer that goes into the growing particles (and is eventually used in the polymerization process) enters the large scale balance as a sink term in the monomer transport equation,
- the heat released at the small scale during polymerization is accounted for at the large scale as a source term in the heat transport equation.

We remark that in the current literature the dominant monomer transport mechanism at both scales is diffusion This is certainly appropriate for those processes using gaseous monomers, however in more recent procedures the conditions of pressure and temperature are such that the monomer is in the liquid state, as it is the case of the SPHERIPOL PROCESS, developed by the company Montell Italia SpA.

When operating with liquids one may argue whether diffusion (based on concentration gradient) is still the real transport mechanism, or if the motion of the monomer within the porous aggregate is better described by Darcy's law (i.e. letting pressure gradient play the major role). We will return this question in Sect. 9.

We are now ready to write down the mathematical model. We will perform this task in Sections 3-5, writing all equations in their general form. By means of suitable rescaling, based on experimental information, we will introduce considerable simplifications, obtaining the final (and physically relevant) version of the model in Sect. 6. The latter is an original contribution of the present paper.

1.3 Governing Equations for the Macroscopic Processes (Heat and Mass Transfer)

Let us first state our assumption on the porosity of the agglomerate:

(H1) *the porosity ϵ of the agglomerate is constant.*

Despite its simplicity, this hypothesis has a well motivated physical origin. The assumption of spherical symmetry implies that, while growing, the particles continuously rearrange their position so that adjacent layers of particles keep the ideal packing configuration (rombohedric packing) minimizing porosity (slightly less than 26%, see [6]). Since adjacent layers have similar growth and the 26% porosity is independent of the radii of the spheres (provided that they are basically equal), we may quite reasonably suppose that the macroscopic porosity remains constant during the process.

The flow problem of the macroscopic level is described by the conservation of the catalytic sites and by some kinetic expansion law

$$\frac{\partial \rho}{\partial t} + \left(\frac{\partial}{\partial r} + \frac{2}{r} \right)(\rho v) = 0, \tag{1.1}$$

$$\left(\frac{\partial}{\partial r} + \frac{2}{r} \right) v = \Theta, \tag{1.2}$$

where Θ is linked to the polymerization rate.

The monomer pore concentration $M(r,t)$ obeys the equation

$$\frac{\partial}{\partial t}(\epsilon M) - \left(\frac{\partial}{\partial r} + \frac{2}{r} \right) \left(D \frac{\partial M}{\partial r} - \epsilon M v \right) = -\rho Q, \tag{1.3}$$

where D is the (bulk) diffusivity and $Q(r,t)$ is the monomer absorption rate of each microsphere that at time t has the radial coordinate r (note that ρ is measured in cm^{-3}). Clearly Q comes from the solution of the problem at the small scale.

The thermal balance equation is

$$c \frac{\partial T}{\partial t} - \left(\frac{\partial}{\partial r} + \frac{2}{r} \right) \left(k \frac{\partial T}{\partial r} - cTv \right) = \rho \widehat{Q}, \tag{1.4}$$

where c, k are the thermal coefficients of the agglomerate (that for simplicity we take constant[1]), and $\widehat{Q}$ is the rate of heat release by each microparticle at the considered time and location.

The heat is produced by the polymerization taking place at the surface of the catalytic particle and therefore $\widehat{Q}$ is a coupling term to the microproblem. All the equations above must be satisfied in the moving domain $\Omega = \{(r,t) \mid 0 < r < R(t),\ 0 < t < \theta\}$.

The boundary conditions for (1.2) are

$$v(0,t) = 0 \quad , \quad t > 0 \tag{1.5}$$

(choice of the frame of reference) while the outer radius increases according to the differential equation

$$\dot{R}(t) = v(R(t),t) \quad , \quad R(0) = R_0 > 0, \tag{1.6}$$

with R_0 given.

---

[1] Apart from the elimination of temperature dependence (not critical from the mathematical point of view), this assumption is consistent with $\epsilon =$ const. Note that the coefficient c is the product of specific heat and density.

The initial and boundary conditions for M are

$$M(r,0) = 0, \quad 0 < r < R \tag{1.7}$$

$$\frac{\partial M}{\partial r}(0,t) = 0, \quad 0 < t < \theta \tag{1.8}$$

$$M(R(t)) = M_R(t), \, 0 < t < \theta. \tag{1.9}$$

Condition (1.7) obviously disregards the previous fragmentation phase, but it can be generalized provided sufficiently high order compatibility conditions are justified

(H2) The boundary value M_R belongs to the space $C^{2+\beta}([0,\theta])$ for some $\beta \in (0,1)$, and is such that

$$M_R(0) = 0, \quad \dot{M}_R(0) = 0 \tag{1.10}$$

and

$$0 \leq M_R \leq M^* \tag{1.11}$$

($M^* > 0$, given).

The initial-boundary conditions for (1.4) are

$$T(r,0) = T_0, \qquad\qquad 0 < r < R_0, \tag{1.12}$$

$$\frac{\partial T}{\partial r}(0,t) = 0, \qquad\qquad 0 < t < \theta, \tag{1.13}$$

$$-k\frac{\partial T}{\partial r}(R(t),t) = h(T(R(t),t) - T_0), \, 0 < t < \theta, \tag{1.14}$$

where T_0 is the (constant) temperature of the external gas (of course cooling in the reactor is needed to dissipate the heat produced during the process and we simplify this aspect of the problem). The constant $h > 0$ is the heat transfer coefficient.

Before rescaling the variables to non-dimensional form, let us state also the problem at the small scale.

1.4 Governing Equation for the Microproblem (Polymerization)

Once more we have to say what happens of the porosity of the polymer, shell during its growth. This time the question is much less clear. Polymer is produced at the inner surface of the shell, pushing outwards the already existing shell.

Thus the correct way of attacking this section of the model should be to relate the porosity to the local value of the stress tensor (for an analysis of stresses see [16]). However, there are no hints in the literature on how to formulate such a relationship. In addition, the stress tensor is temperature dependent and this fact increases the complexity of the mechanical behaviour of the microparticles.

An approach presently under investigation, which includes porosity and stress as further unknowns, leads to a very complicated nonlinear hyperbolic system. However such a research, although promising, is still at an early stage and here we proceed on a heuristic basis.

In the absence of any physical suggestion, the model we are discussing is based on the simplifying assumption

(H3) the polymer porosity ϵ_p is constant.

This implies that the polymer density is also constant, so that the velocity field $v_p(y,t)$ satisfies the single equation

$$\left(\frac{\partial}{\partial y} + \frac{2}{y}\right) v_p = 0 \tag{1.15}$$

in the domain $S = \{(y,t) \mid y_0 < y < s(t), \ 0 < t < \theta\}$.

We recall that y_0 is the radius of the catalytic particles and that we omit the macroscopic coordinate locating each particle in the agglomerate. The boundary $y = s(t)$ is a free boundary.

The monomer transport equation within the shell is

$$\frac{\partial(\epsilon_p m)}{\partial t} - \left(\frac{\partial}{\partial y} + \frac{2}{y}\right)\left(d\frac{\partial m}{\partial y} - \epsilon_p m v_p\right) = 0, \quad \text{in } S, \tag{1.16}$$

d representing the (bulk) diffusivity in the polymer. As we shall see, we can simplify this equation as well as the one describing heat balance, but for the moment we keep the complete form for both. The heat transport equation in the polymeric shell is

$$c_p \frac{\partial T_p}{\partial t} - \left(\frac{\partial}{\partial y} + \frac{2}{y}\right)\left(k_p \frac{\partial T_p}{\partial y} - c_p T_p v_p\right) = 0. \tag{1.17}$$

The boundary condition for v_p at the inner surface specifies the polymer production rate and we suppose that

$$v_p(y_0, t) = \lambda m(y_0, t), \tag{1.18}$$

where λ is a temperature dependent positive coefficient (measured in $m^4 \, kg^{-1} sec^{-1}$). The dependence on temperature is essential because polymerization occurs in a given temperature range.

The motion of the free boundary is governed by the equation

$$\dot{s}(t) = v_p(s(t), t), \quad s(0) = s_0, \tag{1.19}$$

and in our approximation we take $s_0 = y_0$ (neglecting fragmentation), but it is clear how to modify the model to include previous fragmentation.

The monomer mass current coming to the active surface $y = y_0$ is proportional to the polymer volume production rate, and hence to v_p:

$$(dm_y - \epsilon_p v_p m)_{y=y_0} = \nu v_p(y_0, t), \quad 0 < t < \theta \tag{1.20}$$

where ν has the dimensions of a concentration.

Eliminating v_p between (1.18) and (1.20) leads to the nonlinear boundary condition

$$(dm_y - \lambda \epsilon_p m^2)_{y=y_0} = \nu \lambda m(y_0, t) \tag{1.21}$$

which may replace (1.20).

In the extreme case of infinitely fast reaction we may replace (1.18) and (1.20) by

$$m(y_0, t) = 0, \qquad 0 < t < \theta, \tag{1.22}$$

$$dm_y(y_0, t) = \nu v_p(y_0, t), \, 0 < t < \theta. \tag{1.23}$$

In this presentation we keep the conditions (1.18), (1.19), or (1.20).

At the other boundary (i.e. the free boundary) we have just one condition, expressing the boundary monomer pore concentration on the large scale, namely

$$m(s(t), t; r) = KM(r, t), \tag{1.24}$$

where $K \in (0, 1]$ is a partition factor and we have emphasized the presence of the variable r (typically $0.3)^2$

Remark 1. Differently from most of the free boundary problems for second order p.d.e.'s, in this problem the additional condition is not prescribed on the unknown moving boundary, but on the fixed boundary $y = y_0$.

It we take $s(0) = y_0$, there is no need of any initial condition for m.

If we suppose that all coefficients in the equations above are given constants and that $M(r, t)$ is known in (1.24), the problem becomes uncoupled from the one at the large scale. In this conditions global existence and uniqueness have been proved in [2]. Even such a problem is far from being trivial and indeed it turned out that the complete problem with (1.16) in the parabolic form was exceedingly difficult. Therefore (1.16) was replaced by its quasi-stationary version

$$\left(\frac{\partial}{\partial y} + \frac{2}{y}\right)\left(d\frac{\partial m}{\partial y} - \epsilon_p v_p m\right) = 0, \quad \text{in } S \tag{1.25}$$

in [3], where an existence and uniqueness theorem has been proved. As we shall see, such an approximation is generally largely justified.

---

^{2} It is curious enough that the concept of partition factor is already present with precisely the same meaning in early and celebrated papers about electrochemistry of cellular membranes (see [12]).

1.5 Computation of the Coupling Terms. Expansion of the Agglomerate

Some of the unknowns can be eliminated easily. The assumption $\epsilon = \text{constant}$ leads to the determination of the density $\rho(r,t)$ of catalytic sites in terms of the radius $s(t;r)$ of the microspheres. Indeed, the "solid" volume fraction in the aggregate is

$$1 - \epsilon = \rho(r,t)\frac{4}{3}\pi s^3(t;r). \tag{1.26}$$

Moreover, $\epsilon = \text{constant}$ means that the specific volume increase rate of the grains must equal the one of the pores and both are equal to the total specific volume increase rate Θ in eq. (1.2). Therefore we deduce that

$$\left(\frac{\partial}{\partial r} + \frac{2}{r}\right)v = 3\frac{\dot{s}}{s}. \tag{1.27}$$

The monomer absorption rate in equation (1.3) is given by the flux entering by diffusion into the microspheres

$$Q(r,t) = d4\pi s^2(t;r)\frac{\partial m}{\partial y}(s,t;r). \tag{1.28}$$

The heat production rate is proportional to the polymerization rate in a microsphere, which, according to (1.18), is proportional to $m(y_0,t)$, and to the surface of the catalytic particle:

$$\widehat{Q}(r,t) = \mu(T)m(y_0,t;r)4\pi y_0^2, \tag{1.29}$$

where $\mu(T)$ is a known positive coefficient (measured in $\text{m}^3\text{sec}^{-3}$). Here we suppose that all of the heat released is transferred to the outer system, consistently with the simplified scheme we are going to introduce in the next section.

We can simply derive a relationship between the coefficient μ and the coefficient λ. Let q be amount of heat produced by the conversion of the unit mass of monomer into polymer. Then multiplying the polymer volume production rate $4\pi y_0^2\lambda m$ by ρ_p (ρ_p density of the (bulk) polymer) and then by q, we obtain the quantity $\widehat{Q}$, hence

$$\mu = \lambda\rho_p q.$$

We conclude this section by deriving the expression of the macroscopic velocity field in terms of the evolution of the microspheres.

To this end we introduce the Lagrangian coordinate x, expressing the radial coordinate of a microsphere in the agglomerate at time $t = 0$, and we denote by $\sigma(x,t)$ the distance from the center reached at time t. The relationship

$$x = \xi(r, t) \tag{1.30}$$

between Lagrangian and Eulerian coordinates is expressed implicitly by

$$\sigma(\xi, t) = r \tag{1.31}$$

i.e. $\xi = \sigma^{-1}(r, t)$.

Thus in (1.27) we may replace r by σ and v by $\dfrac{\partial}{\partial t}\sigma$ and we use the symbol $s(t; x)$ instead of $s(t; r)$ with obvious meaning. Moreover we note that $\dfrac{\partial}{\partial r} = \dfrac{\partial}{\partial x} \cdot \dfrac{\partial}{\partial r}\sigma^{-1}(r, t)$, i.e. $\dfrac{\partial}{\partial r} = \left(\dfrac{\partial \sigma}{\partial x}\right)^{-1}\dfrac{\partial}{\partial x}$. Summing up, we rewrite equation (1.27) in the form

$$\frac{\partial^2 \sigma}{\partial x \partial t}\left(\frac{\partial \sigma}{\partial x}\right)^{-1} + \frac{2}{\sigma}\frac{\partial \sigma}{\partial t} = 3\frac{\dot{s}(t; x)}{s(t; x)}. \tag{1.32}$$

Equation (1.32) shows that the Lagrangian frame of reference is the most appropriate, since we can write it in the form

$$\frac{\partial}{\partial t}\log\left[\frac{\sigma^2}{x^2}\frac{\partial \sigma}{\partial x}\right] = \frac{\partial}{\partial t}\log\left(\frac{s}{y_0}\right)^3 \tag{1.33}$$

and obtain the integral

$$\sigma^2\frac{\partial \sigma}{\partial x} = x^2\left(\frac{s}{y_0}\right)^3 \tag{1.34}$$

(note: $\sigma(x, 0) = x$ by definition and hence $\dfrac{\partial \sigma}{\partial x}(x, 0) = 1$).

From (1.34) we obtain the desired expression by means of a further integration

$$\sigma(x, t) = \left[3\int_0^x z^2 s^3(t; z)\, dz\right]^{1/3}\frac{1}{y_0}. \tag{1.35}$$

Finally we recover the expansion velocity field differentiating with respect to t

$$v(\sigma(x, t), t) = 3y_0^{-3}(\sigma(x, t))^{-2}\int_0^x z^2 \dot{s}(t, z)s^2(t, z)\, dz \tag{1.36}$$

which is nothing but what we get from direct integration of (1.27), i.e.

$$v(r, t) = 3r^{-2}\int_0^r \frac{\dot{s}}{s}r'^2\, dr'. \tag{1.37}$$

1.6 Rescaling and Simplifications

The data here used are taken from Table 2 of [17].

An appropriate rescaling of the macroscopic level can be performed taking as reference length the typical size R_f of the aggregate after a typical time duration t_f of the process. A realistic selection is $R_f \simeq 2$ mm, $t_f \simeq 4 \cdot 10^3$ sec. Since the ratio R_f/R is about 30, typical average expansion velocity is $V = \dfrac{R_f}{t_f} = 0.5\mu \sec^{-1}$.

So we define the nondimensional quantities

$$r^* = \frac{r}{R_f} \quad , \quad x^* = \frac{x}{R_f} \quad , \quad \sigma^* = \frac{\sigma}{R_f} \quad , \quad v^* = \frac{v}{V}$$

and we rescale M by $M_0 = \sup M_R(t)$: $M^* = \dfrac{M}{M_0}$. The nondimensional version of equation (1.3) is

$$\frac{\partial}{\partial t^*}(\epsilon M^*) - \left(\frac{\partial}{\partial r^*} + \frac{2}{r^*}\right)\left(\tau_M \frac{\partial M^*}{\partial r^*} - \epsilon v^* M^*\right) = -\frac{\rho_0 t_f}{M_0}\rho^* Q, \qquad (1.38)$$

where $\rho^* = \dfrac{\rho}{\rho_0}$, and $(D \simeq 3.2 \cdot 10^{-8}$ m^{2}sec$^{-1})$

$$\tau_M = \frac{t_f}{t_D} \quad , \quad t_D = \frac{R_f^2}{D} \simeq 125\,\sec \;\Rightarrow\; \tau_M \simeq 32, \qquad (1.39)$$

while $0 \le M^* \le 1$. In equation (1.38) we still have to express Q in a convenient way. We will do it later, after considering the scaling of microscopical quantities.

If we write $T^* = \dfrac{T}{T_0}$, we get from (1.4)

$$\frac{\partial T^*}{\partial t^*} - \left(\frac{\partial}{\partial r^*} + \frac{2}{r^*}\right)\left(\tau_T \frac{\partial T^*}{\partial r^*} - V^* T^*\right) = \frac{\rho_0 t_f}{cT_0}\rho^* \widehat{Q} \qquad (1.40)$$

with

$$\tau_T = \frac{t_f k}{cR_f^2} \simeq 57 \qquad (1.41)$$

$(k \simeq 8 \cdot 10^{-2}$Wm^{-1}K^{o-1}, $c \simeq 14 \cdot 10^5$Jm^{-3}K$^{o-1} \Rightarrow \dfrac{k}{c} \simeq 5.7 \cdot 10^{-8}$m^{2}sec$^{-1})$.

Thus we can say that for the monomer transport diffusion is a more important mechanism than convection, and even more important for heat transport.

The equation describing the motion of the outer boundary is

$$\frac{dR^*}{dt^*} = v^*(R^*, t^*) \quad , \quad R^*(0) = \frac{R_0}{R_f} = R_0^*. \tag{1.42}$$

The initial and boundary conditions are

$$M^*(r^*, 0) = 0, \tag{1.43}$$

$$\frac{\partial M^*}{\partial r^*}(0, t^*) = 0, \tag{1.44}$$

$$M^*(R^*(t^*), t^*) = M_R^*(t), \tag{1.45}$$

$$T^*(r^*, 0) = 1, \tag{1.46}$$

$$\frac{\partial T^*}{\partial r^*}(0, t^*) = 0, \tag{1.47}$$

$$-\frac{\partial T^*}{\partial r^*}(R^*(t^*), t^*) = h^*[T^*(R^*(t^*), t^*) - 1], \tag{1.48}$$

where $M_R^* = M_R/M_0$, $h^* = \dfrac{R_f h}{k}$.

The relationship (1.35) between Eulerian and Lagrangian coordinates becomes

$$\sigma^*(x^*, t^*) = \left[3 \int_0^{x^*} z^{*2} s^{*3}(t^*, z^*)\, dz^*\right]^{1/3} \frac{1}{y_0^*}, \tag{1.49}$$

once we rescale the microscopic length variables as follows

$$y^* = \frac{y}{y_f} \quad , \quad s^* = \frac{s}{y_f} \quad , \quad y_0^* = \frac{y_0}{y_f}$$

where $y_f \simeq 30 \cdot y_0$, i.e. $y_f = 0.15\mu$. The non-dimensional velocity field is

$$v^*(\sigma^*, t^*) = 3 y_0^{*-3} \sigma^{*-2} \int_0^{x^*} z^{*2} \frac{ds^*}{dt^*} s^{*2}\, dz. \tag{1.50}$$

Now we rescale the remaining microscopic quantities:

$$m^* = \frac{m}{M_0} \quad , \quad v_p^* = \frac{v_p}{V} \frac{R_f}{y_f} = \frac{v_p}{y_f} t_f.$$

In this way we obtain

$$\left(\frac{\partial}{\partial y^*} + \frac{2}{y^*}\right) v_p^* = 0 \quad , \quad y_0^* < y^* < s^*(t^*) \tag{1.51}$$

$$\frac{\partial m^*}{\partial t^*} - \left(\frac{\partial}{\partial y^*} + \frac{2}{y^*}\right)\left(\tau_m \frac{\partial m^*}{\partial y^*} - \epsilon_p m^* v_p^*\right) = 0 \tag{1.52}$$

with $(d \simeq 10^{-12} \mathrm{m^2 sec^{-1}})$

$$\tau_m = \frac{t_f}{t_d} \quad , \quad t_d^{-1} = \frac{d}{y_f^2} \simeq 45\text{sec}^{-1} \Rightarrow \tau_m \simeq 2 \cdot 10^5 \qquad (1.53)$$

showing that diffusion is absolutely dominant and quasi-steady. The boundary conditions are

$$v_p^*(y_0^*, t^*) = \lambda^*(T^*)\zeta m^*(y_0^*, t^*) \qquad (1.54)$$

with

$$\lambda^*(T^*) = \lambda(T)/\lambda(T_0) \quad , \quad \zeta = M_0 \frac{t_f}{y_f}\lambda(T_0) \qquad (1.55)$$

and

$$\frac{ds^*}{dt^*} = v_p^*(s^*(t^*), t^*), \qquad (1.56)$$

$$\left(\tau_m \frac{\partial m^*}{\partial y^*} - \epsilon_p m^* v_p^*\right)_{y^*=y_0^*} = \nu^* v_p^*(y_0^*, t^*) \qquad (1.57)$$

with

$$\nu^* = \frac{\nu}{M_0} \quad , \quad m^*(s^*, t^*) = KM^*. \qquad (1.58)$$

Putting $T_p^* = \dfrac{T_p}{T_0}$, the equation for the temperature is

$$\frac{\partial T_p^*}{\partial t^*} - \left(\frac{\partial}{\partial y^*} + \frac{2}{y^*}\right)\left(\tau_{T_p} \frac{\partial T_p^*}{\partial y^*} - T_p^* v_p^*\right) = 0 \qquad (1.59)$$

with

$$\tau_{T_p} = t_f \frac{k_p}{c_p y_f^2} \simeq 1.5 \cdot 10^{11}$$

($k_p = 0, 15\text{Wm}-1\text{K}^{o-1}$, $c_p = 1.8 \cdot 10^6 \text{Jm}^{-3}\text{K}^{o-1} \Rightarrow \dfrac{k_p}{c_p} \simeq 8.3 \cdot 10^{-7} \text{ m}^2\text{sec}^{-1}$)
showing that heat flux in the spherules is (practically rigorously) divergence free and the heat flowing out of the outer surface is precisely the one produced at the same time over the catalyst surface.

In other words

$$y^{*2} \frac{\partial T_p^*}{\partial y^*} = \text{constant.} \qquad (1.60)$$

Coming back to the determination of the velocity field within the polymer shell, equation (1.51) says that the product $y^2 v_p$ is independent of y, and therefore, using (1.54),

$$y^{*2} v_p^{*2}(y^*, t^*) = y_0^{*2} \lambda^*(T^*)\zeta m^*(y_0^*, t^*) \qquad (1.61)$$

which, via (1.56), specifies the motion of the outer surface

$$s^{*2}(t^*)\frac{ds^*}{dt^*} = y_0^{*2}\lambda^*(T^*)\zeta m^*(y_0^*, t^*).$$
(1.62)

We now turn our attention to the expression of the right-hand sides of (1.38), (1.40).

First of all we remark that dividing both members of (1.26) by the corresponding quantities at $t = 0$, we obtain

$$\rho^* s^{*3} = 1.$$
(1.63)

Remembering (1.28) and (1.53) we see that

$$t_f \frac{\rho_0 \rho^*}{M_0} Q = 3(1 - \epsilon)y_0^{*-3} s^{*-1} \tau_m \frac{\partial m^*}{\partial y^*}\bigg|_{y^* = y_0^*}$$

and in connection with (1.57)

$$t_f \frac{\rho_0 \rho^*}{M_0} Q = 3(1 - \epsilon)y_0^{*-3} s^{*-1} \zeta \lambda^*(T^*)[m^*(\nu^* + \epsilon_p m^*)]_{y^* = y_0^*} \equiv S^*.$$
(1.64)

Similarly we obtain

$$t_f \frac{\rho_0 \rho^*}{cT_0} \widehat{Q} = (1 - \epsilon)\frac{M_0 \mu_0}{cT_0} \tau_m y_0^{*-1} s^{*-3} \mu^*(T^*)m^*(y_0^*, t^*; \sigma^*) \equiv \widehat{S}^*$$
(1.65)

where $\mu_0 = \mu(T_0)$, $\mu^* = \mu/\mu_0$.

We may write $\widehat{S}^*$ in the more expressive form

$$\widehat{S}^* = 3(1 - \epsilon)\omega s^{*-3} \mu^*(T^*)m^*$$
(1.66)

where

$$\omega = \frac{M_0 \mu_0 4\pi y_0^2 t_f}{cT_0(4\pi/3)y_0^3}$$
(1.67)

is the ratio between two thermal energies: the polymerization heat produced during the time t_f corresponding to concentration M_0, temperature T_0 (= 343°K), and the heat content at temperature T_0 of a sphere of radius y_0 with specific heat c.

In order to get an estimate of ω, we start from the consideration that after a time t_f we expect $s^* = O(1)$, and the same for μ^*, m^*. If we simplify (1.40) to

$$-\left(\frac{\partial}{\partial r^*} + \frac{2}{r^*}\right)\frac{\partial T^*}{\partial r^*} = \frac{1}{\tau_T}\widehat{S}^* \simeq \frac{1}{\tau_T}2\omega \text{ at time } t_f$$

this amounts to saying $-r^{*2}\frac{\partial T^*}{\partial r^*} \simeq \int_0^{r^*} z^2 dz \frac{2\omega}{\tau_T} = \frac{r^{*3}}{3}\frac{2\omega}{\tau_T}$, so that the temperature difference between the center ($r^* = 0$) and the boundary

$(r^* = R^* = O(1))$ is of order $\dfrac{\omega}{3\tau_T}$. Such a difference has typical values around $5^\circ K$. Hence $\omega \simeq 15\tau_T \simeq 8.5 \cdot 10^2$.

Putting this estimate in (1.67) gives $M_0\mu_0 \simeq 1.7 \cdot 10^{-1}$ $\mathrm{Jm^{-2}sec^{-1}}$ and finally $\mu_0 \simeq 6 \cdot 10^{-3}\mathrm{m^3sec^{-3}}$ (for $M_0 \simeq 30\mathrm{Kg\,m^{-3}}$).

1.7 Summary of the Simplified Model and Sketch of the Proof

In this section we drop all "$*$" from the symbols in order to simplify notation. Of course the reader must be careful and remember that all quantities used here are nondimensional. We keep the complete form of transport equations at the macroscopic scale, although we have seen that diffusion is prevailing (more neatly for heat transport):

- *monomer transport in the agglomerate*

$$\frac{\partial}{\partial t}(\epsilon M) - \left(\frac{\partial}{\partial r} + \frac{2}{r}\right)\left(\tau_M \frac{\partial M}{\partial r} - \epsilon v M\right) = -S, \ \tau_M \simeq 30 \tag{1.68}$$

- *heat transport in the agglomerate*

$$\frac{\partial T}{\partial t} - \left(\frac{\partial}{\partial r} + \frac{2}{r}\right)\left(\tau_T \frac{\partial T}{\partial r} - vT\right) = \widehat{S}, \ \tau_T \simeq 60 \tag{1.69}$$

- *motion of the outer boundary*

$$\frac{dR}{dt} = v(R(t), t) \quad , \quad R(0) = R_0 \tag{1.70}$$

- *initial-boundary conditions for M, T*

$$M(r, 0) = 0, \tag{1.71}$$

$$\frac{\partial M}{\partial r} = 0 \quad , \quad r = 0, \tag{1.72}$$

$$M(R(t), t) = M_R(t), \tag{1.73}$$

$$T(r, 0) = 1, \tag{1.74}$$

$$\frac{\partial T}{\partial r} = 0 \quad , \quad r = 0, \tag{1.75}$$

$$-\frac{\partial T}{\partial r} = h(T - 1) \quad , \quad r = R(t) \tag{1.76}$$

- *relationship between Lagrangian and Eulerian coordinates*

$$\sigma(x, t) = \left[3 \int_0^x z^2 s^3(t; z)\, dz\right]^{1/3} \tag{1.77}$$

- *expansion velocity field*

$$v(\sigma(x,t),t) = 3\sigma^{-2} \int_0^x z^2 \frac{ds}{dt} s^2 \, dz, \tag{1.78}$$

- *monomer transport in polymeric shells*

$$\left(\frac{\partial}{\partial y} + \frac{2}{y}\right) \frac{\partial m}{\partial y} = 0, \tag{1.79}$$

- *velocity field in polymeric shells*

$$v_p(y,t) = y^{-2} \lambda^* \zeta m(y_0, t), \tag{1.80}$$

- *growth rate of spherules*

$$\frac{ds}{dt} = v_p(s,t) \quad, \quad s(0) = y_0 \tag{1.81}$$

- *boundary conditions*

$$\left. \frac{\partial m}{\partial y} \right|_{y=y_0} = 0 \tag{1.82}$$

(from (1.57))

$$m(s,t) = KM. \tag{1.83}$$

From (1.60) we can deduce the thermal field in the polymeric shell, since we know the rate of heat production at the inner surface and the temperature at the outside surface.

This is not a relevant information for our purposes. Indeed, if the expected temperature difference between the center and the boundary of the aggregate is about $5^o\mathrm{K}$, the temperature variation across the microspherules is absolutely negligible.

If we use (1.79) in connection with (1.82) we conclude that m is uniform over each polymeric shell, i.e.

$$m = KM \tag{1.84}$$

tanks to (1.83).

This approximation greatly simplifies the small scale problem, since (1.80) now reduces to

$$v_p(y,t) = y^{-2} \lambda^* \zeta K M, \tag{1.85}$$

where λ^*, M depend on macroscopical coordinates.

At this point we have a simple way of integrating (1.81), that is equivalent to

$$\frac{d}{dt}s^3 = 3\lambda^*\zeta KM, \tag{1.86}$$

hence

$$s^3(t) - y_0^3 = 3K \int\limits_0^t \lambda^*\zeta M \, d\tau, \tag{1.87}$$

where the integral on the right-hand side is performed along the trajectory of the particle considered.

Equations (1.77), (1.78) simplify accordingly. In particular (1.78) becomes

$$v(\sigma(x,t),t) = 3\sigma^{-2}K \int\limits_0^x z^2\overline{\lambda}^*(z,t)\zeta\overline{M}(z,t) \, dz \tag{1.88}$$

where $\overline{\lambda}^*(z,t) = \lambda^*(T(\sigma(z,t),t))$, $\overline{M}(z,t) = M(\sigma(z,t),t)$, while (1.87) must used to express σ.

Of course also the expressions (1.64) of S and (1.66) of $\widehat{S}$ can be simplified using (1.84) and (1.87).

In conclusion, we have totally uncoupled the macroscopic problem, which now consists in the system (1.68)-(1.76), where the scalar fields $v, S, \widehat{S}$ are functionals of T, M (see the previous section).

We also remark that

$$R(t) = \sigma(R_0, t) = \left\{3 \int\limits_0^{R_0} z^2 \left[y_0^3 + 3K \int\limits_0^t \overline{\lambda}^*(z,t)\overline{M}(z,t) \, dz\right]\right\}^{1/2}, \tag{1.89}$$

replaces (1.70).

This is a considerably simpler problem than the one studied in the quoted papers [2], [4]. As we have seen, the reduction to the present form is largely justified.

The sketch of the existence proof is as follows: give a pair $(\overline{M}(x,t), \overline{T}(x,t))$ with the Lagrangian coordinate x varying in $[0, R_0]$ and t in some interval $[0, \theta]$. The functions must be $C^{2+\alpha}$ and compatible with the boundary data (1.71)-(1.76) and with the equations (1.68), (1.69) for $r = 0$, $r = R_0$, $t = 0$.

Using $(\overline{M}, \overline{T})$ we can construct the functions $\sigma, v, R, S, \widehat{S}$ and thus solve the system (1.68), (1.69), (1.71)-(1.76), obtaining another pair $(\widetilde{M}, \widetilde{T})$. It is easy to see that we can select the subset X of $C^{2+\alpha}$ in which we choose the pair $(\overline{M}, \overline{T})$ in such a way that $(\widetilde{M}, \widetilde{T})$ belongs to the same set. Also, it is possible to show that the mapping $(\overline{M}, \overline{T}) \rightarrow (\widetilde{M}, \widetilde{T})$ is continuous in the sup norm and even contractive for θ small enough. Thus the mapping has a unique fixed point, solving the original problem.

Remark 2. The coefficients λ^*, μ^* can be taken equal to 1, if we disregard a very short temperature pulse at the beginning of the process (see [17]). Such a pulse is the consequence of the presence of residual monomer in the aggregate (here disregarded) after the fragmentation stage.

1.8 Fragmentation

As we said, fragmentation of the aggregate occurs during a first stage of the process, in which the system is—in a sense—prepared to perform in the way we have described in the previous section. During this stage the catalyst comes into contact with the monomer for the first time and it is crucial to keep the polymerization rate within safe limits. The description of possible damages produced by an excessive reaction rate during the first stage of polymerization are described in [22]. For this reason the monomer temperature is fixed at a lower value than the one used during the main process, thus depressing the factor λ in (1.18).

Based on pictures shown in [22], the aggregate diameter after the first low temperature stage (prepolymerization) increases by a factor 5-6, to a diameter of about 400μ. In view of this fact, if we consider the main section of the process separately it would be sensible to slightly modify the model illustrated in the previous sections, adjusting the initial conditions to the outcome of the fragmentation process. This will be particularly important in order to explain the sudden temperature burst experienced at the start of the second stage of polymerization (see [17]).

Let us summarize the recent work [20], dealing exclusively with the specific aspects of fragmentation and adding some considerations about the thermal field and again dimensional analysis.

The basic assumption is that fragmentation takes place on a front moving inwards through the aggregate. This picture corresponds to the rapid formation of a thin polymeric shell around the catalytic sites. Based upon the pictures of [22] the average growth factor after prepolymerization is between 5 and 6, so we may expect an average thickness of about $0.03\,\mu$, reached after 600 sec.

As in the previous section, here too we keep the assumption that the monomer transport occurs mainly by diffusion.

The front advancement requires the conversion of a finite amount of monomer in a negligible time and can be schematized as an infinitely fast reaction.

If we denote by $\varphi(t)$ the (dimensional) radius of the unpenetrated core, the hypothesis of fast reaction is expressed by

$$M(\varphi(t), t) = 0. \tag{1.90}$$

The quantity $\left| \bar{\rho}_0 \dfrac{d}{dt}\left(\dfrac{4\pi}{3}\varphi^3 \right) \right|$ represents the number of catalytic nuclei involved by the fragmentation process per unit time, where $\bar{\rho}_0$ is the density of catalytic sites in the unpenetrated core. We assume that the value of ρ soon after fragmentation is some given ρ_0. If we neglect the change of porosity and we assume that the radius of the microparticles changes from $\bar{y}_0 = 5 \cdot 10^{-9}$m to e.g. $\bar{y}_0 = 8 \cdot 10^{-9}$m (corresponding to fractional increment δ of about 10% of the final average increase after the prepolymerization stage), we have $\rho_0 \simeq \bar{\rho}_0 \cdot 0.625$ (the exact formula is $(1-\bar{\epsilon})\bar{y}_0^3\bar{\rho}_0 = (1-\epsilon)\bar{y}_0^3(1+\delta)^3\rho_0$, i.e.

$$\rho_0 = \bar{\rho}_0 \frac{1-\bar{\epsilon}}{1-\epsilon}\frac{1}{(1+\delta)^3}).$$

The rate of polymer mass production at the fragmentation front will be $\left| \bar{\rho}_0 \dfrac{d}{dt}\left(\dfrac{4\pi}{3}\varphi^3 \right) \right|$ times the mass m_f of the first polymer layer, i.e.

$$m_f \simeq 900\,\mathrm{Kg\,m}^{-3} \cdot 1.6 \cdot 10^{-24}\mathrm{m}^3 \simeq 1.5 \cdot 10^{-21}\mathrm{Kg}.$$

Since the monomer flow rate per unit surface at the fragmentation front is $D\dfrac{\partial M}{\partial r}\Big|_{r=\varphi(t)}$ (to be multiplied by $4\pi\varphi^2$), we arrive at the following equation, expressing the front advancement

$$m_f\bar{\rho}_0\dot{\varphi} = -D\frac{\partial M}{\partial r}\Big|_{r=\varphi(t)}, \tag{1.91}$$

showing in particular that φ is strictly decreasing.

A natural initial condition for φ is $\varphi(0) = R_0$.

In our picture there is a sudden volume change of the fragmented particles at the front. As a consequence the particles already posses a non-zero expansion velocity at the front. Such a velocity can be found observing that if the front has penetrated by an amount $|d\varphi|$, the thickness of the corresponding layer is multiplied by the fraction $\dfrac{\bar{\rho}_0}{\rho_0} = \dfrac{1-\epsilon}{1-\bar{\epsilon}}(1+\delta)^3$, and the swelling is $\left(\dfrac{\bar{\rho}_0}{\rho_0} - 1 \right)|d\varphi|$. Hence we have the formula

$$v(\varphi(t),t) = -\dot{\varphi}(t)\left(\frac{\bar{\rho}_0}{\rho_0} - 1 \right). \tag{1.92}$$

Clearly condition (1.92) modifies the expression of the expansion velocity field. Integrating (1.27) over the interval (φ, r) we get

$$r^2 v(r,t) + \varphi^2\dot{\varphi}\left(\frac{\bar{\rho}_0}{\rho_0} - 1 \right) = \int_\varphi^r 3\frac{\dot{s}(t;r')}{s(t;r')}r'^2\,dr'. \tag{1.93}$$

The relationship between the Eulerian coordinate r and the Lagrangian coordinate x is now more complicated. Clearly we have $r \equiv x$ in the region

$0 \leq r \leq \varphi(t)$, i.e. in the unreacted core. In the complementary zone we have to integrate equation (1.33) for fixed x over the time interval $(\theta(x), t)$, $\theta(x)$ being the inverse function of $\varphi(t)$. We know that $\sigma(\varphi(t), t) = \varphi(t)$ and that

$$\left. \frac{\partial \sigma}{\partial t} \right|_{x = \varphi(t)} = -\dot{\varphi} \left(\frac{\bar{\rho}_0}{\rho_0} - 1 \right).$$

From $\left(\frac{\partial \sigma}{\partial x} \dot{\varphi} + \frac{\partial \sigma}{\partial t} \right)_{x = \varphi(t)} = \dot{\varphi}(t)$, we deduce that $\left. \frac{\partial \sigma}{\partial x} \right|_{x = \varphi(t)+} = \frac{\bar{\rho}_0}{\rho_0}$ (mak-

ing use of the fact that $\dot{\varphi} \neq 0$). Thus $\frac{\partial \sigma}{\partial x}$ (which is equal to 1 for $x < \varphi(t)$) jumps across the front $x = \varphi(t)$. At this point the integration of (1.33) yields $(\varphi(\theta(x)) = x)$

$$\sigma^2 \frac{\partial \sigma}{\partial x} = x^2 \frac{\bar{\rho}_0}{\rho_0} \left(\frac{s(t; x)}{y_0} \right)^3. \tag{1.94}$$

Now for t fixed we integrate over the interval $(\varphi(t), x)$, obtaining

$$\sigma(x, t) = \left\{ \varphi^3(t) + 3 \frac{\bar{\rho}_0}{\rho_0} \frac{1}{y_0^3} \int_{\varphi(t)}^{x} z^2 s^3(t, z)\, dz \right\}^{1/3}. \tag{1.95}$$

Differentiating with respect to t, we recover the expression (1.93) of the velocity (just perform the substitution $r' = \sigma(z, t)$ after differentiation).

At the same time we can write down the heat released at the fragmentation front as $\bar{\rho}_0 4\pi\varphi^2 |\dot{\varphi}| m_f q$ (remember that $q = \frac{\mu}{\lambda \rho_p}$) and put it equal to the outgoing heat flux

$$m_f \bar{\rho}_0 \dot{\varphi} q = k \left. \frac{\partial T}{\partial r} \right|_{r = \varphi(t)+} - k_c \left. \frac{\partial T}{\partial r} \right|_{r = \varphi(t)-} \tag{1.96}$$

where k_c is the thermal conductivity of the unreacted core. The heat transport equation (1.4) must be coupled with

$$c_c \frac{\partial T}{\partial t} - k_c \left(\frac{\partial}{\partial r} + \frac{2}{r} \right) \frac{\partial T}{\partial r} = 0 \tag{1.97}$$

($c_c = 2 \cdot 10^6\,\mathrm{Jm^{-3}\,{}^\circ K^{-1}}$, $k_c = 0.15\,\mathrm{Wm^{-2}\,{}^\circ K^{-1}}$), through (1.92) and the continuity condition

$$T(\varphi(t)-, t) = T(\varphi(t)+, t). \tag{1.98}$$

We remark that (1.90), (1.91) are precisely the Stefan conditions, well known as phase-change conditions. Let us now choose non-dimensional coordinates. We may normalize $\bar{\rho}_0$ by ρ_0 ($\bar{\rho}_0^* = \bar{\rho}_0/\rho_0 \simeq 1.6$). Time can be rescaled by $\bar{t}_f = 600\,\mathrm{sec}$ and radial coordinate by $\bar{R}_f = 4 \cdot 10^{-4}\,\mathrm{m}$: $r^* = r/\bar{R}_f$, $\varphi^* =$

$\varphi/\overline{R}_f$. Velocity is rescaled by $\overline{v} = \dfrac{\overline{R}_f}{\overline{t}_f} \simeq 0.6\mu/\text{sec}$. For the other variables we choose normalizations similar to the ones of Sect. 6, recalling that during prepolymerization we generally have a different M_0 (now $\overline{M}_0$) and a different T_0 (now $\overline{T}_0 = 293^\circ K$). The new coefficients $\overline{\tau}_M, \overline{\tau}_T$ in equations (1.38), (1.40) (now holding for $\varphi(t) < r < R(t)$) are $\overline{\tau}_M = D\dfrac{\overline{T}_f}{\overline{R}_f^2} \simeq 1.5\cdot 10^{-8}\dfrac{600}{1.6\cdot 10^{-7}} \simeq 60$,

$\overline{\tau}_T = \dfrac{600}{1.6\cdot 10^{-7}}\dfrac{0.1}{1.5\cdot 10^6} \simeq 288.$

In the nondimensional form the above equations take the form

$$M^*(\varphi^*(t^*), t^*) = 0, \tag{1.99}$$

$$\frac{d\varphi^*}{dt^*} = -\chi\frac{\partial M^*}{\partial r^*}\bigg|_{r^*=\varphi^*(t^*)} \quad , \quad \chi = D\frac{\overline{t}_f}{\overline{R}_f^2}\frac{\overline{M}_0}{\rho_0 m_f} \simeq 3\cdot 10^{-3}\overline{M}_0 \tag{1.100}$$

$$v^*(\varphi^*, t) = -\frac{d\varphi^*}{dt^*}\left(\frac{\overline{\rho}_0}{\rho} - 1\right) \tag{1.101}$$

$$\frac{\partial T^*}{\partial r^*}\bigg|_{r^*=\varphi^*+} - \frac{k_c}{k}\frac{\partial T^*}{\partial r^*}\bigg|_{r^*=\varphi^*-} = \widehat{\chi}\frac{d\varphi^*}{dt^*} \quad , \quad \widehat{\chi} = \frac{\rho_0 m_f \overline{R}_f^2}{k\overline{T}_0\overline{t}_f}q \tag{1.102}$$

$(k_c/k \simeq 1.36)$

$$\frac{\partial T^*}{\partial t^*} - \tau_{T_c}\left(\frac{\partial}{\partial r^*} + \frac{2}{r^*}\right)\frac{\partial T^*}{\partial r^*} = 0 \quad , \quad \tau_{T_c} = \frac{k_c}{c_c}\frac{\overline{t}_f}{\overline{R}_f^2} \simeq 300 \quad , \quad 0 < r^* < \varphi^*(t^*). \tag{1.103}$$

The last equation shows that it is a good approximation to take $T^* = T_c^*(t^*)$, spatially constant in the unreacted core, and consequently we may rewrite (1.102) as

$$\frac{\partial T^*}{\partial r^*}\bigg|_{r^*=\varphi^*+} = \widehat{\chi}\frac{d\varphi^*}{dt^*}. \tag{1.104}$$

In this way the thermal problem is well defined in the polymerized region and, using temperature continuity, we just put $T_c^*(t^*) = T^*(\varphi^*(t^*), t^*)$.

The expression for the non-dimensional velocity field is simply derived from (1.93):

$$r^{*2}v^*(r^*, t^*) + \varphi^{*2}\frac{d\varphi^*}{dt^*}\left(\frac{\overline{\rho}_0}{\rho_0} - 1\right) = \int_{\varphi^*}^{r^*} 3\frac{ds^*}{dt^*}\frac{1}{s^*}\xi^2\, d\xi, \tag{1.105}$$

where $s^* = s/\overline{y}_f$ with $\overline{y}_f = 3 \cdot 10^{-8}\,\mathrm{m}$.

Indeed, small-scale normalization of speed is now $v_p^* = \dfrac{v_p}{\overline{v}}\dfrac{\overline{R}_f}{\overline{y}_f} = v_p\dfrac{\overline{t}_f}{\overline{y}_f}$ so that the coefficient ζ in (1.54) changes to

$$\overline{\zeta} = \overline{M}_0 \frac{\overline{t}_f}{\overline{y}_f} \lambda(\overline{T}_0)$$

while λ is normalized by $\lambda(\overline{T}_0)$: $\lambda^* = \lambda(T)/\lambda(\overline{T}_0)$.

These are the only changes to be introduced in (1.85)-(1.87). In particular, the equation replacing (1.88), i.e. the "macroscopic" expression of the expansion velocity field is now

$$\sigma^{*2}v^* + \varphi^{*2}\frac{d\varphi^*}{dt^*}\left(\frac{\overline{\rho}_0}{\rho_0} - 1\right) = \int_{\varphi^*}^{x^*} z^2 3\frac{\overline{\rho}_0}{\rho_0}\frac{1}{y_0^{*3}}\lambda^* \overline{\zeta} k \overline{M}\, dz \qquad (1.106)$$

(using the more convenient form in Lagrangian coordinates).

In summary, the prepolymerization process is described by the equations corresponding to (1.68), (1.69), but with a more marked influence of diffusion ($\overline{\tau}_M \simeq 60$, $\overline{\tau}_T \simeq 300$) and with slightly corrected expressions for $S^*, \widehat{S}^*$, with the boundary conditions (1.73), (1.99) for M, and (1.76), (1.104) for T. Moreover we have the free boundary condition (1.100) and the expression (1.106) for the expansion velocity field. The growth polymer shells is described by equation (1.87), with the corresponding changes.

For the purpose of proving existence it is convenient to introduce at $t = 0$ a small layer of already penetrated aggregate, so that the initial condition for φ is not $\varphi(0) = R_0$, but $\varphi(0) = \varphi_0 < R_0$ and $1 - \dfrac{\varphi_0}{R_0} \ll 1$. Of course in this region some artificial smooth initial condition must be defined.

At this point we confine ourselves to note that if a function $\varphi > 0$ is given in a Lipschitz class in some time interval, then it is possible to find M, T, v, m, v_p, s and by means of

$$\frac{d\widetilde{\varphi}^*}{dt^*} = -\chi\frac{\partial M^*}{\partial r^*}\bigg|_{r^*=\varphi^*} \quad , \quad \widetilde{\varphi}^*(0) = R_0^* \qquad (1.107)$$

it is possible to define a mapping $\varphi^* \to \widetilde{\varphi}^*$, whose fixed points correspond to solutions of the problem. This procedure is basically a simplification of the one used in [20] and we may conclude the the scheme here proposed is well posed.

1.9 Conclusions and Open Questions

We have presented an updated situation about modeling of the Ziegler-Natta type polymerization process with the latest generation of catalysts. In addition to the main process we have analyzed also the preceding step, which is the so-called prepolymerization, including the fragmentation of the catalyst aggregate. The quantitative analysis of the relative influence of the various concurrent phenomena has proven very useful in suggesting approximations that allow to simplify the mathematical model to a great extent without altering its accuracy.

Let us now discuss some question left open and also some future project.

A) Modelling the growth of polymeric shells We recall that the choice of constant porosity of the growing polymeric shell is completely arbitrary, in absence of other data. A more realistic picture of the behaviour of the expanding cells should come from the analysis of the stresses. Such an approach is being investigated (for the case of high pressure reactors) and leads to a nonlinear hyperbolic system of great complexity .

B) Shapes The spherical symmetry of the microparticles and of the aggregate makes it possible to describe the velocity. Of course this information is no longer sufficient if we drop the symmetry assumption. We may conjecture that the motion is driven by a potential, but a mechanical analysis has not been performed.

C) Monomer transport mechanisms At the end of Section 2 we have briefly commented on the use of Fick's law as the basic monomer transport mechanism inside the aggregate. The question arises whether this is appropriate when the monomer in the reactor is liquid and not gaseous. The alternative picture could be to look at the aggregate as a saturated medium with pressure prescribed on its boundary. The corresponding mathematical model would be Darcy's law and pressure would obey an elliptic equation (still under the assumption of constant porosity) with a sink term.

I believe it is worth exploring this possibility. On the small scale the situation is probably different, because of the extremely complex structure of the forming polymeric shells (whose properties have been taken uniform for simplicity). However there is no definite evidence on what the correct monomer displacement mechanism inside the shells could be.

D) Temperature bursts At the beginning of the main process (i.e. soon after fragmentation), a sudden (and very short lived) raise of temperature is observed. As remarked in [17], this is certainly due to the fact that the monomer available in the pores of the aggregate at the moment in which the temperature is increased to values that make the reaction much faster, is rapidly consumed and a large amount of heat is released. The corresponding sudden drop of monomer concentration (caused by the fact that diffusion is not able to provide new material in a sufficiently short time) causes the reaction to slow down almost immediately to a normal rate. This rapid transient should be modelled separately since it involves a time scale which is much shorter than diffusion time (both for monomer and for heat).

E) Alternative approach to fragmentation The pictures shown in [17] point out that at the beginning of the pre-polymerization there is some excess activity of the catalysts, although much more modest than the one at the beginning of the subsequent stage.

Again, this suggest that we should model the very first stage of fragmentation separately.

F) Copolymers Copolymers are produced by a two-stage process in which different monomers are used. The final structure of the product (see the paper [5]) poses still unsolved questions about modelling the second polymerization. This is one of the most stimulating subject of investigation in the large area of polymerization and certainly many efforts will be oriented to this target in the next future.

1.10 Acknowledgements

I wish to thank Dr. S. Mazzullo of former Montell Italia ("G. Natta" R&DC Ferrara, Italy) for his continuous advice and for the many constructive discussions I had with him, particularly concerning technological aspects that have been developed exclusively at Montell.

References

1. D. Andreucci, A. Fasano, R. Ricci, *Modello matematico di replica nel caso limite di distribuzione continua di centri attivi*, in "Simposio Montell 96, S. Mazzullo, G. Cecchi eds., SATE, Ferrara (1996), 154–174.
2. D. Andreucci, A. Fasano, R. Ricci, *On the growth of a polymer layer around a catalytic particle: a free boundary problem*, NoDEA **4** (1997), 511–520.
3. D. Andreucci, A. Fasano, R. Ricci, *Existence of solutions for a continuous multigrain model of polymerization in polymers*, to appear on M3AS.
4. D. Andreucci, R. Ricci, *Mathematical problems in the Ziegler-Natta polymerization process*. In "Complex Flows in Industrial Processes", A. Fasano ed., MSSET Birkhäuser (2000), 215–238.
5. G. Collina, P. Sgarzi, G. Baruzzi, *Recenti sviluppi del concetto di Reactor Granule Technology: Multicatalysts Reactor Granule Technology*, in "Simposio Montell 96", S. Mazzullo, G. Cecchi eds., SATE, Ferrara (1996), 41–59.
6. J. Bear, *Dynamics of Fluids in Porous Media*, America Elsevier, New York, 1972.
7. N.A. Dotson, R. Galván, R.L. Laurence, M. Tirrel, *Polymerization Process Modelling*, VCH Publ. (1996).
8. G. Eder, H. Janeschitz-Kriegl, *Structure development during processing: crystallization*. In Proceedings of Polymers (H.E.H. Meijer ed.), Material Science and Technology **18**, VCH (1997).
9. A. Fasano, *Mathematical models in polymer processing*. MECCANICA 35 (2000), 163-198.

10. A. Fasano, A. Mancini, S. Mazzullo, *Isobaric crystallization of polypropylene*, "Complex Flows in Industrial Processes", A. Fasano ed., MSSET Birkhäuser (2000), 149–189.

11. P. Galli, *Modello di crescita di un polimero e fenomeno della replica: dalla teoria alla conferma sperimentale*, in "Simposio Montell 96", S. Mazzullo, G. Cecchi eds., SATE, Ferrara (1996), 12–40.

12. D.E. Goldman, *Potential, impedance and rectification in membranes*, J. Gen. Physical **27** (1943), 37–60.

13. R.L. Laurence, M.G. Chiovetta, *Heat and mass transfer during olefin polymerization from the gas phase*, Polymer Reaction Eng., K.H. Reichert and W. Geiseler eds., Hanser (Munich), 1983, 73–112.

14. G. Marrucci, G. Ianniuberto, *Molecular theories of polymer viscosity*, "Complex Flows in Industrial Processes", A. Fasano ed., MSSET Birkhäuser (2000), 3–24.

15. S. Mazzullo, G. Cecchi, eds., *Meccanismi di accrescimento su catalizzatori Ziegler-Natta*, "Simposio Montell 96", S. Mazzullo, G. Cecchi eds., SATE, Ferrara (1996).

16. S. Mazzullo, A. Fait, *Natura polimerica dell'addotto* $MgCl_2$ *(EtOH) e modelli di Transizione di fase liquido-solido*, in "Simposio Montell 96", S. Mazzullo, G. Cecchi eds., SATE, Ferrara (1996), 189-218.

17. G. Mei, *Modello polimerico multigrain e double grain*, in "Meccanismi di accrescimento di poliolefine su catalizzatori Ziegler-Natta", Simposio Montell 96, S. Mazzullo and G. Cecchi eds., SATE, Ferrara (1996), 135–153.

18. E.P. Moore, Jr., ed., *Polypropylene Handbook*, Hauser Publ. 1996.

19. E.P. Moore, Jr., *The Rebirth of polypropylene: Supported Catalysts*, Hanser Pub., Munich 1998.

20. S. Pizzi, *The multigrain two-scale model for the Ziegler-Natta polymerization process with fragmentation of the catalytic aggregate*, MECCANICA 35 (2000), 313-323.

21. I.J. Rao, K.R. Rajagopal, *Phenomenologycal modeling of polymer crystallization using the notion of multiple natural configuration*, Interfaces & Free Boundaries **2** (2000), 73–94.

22. G. Pennini, M. Sacchetti, *Prepolimerizzazione e fenomeni di sporcamento (etilene)*, in "Simposio Montell 96", S. Mazzullo, G. Cecchi eds., SATE, Ferrara (1996), 60–77.

23. P. Sgarzi, *Morfologia nascente del polipropilene su catalizzatori sferici* $MgCl_2/TiCl_4$, in "Simposio Montell 96", S. Mazzullo, G. Cecchi eds., SATE, Ferrara (1996), 78–106.

Nucleation

Dedication

2 Classical Kinetic Theory of Nucleation and Coarsening

J. C. Neu[1] and L. L. Bonilla[2]

[1] Department of Mathematics, University of California at Berkeley, Berkeley, CA 94720, USA
[2] Escuela Politécnica Superior, Universidad Carlos III de Madrid, Avda. Universidad 30, 28911 Leganés, Spain

2.1 Introduction

The purpose of this chapter is to explain the classical kinetic theory of nucleation in a context simpler than polymer crystallization. Many theories start by assuming that polymer crystallization is an activated process involving crossing of a free energy barrier [1]. The latter separates two accessible stable states of the system such as monomer solution and crystal. This general setting for activated processes can be used to describe the formation of a crystal from a liquid cooled below its freezing point [2], precipitation and coarsening of binary alloys [3], colloidal crystallization [4], chemical reactions [5], polymer crystallization [1, 6, 7], etc. In all these cases, the theory of homogeneous isothermal nucleation provides a framework to study the processes of formation of nucleii from density fluctuations, and their growth until different nucleii impinge upon each other. In the early stages of these processes, nucleii of solid phase are formed and grow by incorporating particles from the surrounding liquid phase. There is a critical value for the radius of a nucleus that depends on a chemical drive potential, which is proportional to the supersaturation for small values thereof. In this limit, the critical radius is inversely proportional to the supersaturation. At the beginning of the nucleation process, nucleii have small critical radius and new clusters are being created at a non-negligible rate. As the size of existing clusters increases, there are less particles in the liquid phase, the supersaturation decreases and the critical radius increases. Then it is harder for new clusters to spontaneously appear from density fluctuations. What happens is that supercritical clusters (whose radii are larger than the critical one) keep growing at the expense of subcritical clusters, that in turn keep losing particles. The size of the nucleii is still small compared to the average distance between them, so that impingement processes (in which two or more clusters touch and interaction between them dominates their growth) can be ignored. This stage of free deterministic growth is called *coarsening* [9].

A convenient framework to describe nucleation and coarsening is the classical Becker-Döring kinetic theory. We assume that the dominant processes for nucleus growth or shrinking are addition or subtraction of one particle. Nucleation is thus treated as a chain reaction whereby nucleii of n particles are created by adding one particle to a nucleus of $n - 1$ particles, or subtracting one particle from a nucleus with $n + 1$ particles. We can then write rate equations for the number density of nucleii of n particles by using the law of mass action. The kinetic rate constants for the processes of addition and depletion have to be determined by using specific information from the physical process we are trying to model. Typically we impose detailed balance which implies that the ratio of rate constants is proportional to the exponential of the free energy cost of adding one particle to a nucleus of n particles (in units of $k_B T$). This leaves one undetermined rate constant. There are different ways of finding the missing constant. One way is to postulate a microscopic theory for particle interaction and use Statistical Mechanics to determine the free energy of a cluster [8]. A different point of view is to impose that our rate constants should provide a description of coarsening compatible with the macroscopic description in terms of balance equations. We shall illustrate this second point of view and be led to a Smoluchowski equation from which the Lifshitz-Slyozov coarsening theory follows [10].

The structure of this paper is as follows. We present the Becker-Döring kinetic equations for cluster with n particles in Section 2.2. One relation between the two rate constants of this theory follows from detailed balance. The other rate constant has to be determined by comparison with the known macroscopic equation for the growth of cluster radii. In the small supersaturation limit, the Becker-Döring equations can be approximated by a Smoluchowski equation for the distribution function of cluster radii. Its drift term yields the growth of cluster radii in terms of the missing rate constant. To compare with experimental data, we consider the case of coarsening of a binary alloy [3]. In Sections 2.3 to 2.5, we review phase equilibria, macroscopic kinetics of precipitate and matrix atoms and the quasistatic limit of the kinetic equations, respectively. As a result, we find the growth of the radius of a supercritical cluster in terms of macroscopic parameters. Comparison with the results in Section 2.2 yields the sought rate constant; see Section 2.5. Numerical values for all the parameters involved in our theories can be calculated from experimental data as explained in Section 2.6. A discussion of our results constitutes the last Section.

2.2 Kinetics of Clusters

Let us assume that we have two stable phases characterized by different values, c_1 and c_2 of the number density c. Phase 1 is solution and Phase 2 precipitate. Or Phase 1 is the liquid and Phase 2 the crystal phase. Initially all precipitate particles are in Phase 1. Classic Becker-Döring (BD) kinetics treats nucleation as a *chain reaction* whereby nucleii (assumed to be *spherical*) of n precipitate particles are created by adding one particle to a nucleus with $n-1$ particles, or subtracting one particle from a nucleus with $n+1$ particles. This chain reaction scheme is natural for the situation that BD has in mind, in which bulk precipitate phase consists only of precipitate particles so distinction between particles in nucleus or in solution is clear.

Let ρ_n be the number density of nucleii of n particles. The monomer density ρ_1 represents the concentration of precipitate particles in solution and as such it will be identified with the concentration c_∞ of the macroscopic theory in Section 2.5. Consider the reaction

$$n + 1 \rightleftharpoons (n + 1).$$

The forward reaction proceeds at a rate proportional to $\rho_1 \rho_n$ with some rate constant k_a. The backward reaction proceeds at a rate proportional to ρ_{n+1} with rate constant k_d. Hence the net rate of creation of $(n+1)$-clusters from n-clusters per unit volume is the *flux*

$$j_n \equiv k_{a,n}\rho_1\rho_n - k_{d,n+1}\rho_{n+1}. \tag{2.1}$$

The fact that the rate constants depend on cluster size has been explicitly indicated in (2.1). Net rate of creation of n-clusters is due to their creation from $(n-1)$-clusters minus the rate of creation of $(n+1)$-clusters from n-clusters,

$$\dot{\rho}_n = j_{n-1} - j_n \equiv -D_- j_n, \quad n \geq 2. \tag{2.2}$$

This formulation specifies the evolutions of ρ_2, ρ_3, ... with $\rho_1 = c_\infty$ *given*. The number density of precipitate particles in n-clusters is $n\rho_n$ and the density c (equal to the initial concentration of precipitate particles in the solution) of *all* precipitate particles is

$$c \equiv \sum_{n=1}^{\infty} n\,\rho_n, \tag{2.3}$$

or equivalently,

$$c - c_1 = \gamma_\infty + \sum_{n=2}^{\infty} n\,\rho_n, \tag{2.4}$$

where we have defined $\gamma = c - c_1$, $\gamma_\infty = c_\infty - c_1$. There is conservation of all precipitate particles so c or $\gamma_0 = c - c_1$ are constant. (2.4) establishes that the constant initial concentration disturbance γ_0 is sum of the disturbance of precipitate particles, γ_∞, and the number density of particles in all cluster sizes $n \geq 2$. Given the constraint (2.3), the evolution of all ρ_n, including ρ_1, is specified.

A most essential point of BD kinetics is identification of rate constants k_a and k_d. Ideally, this would be based on basic energetics and dynamics at the microscopic level, but a complete realization of this ideal is clearly elusive. Here is what *is* done: *The ratio is determined by detailed balance.* Equilibrium, if achievable, is described by a zero flux. Setting $j_n = 0$ in (2.1) implies

$$\frac{k_{a,n}}{k_{d,n+1}} = \frac{\rho_{n+1}}{\rho_1 \rho_n}.$$ (2.5)

Standard equilibrium physicochemical theory states that

$$\frac{\rho_{n+1}}{\rho_n} = e^{-\frac{\mu_n}{\tau}},$$ (2.6)

where μ_n is the free energy cost of creating an $(n + 1)$-particle nucleus from an n-particle nucleus relative to the state of no nucleus. $\tau = k_B T$ is the temperature measured in units of energy. Clearly, $\mu_n = G_{n+1} - G_n$ (G_n is the free energy of a n-cluster), so (2.6) becomes

$$\frac{\rho_{n+1}}{\rho_n} = e^{-\frac{G_{n+1}-G_n}{\tau}},$$

and (2.5) now reads

$$\rho_1 k_{a,n} = e^{-\frac{G_{n+1}-G_n}{\tau}} k_{d,n+1}.$$

Thus formula (2.1) for the flux is now

$$j_n = k_{d,n+1} \left\{ e^{-\frac{G_{n+1}-G_n}{\tau}} \rho_n - \rho_{n+1} \right\}.$$ (2.7)

The equilibrium considered in the detailed balance argument is achievable only if $G_n \to +\infty$ as $n \to \infty$, whereas in a supersaturated solution, G_n achieves a maximum for finite n and then $G_n \to -\infty$ as $n \to \infty$. The determination of the ratio k_a/k_d is assumed to hold regardless.

What is known about G_n? Microscopic models for G_n (or, equivalently, the cluster partition function $Q_n \equiv \sum_K e^{U b(K)/\tau} = e^{-G_n/\tau}$, where U is the binding energy per pair of particles in the cluster of n particles, $b(K)$ is the number of nearest-neighbor pairs of particles in the cluster K, and the sum is over all translationally inequivalent n-particle clusters) are described in [8, 11, 12]. We would like to follow here a simpler approach, consisting of identifying the resulting expressions for large spherical nucleii with known facts about radius growth in the quasistatic approximation. For nucleii of macroscopic size $n \gg 1$, n can be written in terms of the cluster radius a by

$$n = \frac{4\pi}{3} c_2 a^3,$$ (2.8)

and $G_n \sim G(a)$, where $G(a)$ is the free energy of a nucleus of radius a as determined by continuum theory. For nucleii of only a few particles, this asymptotic correspondence with continuum theory breaks down. But if the critical nucleus has $n \gg 1$ particles, the continuum approximation works for n on the order of critical cluster size.

In the limit $|(G_{n+1} - G_n)/\tau| \ll 1$, formula (2.7) for the flux reduces to

$$j_n = k_{d,n+1} \left\{ -\frac{(G_{n+1} - G_n)\,\rho_n}{\tau} + \rho_n - \rho_{n+1} \right\}$$

$$= -k_{d,n+1} \left\{ \frac{1}{\tau}(D_+G_n)\,\rho_n + D_+\rho_n \right\}, \qquad (2.9)$$

where $D_{\pm}h_n \equiv \pm(h_{n\pm1} - h_n)$. The basic evolution equation (2.2) now reads

$$\dot\rho_n - D_- \left\{ k_{d,n+1} \left(\frac{1}{\tau}(D_+G_n)\,\rho_n + D_+\rho_n \right) \right\} = 0. \qquad (2.10)$$

This equation looks like a spatially discretized Smoluchowski equation. Asymptotic replacement of difference operators D_+, D_- by derivatives is justified if the relative changes in G_n, ρ_n and k_d when n increases by one are small. Here we will follow the simple procedure of formulating the continuum limit an checking its validity a posteriori.

The space-like variable in (2.10) is n. Experimental data usually contain histograms showing the distribution of nucleii in the space of their radii a, so we adopt the radius a as a more convenient space-like variable. The dependent variable should be $\rho = \rho(a,t)$, the distribution of nucleii in space of radius a. Thus $\rho(a,t)\,da$ is the number of nucleii per unit volume with radii in $(a, a+da)$. Conversions $n \to a$, $\rho_n \to \rho$ are now determined. From (2.8), it follows that the change da in a when n increases by 1 is given by

$$1 \sim 4\pi\,c_2 a^2\,da. \qquad (2.11)$$

In the general continuum theory of nucleii, the concentration of precipitate particles inside a nucleus, c_2, is a function of the radius a. But for many experiments, deviations of c_2 from its equilibrium value for a planar interface are negligible, so in (2.11) any term arising from a-dependence of c_2 is dropped. ρ_n is related to $\rho(a,t)$ by

$$\rho_n \sim \rho(a,t)\,da \sim \frac{1}{4\pi c_2}\frac{\rho}{a^2}. \qquad (2.12)$$

Given any sequence h_n with continuum approximation $h(a)$,

$$D_+h_n \sim D_-h_n \sim \frac{h_a}{4\pi c_2\,a^2}. \qquad (2.13)$$

It follows from (2.12) and (2.13) that the continuum limit of (2.10) is

$$\rho_t - \frac{\partial}{\partial a}\left\{ \frac{k_d}{(4\pi c_2 a^2)^2}\left(\frac{\rho}{\tau}\frac{\partial G}{\partial a} + a^2\frac{\partial}{\partial a}\left(\frac{\rho}{a^2}\right) \right) \right\} = 0. \qquad (2.14)$$

(Here k_d is a function of a, to be specified). The constraint (2.4) can be written as

$$\gamma_\infty + c_2 \int_0^\infty \frac{4\pi}{3} a^3 \rho(a, t) \, da = c - c_1 \equiv \gamma_0. \qquad (2.15)$$

As time elapses, it will be seen that the diffusive term in Eq. (2.14) becomes negligible in comparison with the drift term. The latter yields the following equation for radius growth:

$$\dot{a} = -\frac{k_d}{(4\pi c_2 a^2)^2 \tau} \frac{\partial G}{\partial a}. \qquad (2.16)$$

We now present a macroscopic theory that gives an explicit expression for $\dot{a}$ that can be compared to experimental data. Then $k_d(a)$ can be determined and this will specify the limit of $k_{d,n}$ for large n.

2.3 Phase Equilibria of a Binary Material

2.3.1 Phase Equilibria of a Binary Material

Let us consider a medium consisting of two different particles. The more abundant type is called "matrix" , the other "precipitate" . Suppose that we have a uniform mixture at fixed temperature and pressure. Let μ be the chemical potential , i.e., the free energy cost of adding one precipitate particle to a pre-existing solution. It is a function of the number density of precipitate particles, c:

$$\mu = \mu(c). \qquad (2.17)$$

Let us now derive the relationship of the chemical potential, $\mu(c)$, to the bulk free energy density, $g(c)$. We shall add one precipitate particle to a solution of total volume V. Then the free energy changes from $g(c)V$ to $g(c)V + \mu(c)$, but the volume changes from V to $V' \equiv V + \nu(c)$, where $\nu(c)$ is the specific volume of a precipitate particle in a solution of number density c. The number density changes from c to $c' \equiv (cV + 1)/(V + \nu)$, and therefore the new free energy is also expressed as $g(c')V'$. Hence we get the identity

$$g(c)V + \mu(c) = g\left(\frac{cV + 1}{V + \nu}\right)(V + \nu). \qquad (2.18)$$

Since $cV \gg 1$, $\nu \ll V$, this identity reduces to

$$\mu(c) = g'(c) + \nu(c)\{g(c) - cg'(c)\}. \qquad (2.19)$$

One can just as easily consider the chemical potential $\bar{\mu}(c)$, which is the free energy cost of adding one matrix particle to the solution. Adding one matrix particle changes the free energy to $g(c)V + \bar{\mu}(c)$. The volume of the solution changes now to $V' = V + \bar{\nu}(c)$, where $\bar{\nu}(c)$ is the specific volume of a matrix particle in solution. The concentration c of the precipitate changes to $c' \equiv cV/(V + \bar{\nu})$, and therefore the new free energy is $g(c')V'$. Hence we get an identity analogous to (2.18),

$$g(c)V + \bar{\mu}(c) = g\left(\frac{cV}{V + \bar{\nu}}\right)(V + \bar{\nu}). \tag{2.20}$$

Again the conditions $cV \gg 1$, $\bar{\nu} \ll V$ lead to an asymptotic reduction of (2.20),

$$\bar{\mu}(c) = \bar{\nu}(c)\{g(c) - cg'(c)\}. \tag{2.21}$$

2.3.2 Phase Equilibria in a Dilute Solution

Suppose that there are two stable phases characterized by different values, c_1 and c_2 of the number density c, and that these phases occupy adjacent half spaces separated by a planar interface. Imagine that one precipitate particle is removed from phase 1 and dropped in phase 2. The free energy cost is $\mu(c_2) - \mu(c_1)$. In equilibrium, the energy cost is to be zero, therefore

$$[\mu] \equiv \mu(c_2) - \mu(c_1) = 0. \tag{2.22}$$

Similarly, the energy cost of removing a matrix particle from phase 1 and dropping it in phase 2 is also to be zero, so

$$[\bar{\mu}] = 0. \tag{2.23}$$

The possible existence of multiple phases is determined by the structure of $g(c)$. Consider a *dilute* solution with c much smaller than the total atomic density. In this case, the specific volumes ν and $\bar{\nu}$ of precipitate and matrix particles should be nearly constants independent of c: "crowding" effects should be insignificant. In this case, (2.21) and (2.23) imply

$$[g(c) - c\,g'(c)] \equiv [g - cg'] = 0. \tag{2.24}$$

Given this result, it now follows from (2.19) and (2.22) that

$$[g'] = 0.$$

Hence,

$$g'(c_1) = g'(c_2) = M \quad \text{(common value)}, \tag{2.25}$$

and

$$[g] - [c]\,M = 0. \tag{2.26}$$

Given $g(c)$ with $g''(c) < 0$ in some interval of c, and $g''(c) > 0$ outside, one discerns the standard geometrical construction of the solution to (2.25) and (2.26) for c_1 and c_2. This is depicted in Figure 2.1.

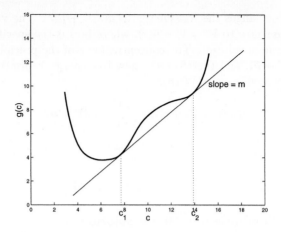

Fig. 2.1. Geometrical construction of stable phase equilibria.

2.3.3 Critical Nucleus

Consider a spherical nucleus of phase 2 surrounded by phase 1. The energy cost of adding one precipitate particle to this nucleus is

$$[\mu] + 8\pi\sigma r\, dr.$$

Here r is the initial radius of the nucleus, dr is the change in radius due to adding one precipitate particle, and σ is the surface tension. One has

$$4\pi r^2 dr = \nu$$

so the energy cost can be written as

$$[\mu] + \frac{2\sigma\nu}{r}\,.$$

For a nucleus in equilibrium, this energy cost is zero, therefore

$$[\mu] = -\frac{2\sigma\nu}{r}\,. \tag{2.27}$$

Now we add one matrix particle to the nucleus. No energy cost for this process implies

$$[\bar{\mu}] = -\frac{2\sigma\bar{\nu}}{r}\,. \tag{2.28}$$

After substituting for $\bar{\mu}$ from (2.21), this equation becomes

$$[g - c\,g'] = -\frac{2\sigma}{r}\,. \tag{2.29}$$

Now substitute (2.19) for μ and (2.29) for $2\sigma/r$ in (2.27) to get

$$[g'] = 0, \tag{2.30}$$

which is the same as in the case of a planar interface, (2.25). Given the concentration c_1 of phase 1, this equation determines the concentration c_2 inside the nucleus, and then the Gibbs-Thomson relation (2.27) determines the radius of the nucleus, r. This determination simplifies when the concentrations c_1 and c_2 are near "planar" values and γ_1 and γ_2 are deviations from the planar values. (2.30) together with (2.24) for the planar case imply

$$[g'' \gamma] = 0. \tag{2.31}$$

Let us denote the common values of $g''(c_1) \gamma_1$ and $g''(c_2) \gamma_2$ by m. The variation of $[g - c\,g']$ in (2.29) is

$$-[c\,g'' \gamma] = -[c]\,m.$$

Hence (2.29) gives

$$[c]\,m = \frac{2\sigma}{r}. \tag{2.32}$$

2.4 Macroscopic Kinetics

2.4.1 Balance Equations and Jump Conditions

Let $c(x,t)$, $\bar{c}(x,t)$ denote the macroscopic number densities of precipitate and matrix atoms. Local volume fractions of precipitate and matrix atoms are νc and $\bar{\nu}\bar{c}$, respectively. Since matrix and precipitate atoms fill space leaving no gaps, we have the *space filling condition*

$$\nu c + \bar{\nu}\bar{c} = 1. \tag{2.33}$$

In conventional kinetics, the density of precipitate is locally conserved, with a flux proportional to the gradient of the precipitate chemical potential $\mu(c)$,

$$c_t + \nabla \cdot (-\delta \nabla \mu) = 0,$$

or

$$c_t = \nabla \cdot (D \nabla c), \quad D = \delta(c)\,\mu'(c). \tag{2.34}$$

Here $\delta(c)$ is a positive mobility coefficient and $-\delta\nabla\mu = -D\nabla c$ is the flux of c. This flux is formally a diffusion with diffusion coefficient $D = \delta(c)\,\mu'(c)$. Given $\mu(c)$ as in (2.19), and provided ν and $\bar{\nu}$ *do not depend on* c (dilute solution),

$$D = \delta(c)\,\mu'(c) = \delta(1 - \nu c)\,g''(c) = \delta\bar{\nu}\,\bar{c}\,g''(c). \tag{2.35}$$

In the last equality, the space filling condition has been used to replace $1 - \nu c$ by $\bar{\nu}\,\bar{c}$. For *stable* bulk phases, D must be positive, and (2.35) then implies $g''(c) > 0$. The description of matrix transport is essentially the same. The flux of matrix concentration $\bar{c}$ is

$$-\bar{\delta}\,\nabla\bar{\mu} = -\bar{\delta}\,\bar{\mu}'\,\nabla\bar{c} = -\overline{D}\,\nabla\bar{c}$$

where $\bar{\mu}(\bar{c})$ is the chemical potential of the matrix atoms as a function of the matrix concentration $\bar{c}$, $\bar{\delta}(c)$ is the mobility of the matrix atoms, and $\overline{D}$ is the matrix diffusion coefficient given by

$$\overline{D} = \bar{\delta}\,\bar{\mu}'.$$

Let $\bar{g}(\bar{c})$ be the free energy density *as a function of* $\bar{c}$. The space filling condition and $g(c) = \bar{g}(\bar{c})$ imply that

$$\bar{\mu} = \bar{g}'(\bar{c}) + \nu\left\{\bar{g}(\bar{c}) - \bar{c}\,\bar{g}'(\bar{c})\right\},$$

which is totally symmetric to (2.19). Then the diffusion coefficient $\overline{D}$ is related to $\bar{g}(\bar{c})$ by a formula symmetric to (2.35),

$$\overline{D} = \bar{\delta}\nu c\,\bar{g}''(\bar{c}). \tag{2.36}$$

The space filling condition leads to a relation between the mobilities δ and $\bar{\delta}$. The linear combination $\nu c + \bar{\nu}\bar{c}$ is locally conserved with flux $-\nu D\nabla c - \bar{\nu}\overline{D}\nabla\bar{c}$. But $\nu c + \bar{\nu}\bar{c} \equiv 1$, so this flux is divergence free. Let C be any closed surface. Use of divergence theorem yields

$$\int_C (\nu\,Dc_n + \bar{\nu}\,\overline{D}\bar{c}_n)\,da = 0.$$

The space filling condition implies $\nu c_n = -\bar{\nu}\bar{c}_n$, so that we get $\int_C \nu\,(D - \overline{D})\,c_n\,da = 0$. This holds for all concentration fields c and closed surfaces C. Hence,

$$D = \overline{D} \tag{2.37}$$

and by (2.35) and (2.36),

$$\delta(c)\,\bar{\nu}\,\bar{c}\,g''(c) = \bar{\delta}(\bar{c})\,\nu c\,\bar{g}''(\bar{c}). \tag{2.38}$$

Now,

$$g(c) = \bar{g}(\bar{c}) = \bar{g}\left(\frac{1}{\bar{\nu}}(1 - c\nu)\right),$$

therefore, provided again that ν and $\bar{\nu}$ *do not depend on c*,

$$g''(c) = \left(\frac{\nu}{\bar{\nu}}\right)^2 \bar{g}''(\bar{c}),$$

and (2.38) becomes

$$\delta \nu \bar{c} = \bar{\delta} \bar{\nu} c. \tag{2.39}$$

This is the relation between mobilities.

The integral form of Equation (2.34) informs the upcoming discussion of boundary conditions on a phase interface. Let $R = R(t)$ be a time sequence of closed regions in which $c = c(x, t)$ is a smooth solution of (2.34). The number of precipitate particles in R is

$$N = \int_R c \, dx,$$

and the time rate of change is

$$\dot{N} = \int_R c_t \, dx + \int_{\partial R} U c \, da,$$

where U is the outward normal velocity of ∂R. Using (2.34) to substitute for c_t above, and then the divergence theorem, we obtain

$$\dot{N} = \int_{\partial R} (U c + D c_n) \, da.$$

The interpretation of this equation is clear: The influx of precipitate atoms per unit area on ∂R is

$$U c + D c_n. \tag{2.40}$$

Suppose now that there is a region R_1 of matrix surrounding a region R_2 of precipitate. The influx of precipitate atoms per unit area on interface C is given by (2.40) with c, c_n evaluated on the precipitate side of C. The outflux of precipitate atoms from the surrounding matrix into R_2 is also given by (2.40) with c, c_n evaluated on the *matrix* side of C. Since precipitate atoms do not accumulate on C to form a surface density, the following jump condition holds

$$[U c + D c_n] = 0 \quad \text{on } C. \tag{2.41}$$

In summary, conservation of precipitate is expressed by the diffusion equation (2.34) and the associated jump condition (2.41). The matrix density $\bar{c}$ satisfies the diffusion equation and jump condition with the same diffusion coefficient D. The space filling condition (2.33) is automatically upheld by this kinetics. In addition there are "thermodynamic" jump conditions

$$[g'] = 0, \tag{2.42}$$
$$[g - cg'] = -2\sigma\kappa, \tag{2.43}$$

expressing local equilibrium about the phase interface. These are in fact Equations (2.29), (2.30) with $1/r$ replaced by the mean curvature κ. Equations (2.34), (2.41), (2.42) and (2.43) constitute a free boundary problem for the evolution of precipitate concentration c and the phase interfaces.

2.4.2 Time Evolution of Gibbs Energy

The total Gibbs free energy is

$$G = G_1 + G_2 + \sigma S, \tag{2.44}$$

where

$$G_1 \equiv \int_{R_1} g \, dx, \quad G_2 \equiv \int_{R_2} g \, dx \tag{2.45}$$

are free energies of bulk matrix and precipitate phases, and S is the surface area of the phase interface, and σ the surface tension. The time evolution of G under the kinetics of the free boundary problem (2.34), (2.41), (2.42) and (2.43) is examined here. To compute the rates of change of G_1 and G_2, it is useful to formulate a transport equation for the free energy density $g(c)$:

$$g_t = g' \, c_t = g' \, \nabla \cdot (D \nabla c) = \nabla \cdot (D \nabla g) - D \, g'' \, |\nabla c|^2$$

or

$$g_t - \nabla \cdot (D \nabla g) = -D \, g'' |\nabla c|^2. \tag{2.46}$$

In stable bulk phases, $g'' > 0$, therefore the source density in (2.46) is generally negative. Let us now look at the rate of G_2.

$$\dot{G}_2 = \int_{R_2} g_t \, dx + \int_C U \, g \, da, \tag{2.47}$$

where U is the normal velocity of the phase interface, positive if outward from precipitate. Inserting g_t from (2.46) and using the divergence theorem, (2.47) becomes

$$\dot{G}_2 = \int_C (U \, g + D \, g_n)\big|_2 \, da - \int_{R_2} D g'' |\nabla c|^2 \, dx. \tag{2.48}$$

In the surface integral, the subscript 2 means evaluation on precipitate side of interface. Similarly,

$$\dot{G}_1 = -\int_C (U \, g + D \, g_n)\big|_1 \, da - \int_{R_1} D g'' |\nabla c|^2 \, dx, \tag{2.49}$$

where subscript 1 means evaluation on matrix side of interface. The rate of change of the surface energy σS in (2.44) is given by the standard formula of differential geometry,

$$\sigma \dot{S} = 2\sigma \int_C \kappa\, U\, da. \tag{2.50}$$

Adding Equations (2.48) to (2.50), we obtain the rate of change of the total free energy

$$\dot{G} = - \int_{R_1+R_2} D g'' |\nabla c|^2\, dx$$
$$+ \int_C \{[U\, g + D\, g_n] + 2\sigma\kappa U\}\, da, \tag{2.51}$$

Here the jump $[\dots]$ denotes values on precipitate side minus values on matrix side. Using the continuity condition $[g'] = 0$, it follows that

$$[Ug + Dg_n] = U\,[g] + g'\,[D\, c_n].$$

In the right hand side, g' denotes a well defined value on the phase interface. By conservation jump condition (2.41), $[D\, c_n] = -U\,[c]$, hence

$$[Ug + Dg_n] = U\,[g - cg']. \tag{2.52}$$

By the thermodynamic jump condition (2.43), $[g - cg'] = -2\sigma\kappa$, so finally,

$$[Ug + Dg_n] = -2\sigma U\kappa,$$

and the energy rate formula (2.51) reduces to

$$\dot{G} = - \int_{R_1+R_2} D g'' |\nabla c|^2\, dx. \tag{2.53}$$

The integral is negative definite. Notice that surface integral contributions to $\dot{G}$ cancel. This point is examined by direct physical argument to see what it really means.

Recall that influx of precipitate atoms into precipitate phase per unit area is

$$U\, c + D\, c_n.$$

The free energy of each precipitate atom changes by an amount $[\mu] = -2\sigma\kappa\nu$, according to (2.27), as it crosses from matrix to precipitate. Hence there is a contribution to $\dot{G}$ of

$$-2\sigma\,(U\nu c + D\nu c_n)\,\kappa \tag{2.54}$$

per unit area of phase interface due to crossing of precipitate atoms. Similarly, influx of matrix atoms into precipitate phase per unit area is

$$U\,\bar{c} + D\,\bar{c}_n$$

and change in free energy for each matrix atom crossing into precipitate is $[\bar{\mu}] = -2\sigma\kappa\bar{\nu}$, according to (2.28). Hence, crossing of matrix atoms gives another surface contribution to $\dot{G}$, of

$$-2\sigma \left(U\bar{\nu}\bar{c} + D\bar{\nu}\bar{c}_n \right) \kappa \tag{2.55}$$

per unit area. Adding (2.54) and (2.55) yields surface contribution to $\dot{G}$ due to crossing of both types of atoms,

$$-2\sigma \left\{ U \left(\nu c + \bar{\nu}\bar{c} \right) + D \left(\nu c_n + \bar{\nu}\bar{c}_n \right) \right\} \kappa = -2\sigma U\kappa$$

per unit area. Here the space-filling constraint has been used. From (2.50) it is seen that $2\sigma U\kappa$ can be identified as a rate of change of surface energy per unit area. Hence, the total rate of free energy production per unit area of phase interface is

$$-2\sigma U\kappa + 2\sigma U\kappa = 0.$$

2.5 Quasistatic Nuclei

An isolated region R_2 of precipitate phase, called a *nucleus* is assumed spherical, and concentration field c is assumed spherically symmetric. The kinetics is quasistatic if the time derivative in the diffusion equation (2.34) is negligible. Here kinetics is analyzed under the quasistatic assumption and regimes of validity are determined a posteriori by the criterion

$$\frac{Ua}{D} = \frac{a\dot{a}}{D} \ll 1. \tag{2.56}$$

Here a is the radius of the nucleus, and the normal velocity $U = \dot{a}$. a^2/D represents the characteristic time of diffusive transport in the matrix phase surrounding the nucleus. The characteristic time associated with the kinetics of the radius is $a/U = a/\dot{a}$. Kinetics is quasistatic if the time scale of the radius is much longer than the diffusion time in the surrounding matrix, as in (2.56).

Under assumptions of radial symmetry and quasistatic kinetics, the diffusion equation (2.34) reduces to

$$\partial_r (r^2 Dc_r) = 0. \tag{2.57}$$

The conservation jump condition (2.41) reads

$$\dot{a} [c] = -[Dc_r]. \tag{2.58}$$

The thermodynamic jump conditions (2.42) and (2.43) read

$$[g'] = 0, \tag{2.59}$$

$$[g - cg'] = -\frac{2\sigma}{a}. \tag{2.60}$$

Given the value c_∞ of c as $r \to \infty$, Equations (2.57) to (2.60) determine an ordinary differential equation (ODE) for the nuclear radius $a(t)$. The first integral of (2.57) is

$$r^2 Dc_r = Q, \quad \text{or} \quad Dc_r = \frac{Q}{r^2}. \tag{2.61}$$

Here Q is a function of time on $r < a$ or $r > a$, but with a possible jump at $r = a$. Regularity of c at $r = 0$ forces $Q \equiv 0$ on $r < a$. From (2.61) it is now evident that

$$[Dc_r] = -\frac{Q}{a^2},$$

where Q now refers to the value on $r > a$. The conservation jump condition (2.58) now reads

$$\dot{a}[c] = \frac{Q}{a^2}. \tag{2.62}$$

This is a differential equation for $a(t)$, once the dependence of $[c]$ and Q upon the radius are determined. The two thermodynamic jump conditions (2.59) and (2.60) determine $c(a-)$ and $c(a+)$, hence $[c]$ as a function of a. Figure 2.2 shows the graphical construction of (2.59) and (2.60).

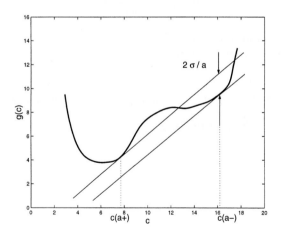

Fig. 2.2. Geometrical construction of $c(a-)$ and $c(a+)$.

To determine Q, integration of (2.61) in $r > a$ is required. Let $h(c)$ be a function such that $h'(c) = D(c)$. (2.61) now reads

$$\partial_r h(c) = \frac{Q}{r^2},$$

and integration from $r = a$ to $r = \infty$ gives

$$h(c_\infty) - h(c_+) = \frac{Q}{a} \quad \text{or} \quad Q = a\left\{h(c_\infty) - h(c_+)\right\}.$$

Here c_+ means $c(a+)$. (2.62) now reads

$$\dot{a} = \frac{h(c_\infty) - h(c_+)}{a\,[c]}. \tag{2.63}$$

Since c_+ and $[c]$ are definite functions of a as determined by the thermodynamic jump conditions (2.59) and (2.60), (2.63) is the required ODE for $a(t)$.

2.5.1 Small Supersaturation

A standard limit called *small supersaturation* is realized when the concentration is close to planar equilibrium values $c = c_1$ in matrix phase $r > a$ and $c = c_2$ in precipitate phase $r < a$. Let γ denote disturbance of c from planar equilibrium values: $\gamma = c - c_1$ in matrix phase, $\gamma = c - c_2$ in precipitate phase. For $|\gamma| \ll c_1, c_2$, (2.63) reduces to

$$\dot{a} \sim \frac{D_1\left(\gamma_\infty - \gamma_+\right)}{a\,[c]}. \tag{2.64}$$

Here $D_1 \equiv D(c_1)$ and $[c] = c_2 - c_1$. Also $\gamma_\infty = \gamma(r = \infty)$ and $\gamma_+ = \gamma(a+)$. γ_+ is determined from asymptotic limit of (2.59) and (2.60): Reduction of (2.59) is

$$[g''\gamma] = 0 \implies g''(c_1)\gamma_+ = g''(c_2)\gamma_- = m \text{ (common value)}, \tag{2.65}$$

and given m, reduction of (2.60) is

$$[c]\,m = \frac{2\sigma}{a} \quad \text{or} \quad g_1''\,[c]\,\gamma_+ = \frac{2\sigma}{a}. \tag{2.66}$$

Substituting this result for γ_+ into (2.64) gives the reduced ODE

$$\dot{a} \sim \frac{D_1}{[c]\,g_1''}\frac{1}{a}\left(g_1''\,\gamma_\infty - \frac{2\sigma}{a\,[c]}\right). \tag{2.67}$$

This equation indicates that clusters whose radii are smaller than the critical value

$$a_c \equiv \frac{2\sigma}{g_1''\,\gamma_\infty\,[c]}, \tag{2.68}$$

shrink and disappear. Supercritical clusters of radius larger than the critical radius $a = a_c$ grow steadily according to (2.68).

Is this small supersaturation kinetics consistent with the quasistatic criterion (2.56)? Natural unit of a is a_c given by (2.68), which is the standard formula for critical radius in small supersaturation limit. Given $a = O(a_c)$, an order of magnitude estimate of $\dot{a}$ based on (2.67) is

$$\dot{a} = O\left(\frac{D_1 \gamma_\infty}{a_c [c]}\right)$$

which can be rearranged as

$$\frac{a \dot{a}}{D_1} = O\left(\frac{\gamma_\infty}{[c]}\right) \tag{2.69}$$

The analysis here is based on $|\gamma| \ll c_1, c_2$. But for quasistatic kinetics, we need

$$|\gamma_\infty| \ll [c] = c_2 - c_1. \tag{2.70}$$

The reduced ODE (2.67) indicates natural units of γ, space and time. The limit (2.70) is embodied by measuring $\gamma \equiv c - c_e$ in units of ϵc_e, where $\epsilon > 0$ is a gauge parameter and limit $\epsilon \to 0+$ is considered. Given this unit of γ, the order of magnitude of a_c in (2.68) is l/ϵ, $l \equiv \sigma/([c]g_1'' c_e)$. l/ϵ is adopted as unit of length. Finally, the unit of time which gives $\dot{a}$ as in (2.69) is τ_e/ϵ^3, $\tau_e \equiv (l^2/D_1)([c]/c_e)$. This system of units is summarized in the following Table:

Variable	Unit
$\gamma \equiv c - c_e$	ϵc_e
x	$\dfrac{l}{\epsilon}, \; l \equiv \dfrac{\sigma}{[c]g_1'' c_e}$
t	$\dfrac{\tau_e}{\epsilon^3}, \; \tau_e \equiv \dfrac{l^2[c]}{D_1 c_e}$
G	$\dfrac{\sigma l^2}{\epsilon^2}$

Scaling Table for small supersaturation.

Given these units, the nondimensional version of (2.67) is

$$\dot{a} = \frac{1}{a}\left(\gamma_\infty - \frac{2}{a}\right). \tag{2.71}$$

2.5.2 Quasistatic Energetics

Given evolution of precipitate concentration field c, corresponding changes in total free energy of medium are quantified by the rate formula (2.53). Now suppose concentration field c corresponds to a nucleus undergoing quasistatic evolution as set by criterion (2.56). It seems reasonable to approximate the integral in (2.53) using the quasistatic approximation to c which satisfies Equations (2.57) to (2.60). The rate formula (2.53) reduces to

$$\dot{G} = -\int_0^\infty Dg'' c_r^2 4\pi r^2 \, dr.$$

¿From (2.61), $r^2 Dc_r = Q$ for $r > a$, and $r^2 Dc_r = 0$ for $r < a$, so this reduces to

$$\dot{G} = -4\pi Q \int_{a+}^\infty g'' c_r \, dr = -4\pi Q \left\{ g'(c_\infty) - g'(c_+) \right\}.$$

Substituting for Q from (2.62),

$$\dot{G} = -[c] \left\{ g'(c_\infty) - g'(c_+) \right\} (4\pi a^2 \dot{a}). \tag{2.72}$$

In the right hand side, $[c]$ and $g'(c_+)$ are definite functions of radius a, determined by thermodynamic jump conditions (2.59) and (2.60). Hence, in quasistatic evolution, G is effectively a function of the radius a. ¿From (2.72), it is seen that $G(a)$ obeys the differential relation

$$dG = -[c] \left\{ g'(c_\infty) - g'(c_+) \right\} (4\pi a^2 da). \tag{2.73}$$

In conventional descriptions of nucleus energetics, the Gibbs free energy cost of a nucleus regarded as a function of the instantaneous radius a, independent of past history. The specific formula is

$$G = 4\pi \sigma a^2 - g_b \frac{4\pi}{3} a^3 \tag{2.74}$$

or in differential form,

$$dG = 8\pi \sigma a \, da - g_b 4\pi a^2 \, da. \tag{2.75}$$

Here σ is surface tension and $4\pi \sigma a^2$ represents surface energy. g_b is a constant with units of energy density, sometimes called "chemical driving force". It is the free energy released per unit volume of increase of precipitate phase. In its present form (2.73) does not have obvious correspondence to (2.75). A correspondence is brought out by reformulation of (2.73) with help of thermodynamic jump conditions (2.59) and (2.60). By (2.59), it follows that $g'(c_-) = g'(c_+) = $ common value m, and hence (2.60) gives

$$[g] - [c] m = -\frac{2\sigma}{a},$$

or, equivalently,

$$[c] g'(c_+) = [g] + \frac{2\sigma}{a}.$$

Hence (2.73) becomes

$$dG = 8\pi \sigma \, a \, da + \{ [g] - [c] g'(c_\infty) \} (4\pi a^2 da). \tag{2.76}$$

The first term on RHS is differential of surface energy, same as in (2.75). An apparent correspondence between (2.75) and (2.76) is completed by identifying the chemical driving force g_b,

$$g_b \equiv -[g] + [c]\, g'(c_\infty). \tag{2.77}$$

Recalling that $[g]$ and $[c]$ are functions of radius a as determined by thermodynamic jump conditions, it is evident that the chemical driving force (2.77) is generally a function of cluster radius, and *not a constant* as in conventional nucleus energetics.

We can now write Eq. (2.62) for $\dot{a}$ in terms of $G_a = -\{g'(c_\infty) - g'(c_+)\}\, 4\pi a^2[c]$ given by (2.73). The result is

$$\dot{a} = -\frac{\int_{c_+}^{c_\infty} D(c)\, dc}{g'(c_\infty) - g'(c_+)} \frac{G_a}{4\pi[c]^2 a^3}. \tag{2.78}$$

2.5.3 Energetics at Small Supersaturation

In the small supersaturation limit, $\int_{c_+}^{c_\infty} D(c)\, dc \sim D(c_1)\,(c_+ - c_\infty) = D_1\,(\gamma_+ - \gamma_\infty)$, $g'(c_\infty) - g'(c_+) \sim g_1''\,(\gamma_+ - \gamma_\infty)$ and (2.78) reduces to

$$\dot{a} = -\frac{G_a D_1}{4\pi g_1''[c]^2 a^3}. \tag{2.79}$$

Inserting the approximate chemical driving force $g_b = [c]g'(c_\infty) - [g] \sim g_1''\gamma_\infty$ from (2.77) into the free energy (2.74), we obtain

$$G \sim 4\pi\sigma a^2 - [c]\, g_1'' \gamma_\infty \left(\frac{4\pi}{3} a^3 \right). \tag{2.80}$$

Notice that the asymptotic chemical driving force,

$$g_b \sim [c]\, g_1'' \gamma_\infty, \tag{2.81}$$

is in fact a constant independent of radius a, as in the conventional wisdom. An alternative expression for g_b is useful: In certain experiments surface tension σ and critical radius a_c are measured observables, so it is convenient to represent g_b in terms of σ and a_c,

$$g_b = \frac{2\sigma}{a_c}. \tag{2.82}$$

This formula follows from the condition $G'(a_c) = 0$. The natural unit of free energy is σl^2. This is entered into the last column of the scaling table. Then the dimensionless version of the free energy formula (2.80) is

$$G = 4\pi\, a^2 - \gamma_\infty \frac{4\pi}{3} a^3. \tag{2.83}$$

2.5.4 Identification of Rate Constant in BD Kinetics

We can now compare Eq. (2.16) for the growth of a (large) cluster radius in BD kinetics with the corresponding equations (2.78) and (2.79) obtained from our macroscopic description. We find

$$k_d = \frac{\tau \int_{c_+}^{c_\infty} D(c)dc}{g'(c_\infty) - g'(c_+)} \left(\frac{c_2}{[c]}\right)^2 4\pi a \tag{2.84}$$

$$\sim \frac{D_1\tau}{g_1''} \left(\frac{c_2}{[c]}\right)^2 4\pi a, \tag{2.85}$$

in the small supersaturation limit.

With this determination of k_d, the continuum limit equation (2.14) of cluster kinetics is completely specified. It is easy to check that the corresponding microscopic rate constant determined by Penrose et al [11, 12] is also proportional to a for large clusters. One might wonder about the micromolecular basis of (2.85). While that requires further work, here is a curious observation which might become relevant to this question: Recall the *mobility* δ defined in the formulation of the macroscopic transport theory of Section 2.4. It is related to the diffusion D by (2.35),

$$D = \delta \left(1 - \nu c\right) g''.$$

Hence the ratio D_1/g_1'' in (2.85) is given by

$$\frac{D_1}{g_1''} = \delta_1 \left(1 - \nu c_1\right). \tag{2.86}$$

In the next Section, we show that the volume fraction νc_1 of precipitate in matrix phase is small, $\nu c_1 \sim (0.04521 \text{ nm}^3) (2.24 \text{ nm}^{-3}) \approx 0.10$, for coarsening experimental data in binary alloys [3]. Then $D_1/g_1'' \approx \delta_1$ and formula (2.85) for k_d reduces to

$$k_d \approx \delta_1\tau \left(\frac{c_2}{[c]}\right)^2 4\pi a.$$

Let us notice that this rate constant k_d is linear in the cluster radius so that it scales as $n^{\frac{1}{3}}$ with cluster size. This contrasts with the usual Turnbull-Fisher rate constant that scales as $n^{\frac{2}{3}}$ [6], but it agrees with the microscopic considerations of Penrose et al [11, 12]. The scaling $n^{\frac{1}{3}}$ has been shown to yield the Lifshitz-Slyozov distribution function for cluster radii [10]. The latter is a roughly adequate description of coarsening [3, 4].

2.6 Material and Energy Parameters of Kinetic Theory Determined from Xiao-Haasen Data

2.6.1 Small Supersaturation

The nucleation in Xiao-Haasen's (XH) paper [3] takes place under low supersaturation: The initial sample has uniform composition, with mole fraction $\chi \equiv 0.12$, or 12 % of Al in a Ni matrix. *Equilibrium* mole fraction of Al in matrix phase at annealing temperature of 773 K is $\chi_1 = 0.101$, or 10.1 %. Equilibrium mole fraction of Al in precipitate phase is $\chi_2 = 0.230$, or 23 %. Hence, *supersaturation* as a function of equilibrium concentration has initial value

$$\frac{\chi - \chi_1}{\chi_1} \approx \frac{0.120 - 0.101}{0.101} \approx 0.19.$$

Mole fractions are converted into number densities: XH report a molar volume of precipitate phase $V_m \approx 27.16 \times 10^{-6}$ m^3. Conversion to an atomic volume by Avogadro's number gives

$$\nu_m \equiv \frac{V_m}{N_A} \approx 4.51 \times 10^{-29}\,\mathrm{m}^3$$

or $\nu_m \approx 0.0451$ nm^3.

XH also report a lattice constant of $a \approx 0.356$ nm for the precipitate phase, and atomic volume corresponding to this lattice constant is $\nu_m = a^3 \approx 0.0451$ nm^3. It is clear that the molar volume V_m was derived from the lattice constant. A lattice constant for the matrix phase is not reported explicitly, so it is presumably close to the value $a \approx 0.356$ nm of the precipitate phase. It seems there is an implicit assumption: Local structure of alloy in a lattice, with sites that can be occupied by Al or Ni atoms. In this case, atomic volumes of Al and Ni are de-facto the same, i.e.,

$$\nu_m = \bar{\nu}_m \approx 0.0451\,\mathrm{nm}^3.$$

Now number densities of Al and Ni easily follow. For instance, c_1, the equilibrium number density of Al in matrix phase is

$$c_1 = \frac{\chi_1}{\nu} \approx \frac{0.101}{0.0451}\,\mathrm{nm}^{-3} \approx 2.24\,\mathrm{nm}^{-3}.$$

Table 1 gives initial concentration of Al in matrix phase, and equilibrium concentrations c_1 and c_2 of Al in matrix and precipitate phases.

c	c_1	c_2
2.66	2.24	5.10

Table 1: Number densities (nm^{-3}).

2.6.2 Nucleation Energetics

In XH, the nucleus energy takes the classic form

$$G = 4\pi\sigma a^2 - g_b \left(\frac{4\pi}{3} a^3\right). \tag{2.87}$$

Here σ is surface tension. A value $\sigma \approx 0.014$ J m^{-2} = 1.4×10^{-20} J nm^{-2} is deduced from interpreting coarsening data with the Lifshitz-Slyozov (LS) theory. XH deals with chemical driving force g_b in two ways:

(i) *"Experimental"*. The distribution of nuclei in the space of their radii goes through a transient phase with two peaks, separated by a local minimum at about 1.2 nm. XH conjecture that the initial radius a_c is in fact this 1.2 nm. Estimate of g_b now follows from (2.81),

$$g_b \sim \frac{2\sigma}{a_c} \approx 2.33 \times 10^{-20} \text{ J nm}^{-3}.$$

Given an experimental estimate of the critical radius, $a_c \approx 1.2$ nm at outset, one can estimate the number of atoms in the critical nucleus, both Al and entrained Ni:

$$\frac{1}{\nu} \frac{4\pi}{3} a_c^3 \approx 160.$$

Of these, 23% are Al, so there are

$$n_c = 0.23 \times 160 = 37$$

Al atoms in the critical nucleus. It seems that the critical nucleus is "just big enough" so energetics based on continuum theory applies.

(ii) *"Theory"*. In standard theories, g_b is computed from both thermodynamic properties of precipitate and bulk phases. As such, it comes out as a constant independent of nucleus radius a. These derivations do not face up to the fine points of the real situation, summarized in the formula (2.77) for g_b. So our approach is to stick with the determination of g_b based on σ and a_c,

$$g_b \approx \frac{2\sigma}{a_c} \approx 2.33 \times 10^{-20} \text{ J nm}^{-3},$$

and then see what can be said about $g(c)$. In (2.81) one can determine g_1'' because all the other quantities are known. In fact, one gets

$$g_1'' = \frac{g_b}{[c]\gamma_\infty} \approx \frac{2.33 \times 10^{-20} \text{ J nm}^{-3}}{(5.10 - 2.24) \text{ nm}^{-3} (2.66 - 2.24) \text{ nm}^{-3}} \approx 1.94 \times 10^{-20} \text{ J nm}^{-3}.$$

The annealing temperature of 773 K defines a basic unit of energy,

$$\tau = (773 \text{ K}) (1.38 \times 10^{-23} \text{ J/K}) \approx 1.07 \times 10^{-20} \text{ J}.$$

One now has

$$\frac{g_1''}{\tau} \approx 1.81\,\text{nm}^3.$$

This is just one number imposed upon free energy function $g(c)$ by the XH data, but it is sufficient to establish the

Nonideal character of Al solution in matrix phase.

Suppose the solution is ideal. Then the chemical potential of an Al particle in matrix phase is given by

$$\mu(c) = \mu_1 + \tau \ln \frac{c}{c_1}, \tag{2.88}$$

where μ_1 is the chemical potential when $c = c_1$ is the planar solvability. Now the relation between $\mu(c)$ and $g(c)$ is given by (2.19), which is repeated here for easy reference,

$$\mu(c) = g'(c) + \nu\,(g - c\,g'). \tag{2.89}$$

¿From this equation is evident that $g''(c)$ gives information about $\mu'(c)$. In fact, differentiation of (2.89) yields

$$\mu'(c) = (1 - \nu c)\,g''(c) \implies \mu_1' = (1 - \nu c_1)\,g_1''. \tag{2.90}$$

Numerical value of μ_1'/τ based upon previous value of g_1'' turns out to be

$$\frac{\mu_1'}{\tau} \approx \{1 - (0.0451\,\text{nm}^3)\,(2.24\,\text{nm}^{-3})\}\,(1.81\,\text{nm}^3) \approx 1.63\,\text{nm}^3.$$

If the ideal solution formula (2.88) were correct, one would get

$$\frac{\mu_1'}{\tau} = \frac{1}{c_1} \approx \frac{1}{2.24\,\text{nm}^{-3}} \approx 0.45\,\text{nm}^3,$$

which is $1/274$ of value that follows from XH parameters. That the solution of Al in Ni phase is not ideal was already known to XH. They in fact considered that our chemical driving force g_b is sum of two terms: (i) a chemical driving force estimated from the activity of Al component at the concentrations χ and χ_1, and (ii) and the elastic strain energy per unit volume. With the corresponding expressions, they obtained a value for the critical radius, $a_c \approx 1.7$ nm, which is not too far from the experimental value, $a_c \approx 1.2$ nm [3].

The experimentally derived values of surface tension σ and chemical driving force g_b in (2.87) set important parameters for macroscopic nucleation theory, namely: Energy barrier for nucleation, and typical free energy cost to add one Al particle to a nucleus.

Energy barrier is given by

$$G_{\mathrm{nuc}} = G(a_c) = \frac{4\pi}{3}\,\sigma a_c^2 \approx \frac{4\pi}{3}\,(1.4\times10^{-20}\,\mathrm{J}\ \mathrm{nm}^{-2})\,(1.2\,\mathrm{nm})^2 \approx 8.4\times10^{-20}\,\mathrm{J}..$$

Energy barrier in units of thermal energy is

$$\frac{G_{\mathrm{nuc}}}{\tau} \approx 7.8.$$

A reasonable looking number. Notice that exponential

$$e^{-\frac{G_{\mathrm{nuc}}}{\tau}} \approx 3.7 \times 10^{-4},$$

which appears in the nucleation rate is not too small. This makes anthropomorphic sense: In XH experiment, nucleation kinetics unfolds in hours and days time scales, not unduly taxing to humans. Evidently, the annealing temperature is tuned so as to achieve a "reasonable" nucleation rate.

Free energy cost to add one particle.
The number n of Al particles in nucleus is related to radius a by

$$n = \frac{4\pi}{3}\,c_2 a^3 \implies a = \left(\frac{3n}{4\pi c_2}\right)^{\frac{1}{3}}. \tag{2.91}$$

Substituting (2.91) for a in (2.87) gives nucleus energy as a function of n,

$$G_n = (36\pi)^{\frac{1}{3}}\,\sigma c_2^{-\frac{2}{3}}\,n^{\frac{2}{3}} - g_b\,c_2^{-1}\,n. \tag{2.92}$$

Free energy cost to add one particle to nucleus of n particles, in units of thermal energy τ, is

$$\frac{\mu_n}{\tau} \equiv \frac{G_{n+1} - G_n}{\tau} = (36\pi)^{\frac{1}{3}}\,\sigma c_2^{-\frac{2}{3}}\tau^{-1}\left\{(n+1)^{\frac{2}{3}} - n^{\frac{2}{3}}\right\} - g_b\,c_2^{-1}\,\tau^{-1}.$$

Since this formula is based on continuum theory, its validity requires $n \gg 1$, in which case it reduces to

$$\frac{\mu_n}{\tau} \sim \frac{2}{3}(36\pi)^{\frac{1}{3}}\,\sigma c_2^{-\frac{2}{3}}\,\tau^{-1}n^{-\frac{1}{3}} - g_b\,c_2^{-1}\,\tau^{-1}. \tag{2.93}$$

Substituting XH parameter values in RHS,

$$\frac{\mu_n}{\tau} \sim 1.43\,n^{-\frac{1}{3}} - 0.42. \tag{2.94}$$

In the limit $n \to \infty$, we get $|\mu_n/\tau| \sim 0.42$. While less than 1, one would not call this value "small compared to one". For $n = 37$, corresponding to a critical nucleus, of course one gets $\mu_n/\tau = 0$. Hence there will be a range of n about $n = n_c = 37$ in which $|\mu_n/\tau| \ll 1$. In particular, $|\mu_n/\tau| < 0.2$ in the rather generous interval $12 < n < 300$. In this range of n, asymptotic reduction of discrete kinetic models such as Becker-Döring to a Smoluchowski partial differential equation (PDE) should be reasonable.

2.7 Discussion

In coarsening experiments, one starts from a situation of equilibrium at high temperature in which most clusters are monomers. Then the temperature is lowered to a value below the critical temperature, and kept there. Clusters are nucleated and grow, supersaturation changes with time so that nucleation of new clusters becomes unlikely, and the coarsening of clusters proceeds. As explained by Penrose et al [11, 12], this process is reasonably well described by the Becker-Döring model (better than by the Lifshitz-Slyozov distribution function), provided the volume fraction of precipitate is small. Let us describe the nucleation and coarsening processes in typical experiments such as XH and which parts thereof are mathematically understood.

The nucleation process described by the BD equations starts at $t = 0$ with some initial value of $\rho_1 = c_\infty$ and no supercritical clusters. According to the XH data, the energy barrier corresponding to the initial value of the critical nucleus is relatively high, $G_{\mathrm{nuc}}/\tau \approx 7.8$, so that we may consider the clusters below critical size ($n < n_c$) to be in a quasistationary state. The flux across the energy barrier is then uniform and it supplies the source for coarsening of clusters larger than the critical size. As explained in Section 2.6, there is a range of sizes (about the critical size) for which we may approximate the discrete BD kinetics by a continuum Smoluchowski equation for the distribution function ρ. The latter will describe the coarsening process and it should be approximately solved with a boundary condition obtained by matching to the solution of the BD equations for $n < n_c$. For $t > 0$, supercritical clusters are created at the rate j per unit volume given in Eq. (2.100) below, and ρ_1 starts to decrease. A small change of ρ_1, $O(1/n_c) = O(\epsilon^3 l^3/c_2)$, induces an $O(1)$ relative change of j. There is a transient situation during which ρ becomes a bimodal distribution function with peaks at sub and supercritical sizes. As time evolves, the supercritical peak increases at the expense of the subcritical peak, which disappears given enough time. Then the resulting unimodal distribution evolves toward a function with the LS scaling.

Currently it is known that the LS distribution function [9] is a solution of the Smoluchowski equation for a very special boundary condition at small cluster size [10]. Although the stability properties of the LS distribution function are not completely elucidated, it seems clear that the Smoluchowski equation may have other stable solutions that may match the quasistationary distribution at small cluster sizes. The appropriate solution of the Smoluchowski equation should then describe the transient stage of coarsening. As the time advances, the peak of the distribution function at subcritical sizes decreases and disappears while the peak at supercritical sizes takes over. The latter should have the LS scaling to explain experimental [3] and numerical data [11, 12]. To carry out an asymptotic analysis of nucleation and coarsening providing the same qualitative description sketched here is a challenging future task.

2.8 Acknowledgements

The present work was financed through the Spanish DGES grant PB98-0142-C04-01 and carried out during J. Neu's sabbatical stay at Universidad Carlos III supported by the Spanish Ministry of Education.

2.9 Appendix: BD Kinetics for $n < n_c$

The quasistationary state is a solution of the BD equations characterized by uniform flux,

$$j_n = k_{d,n+1} \left\{ e^{-\frac{G_{n+1}-G_n}{\tau}} \rho_n - \rho_{n+1} \right\} \equiv j, \qquad (2.95)$$

for $n < n_c = c_2 4\pi a_c^3/3$. At high temperature, before the experiment starts, we have the following equilibrium solution

$$\rho_{eq,n} = \frac{c\, e^{-\frac{G_n}{\tau}}}{\sum_{l=1}^{\infty} l\, e^{-\frac{G_l}{\tau}}} \sim c\, e^{-\frac{G_n-G_1}{\tau}}. \qquad (2.96)$$

To write the above approximation, we have assumed that $G_2/\tau \gg G_1/\tau \gg 1$ and that G_n increases with n. After nucleation and coarsening start, we shall assume that ρ_n is close to its equilibrium value [given by the approximate expression (2.96) at the correct temperature τ], as $n \ll n_c$. Thus $\rho_n = O(c\, e^{-(G_n-G_1)/\tau})$ if $n < n_c$, and much smaller than this order if $n > n_c$. This means that $e^{\frac{G_n}{\tau}} \rho_n/c = O(e^{G_1/\tau})$ if $n < n_c$, and that $e^{\frac{G_n}{\tau}} \rho_n/c = o(e^{G_1/\tau})$ if $n \gg n_c$.

Equation (2.95) can be written as

$$e^{\frac{G_{n+1}}{\tau}} \rho_{n+1} - e^{\frac{G_n}{\tau}} \rho_n = -\frac{j}{k_{d,n+1}} e^{\frac{G_{n+1}}{\tau}},$$

and therefore easily integrated under the condition $e^{\frac{G_n}{\tau}} \rho_n \to 0$ as $n \to \infty$:

$$e^{\frac{G_n}{\tau}} \rho_n = j \sum_{l=n}^{\infty} \frac{e^{\frac{G_{l+1}}{\tau}}}{k_{d,l+1}}. \qquad (2.97)$$

The terms in this sum are largest for $l \sim n_c$, at which G_l is maximum. For such integers, the continuum approximation holds, and we can write

$$\frac{e^{\frac{G_{l+1}}{\tau}}}{k_{d,l+1}} \sim \frac{e^{\frac{G_{n_c}}{\tau}}}{k_{d,n_c}} \sum_{l=n}^{\infty} e^{-\frac{4\pi\sigma}{\tau}(a_l-a_c)^2}. \qquad (2.98)$$

We have used $G_{l+1} - G_{n_c} \sim -4\pi\sigma(a_l - a_c)^2$, for $l+1$ close to n_c. We now approximate $a_l = a_c + x\sqrt{\sigma/\tau}$ in (2.97), and $1 \sim 4\pi c_2 a_c^2 da_l = 4\pi c_2 a_c^2 \sqrt{\tau/\sigma}\, dx$, so that (2.97) becomes

$$e^{\frac{G_n}{\tau}} \rho_n \sim \frac{j}{D_1} \frac{[c]}{c_2} \frac{\sqrt{\frac{\sigma}{\tau}}}{\gamma_\infty} e^{\frac{G_{n_c}}{\tau}} 2 \int_{\sqrt{\frac{\sigma}{\tau}}(a-a_c)}^{\infty} e^{-4\pi x^2} dx. \qquad (2.99)$$

The equilibrium solution of the BD equations is (2.96). If we impose that $\rho_n \sim \rho_{eq,n}$ as $n \ll n_c$, (2.96) and (2.99) yield

$$j \sim \sqrt{\frac{\tau}{\sigma}} \frac{cD_1\gamma_\infty c_2}{[c]} e^{-\frac{G_{n_c}-G_1}{\tau}}. \qquad (2.100)$$

This constant flux is exponentially small because $(G_{n_c} - G_1)/\tau \sim G(a_c)/\tau \gg 1$. Notice that it is also proportional to the supersaturation γ_∞. It is clear that a small change in the supersaturation, $\delta\gamma = O(\gamma_\infty/n_c)$, produces an $O(1)$ change in n_c and in G_{n_c}, $\delta n_c = -3n_c\delta\gamma/\gamma_\infty$ and $\delta G_{n_c} = -2G_{n_c}\delta\gamma/\gamma_\infty$, and hence a significant relative change of j in (2.100):

$$\frac{\delta j}{j} \sim \exp\left(\frac{g_1''[c]n_c}{\tau}\delta\gamma\right). \qquad (2.101)$$

Notice that, in the continuum limit, the flux j_n becomes

$$\begin{aligned}
j_n &\sim -\frac{D_1\tau}{g_1''[c]^2} \frac{e^{-\frac{G(a)}{\tau}}}{4\pi a} \frac{\partial}{\partial a}\left(e^{\frac{G(a)}{\tau}} \frac{\rho}{a^2}\right) \\
&= \frac{D_1\tau}{g_1''[c]^2} \frac{1}{4\pi a}\left(\frac{\rho}{\tau a^2}\frac{\partial G}{\partial a} + \frac{\partial}{\partial a}\left(\frac{\rho}{a^2}\right)\right). \qquad (2.102)
\end{aligned}$$

The relation between drift and diffusion coefficients here is G_a/τ in agreement with the formulas provided by Nonequilibrium Thermodynamics; see Ref. [6].

References

1. A. Ziabicki, *Generalized theory of nucleation kinetics. I. General formulations.* J. Chem. Phys. **48**, 4368-4374 (1968).
2. I. V. Markov, *Crystal growth for beginners.* (World Sci., Singapore 1995).
3. S. Q. Xiao and P. Haasen, *HREM investigation of homogeneous decomposition in a Ni-12 at.% Al alloy.* Acta metall. mater. **39**, 651-659 (1991).
4. U. Gasser, E.R. Weeks, A. Schofield, P.N. Pursey and D. A. Weitz, *Real-space imaging of nucleation and growth in colloidal crystallization.* Science **292**, 258-262 (2001).
5. I. Pagonabarraga, A. Pérez-Madrid and J. M. Rubí, *Fluctuating hydrodynamics approach to chemical reactions.* Physica A **237**, 205-219 (1997).
6. D. Reguera, J. M. Rubí and L. L. Bonilla, contribution to this book.
7. A. Ziabicki, contribution to this book.
8. J. Lebowitz and O. Penrose, *Cluster and Percolation inequalities for lattice systems with interactions.* J. Stat. Phys. **16**, 321-337 (1977).
9. I. M. Lifshitz and V. V. Slyozov, *The kinetics of precipitation from supersaturated solid solutions.* J. Phys. Chem. Solids **19**, 35-50 (1961).

10. J. J. L. Velázquez, *The Becker-Döring equations and the Lifshitz-Slyozov theory of coarsening.* J. Stat. Phys. **92**, 195-236 (1998).
11. O. Penrose and A. Buhagiar, *Kinetics of nucleation in a lattice gas model: microscopic theory and simulation compared.* J. Stat. Phys. **30**, 219-241 (1983).
12. O. Penrose, J. L. Lebowitz, J. Marro, M. Kalos and J. Tobochnik , *Kinetics of a first order phase transition: computer simulations and theory.* J. Stat. Phys. **34**, 399-426 (1984).

3 Multidimensional Theory of Crystal Nucleation

Andrzej Ziabicki

Polish Academy of Sciences, Institute of Fundamental Technological Research
21 Swietokrzyska St.,00-049 Warsaw, Poland.

3.1 Introduction

The concept of nucleation as a first step in phase transitions, was originally developed in nineteen twenties and thirties by Volmer and Weber [1], Kaischew and Stransky [2], Becker and Döring [3], Zeldowich [4], Frenkel [5], Turnbull and Fisher [6], and others. With some modifications, the original theory has been applied to crystallization of polymers by Lauritzen and Hoffman [7], Mandelkern [8], Frank and Tosi [9]. The original theory concerned one-dimensional growth of molecular (atomic) clusters and could not explain more complex processes, like crystallization in oriented systems, effects of potential fields, anisotropic growth, etc.

The classical, one-dimensional nucleation theory in its original form, appears inadequate for description of polymer crystallization. Large size of crystallizing molecules and numerous internal degrees of freedom make growth of a crystalline cluster dependent not only on its *volume* (as assumed in the classical theory), but also on *shape*, and *internal structure*. Characteristics of cluster *orientation* and/or *position* appear as additional degrees of freedom when crystallization takes place in an external potential field. Long relaxation times make nucleation slow, which accounts for effects of history of external conditions (temperature, pressure, etc).

The present author formulated a more general model of nucleation, including effects of cluster shape, orientation, position and internal structure [10]. The model is a natural extension of the classical theory. Although no specific materials or transitions are implied, motivation for the new model was associated with orientation and stress controlled crystallization of polymers. The model predicts new physical effects, some of which have already been observed (selective crystallization). However, also the new model does not pretend to solve all problems encountered in polymer crystallization, in particular complex crystallization morphology.

In this review, we will start with brief presentation of the classical, one-dimensional nucleation theory. Modifications of the classical theory will be discussed, followed by presentation of the generalized, multidimensional model. Example applications will conclude the review.

3.2 One-dimensional Theory of Nucleation

Fundamental processes of the classical nucleation theory are reversible, bi-molecular reactions of aggregation/dissociation. In a general case, a $(g - k)$-element cluster of the new phase reacts with a k-element cluster forming a cluster of size g

$$\beta_{g-k} + \beta_k \overset{k^+}{\underset{k^-}{\leftrightarrow}} \beta_g \tag{3.1}$$

k^+ and k^- denote reaction rate coefficients, respectively, for addition and dissociation. Considering concentration of clusters small, compared to unaggregated kinetic units, β_1

$$[\beta_k] \ll [\beta_1] ; \quad k > 1 \tag{3.2}$$

the theory usually considers only reactions involving single kinetic elements, i.e. those with $k = 1$

$$\beta_{g-1} + \beta_1 \overset{k^+}{\underset{k^-}{\leftrightarrow}} \beta_g \tag{3.3}$$

In the classical theory, cluster is characterized by a single variable - volume v - approximately proportional to the number of single kinetic elements contained in the cluster, g

$$v \cong g \cdot v_0 \tag{3.4}$$

v_0 is molecular volume of a single kinetic element, β_1 . Free energy of aggregation, ΔF , includes a volume contribution, proportional to v, and surface contribution, proportional to $v^{2/3}$

$$\Delta F(g) = (v - v_0)\Delta f + (v^{2/3} - v_0^{2/3})C\sigma \tag{3.5}$$

Δf is volume free energy density controlled by temperature. In first approximation, Δf is proportional to undercooling, and changes sign in the equilibrium melting temperature, T_m

$$\Delta f = \Delta h(T - T_m)/T_m \tag{3.6}$$

Δh denotes density of the heat of melting. Surface contribution, proportional to surface energy density (interface tension), σ , is always positive. The constant C depends on cluster shape: for spherical clusters $C = (36\pi)^{1/3}$, for cubic clusters $C = 6$.

$$\Delta F(v) = (v - v_0)\Delta f + (v^{2/3} - v_0^{2/3})C\sigma \tag{3.7}$$

Critical crystallization (melting) temperature T_m is obtained from the free energy density in the thermodynamic limit $(v \to \infty)$

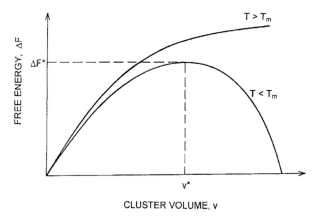

Fig. 3.1. Free energy of aggregation, ΔF, as a function of cluster volume, v (schematic).

$$\lim_{v \to \infty} \frac{\Delta F(v, T)}{v} = 0 \quad \Longleftrightarrow \quad \Delta f(T) = 0 \qquad (3.8)$$

Above T_m, free energy of aggregation, $\Delta F(v)$, is positive and monotonically increases with cluster volume (Fig. 1). Phase transition does not take place, and equilibrium, Boltzmann-type cluster distribution is obtained. Below T_m, free energy of small clusters is positive, passes through a maximum and then decreases to minus infinity. Critical cluster size (volume, v^* or number of units, g^*) corresponding to maximum free energy plays a crucial role. Small, subcritical clusters (*embryos*) are unstable as their growth requires increase of free energy. On the other hand, free energy of large, supercritical clusters (*nuclei*) decreases with increasing volume, making their growth spontaneous. In temperatures higher than T_m, free energy monotonically increases with volume, and all clusters, however large, are unstable.

The classical theory defines nucleation in a supercooled system ($T < T_m$) as a *motion of clusters in the space of volumes, v*. Unstable, subcritical clusters passing into the stability region must overcome the potential barrier, ΔF^* (cf. Figure 3.1). The process of cluster growth is thermally activated, and therefore is also called *thermal nucleation.* To describe kinetics of thermal nucleation, Turnbull and Fisher [6] introduced *continuous, one-dimensional space of cluster volumes, $v \in \{v_0, \infty\}$,* replacing discrete space of cluster sizes, $g \in \{1, \infty\}$, corresponding to equations (3). The continualized model defines normalized density of clusters

$$w(v, t) = \frac{1}{N_0} \frac{dN(v)}{dv} \ ; \quad \int w(v, t) dv = 1 \qquad (3.9)$$

N_0 is total number of kinetic units in the system. Cluster density, w, satisfies the condition of conservation (continuity)

$$\frac{\partial w}{\partial t} + \frac{\partial}{\partial v} j(v) = 0 \tag{3.10}$$

The flux j (flux of growth in the space of volumes) includes a diffusional and a thermo- dynamic contribution. The latter is associated with gradient of the free energy of aggregation with respect to cluster volume

$$j(v) = -D_{gr} \left[\frac{\partial w}{\partial v} + \frac{w}{kT} \frac{\partial \Delta F}{\partial v} \right] \tag{3.11}$$

The "growth diffusion coefficient", D_{gr}, is associated with molecular mobility of aggregating molecules included in the reaction rate coefficients, k^+, k^-.

Critical cluster volume, v^*, is obtained from the condition of maximum free energy

$$\frac{\partial \Delta F}{\partial v} = 0 \tag{3.12}$$

which yields

$$v^* = \frac{-8C^3 \sigma^3}{27(\Delta f)^3} \tag{3.13}$$

The sign $(-)$ in Eq. (3.13) results from the fact, that denominator - volume density of the free energy - is negative. Critical cluster size, v^*, determines maximum free energy of cluster formation (potential barrier for nucleation)

$$\Delta F^* = \Delta F(v = v^*) = \frac{4C^3 \sigma^3}{27(\Delta f)^2} \tag{3.14}$$

and thermal nucleation rate, i.e. flux of clusters entering the stability region reads

$$\dot{N}_{th} = N_0 \cdot j(v^*) = -D_{gr} \frac{\partial w}{\partial v} \bigg|_{v=v^*} \tag{3.15}$$

Assumption of zero flux $(j = 0)$ in Eq. (3.11) yields *equilibrium cluster distribution*, realizable above the melting temperature, T_m

$$w_{eq}(v) = w_1 e^{-\Delta F/kT} \tag{3.16}$$

In the conditions of undercooling $(T < T_m)$, free energy of aggregation, ΔF, and the distribution (16) become divergent. Instead of *equilibrium-*, one can ask about *steady-state distribution* characterized by constant, non-zero flux $(j > 0)$

$$w_{st}(v) = w_1 \frac{\int\limits_v^V \dfrac{e^{\Delta F/kT}}{D_{gr}(v)} dv}{\int\limits_{v_0}^V \dfrac{e^{\Delta F/kT}}{D_{gr}(v)} dv} e^{-\Delta F/kT} \tag{3.17}$$

The distributions (16) and (17) satisfy the boundary condition

$$w(v = v_0) = w_1 \tag{3.18}$$

Existence of the steady state requires that cluster density disappears at some finite limit, V , much larger than the critical cluster volume, v^*

$$w(v = V \gg v^*) = 0 \tag{3.19}$$

Steady-state flux assumes the form

$$j_{st} = -D_{gr} \frac{\partial w}{\partial v}\bigg|_{v=v^*} = \frac{w_1}{\int\limits_{v_0}^{V} \dfrac{e^{\Delta F/kT}}{D_{gr}(v)} dv} \tag{3.20}$$

The integral in Eq.(3.20) is often approximated by its largest term which reduces steady-state nucleation rate to

$$\dot{N}_{st} = N_0 \cdot j_{st}(v^*) \cong \text{const. } D_{gr}(v^*) e^{-\Delta F^*/kT} \tag{3.21}$$

Classical theory of nucleation does not define shape of clusters or aggregating kinetic elements, β_1 . However, restriction of cluster configuration to a single variable - volume - implies that aggregating molecules and the resulting clusters are spherical and isotropic.

The classical model of nucleation has been successfully applied to simple transitions, such as condensation of gases, crystallization of metals, etc. The same model with various modifications has also been used for crystallization of low molecular liquids and polymers, although it cannot explain many effects encountered in real crystallization conditions.

3.3 Modifications of the Classical Theory

3.3.1 Athermal Nucleation

The flux of clusters passing through potential barrier ΔF^* from the region of embryos to the region of stable clusters (Eq. (3.21)), determines *thermal nucleation* . In late forties, Fisher, Hollomon and Turnbull [11] suggested another mechanism, *athermal nucleation* . I am using this term in its original meaning, exactly as defined by Fisher et al. [11]. In polymer literature, the term "athermal nucleation" is sometimes used to denote heterogeneous nucleation, or nucleation on pre-determined (seeded) nuclei. Ziabicki [12] presented simple derivation of both nucleation mechanisms for the one-dimensional model of nucleation. The *number of nuclei* (i.e. stable clusters) at the instant t is determined by integral of cluster density, taken over the region of stability $(v \geq v^*)$

$$N(t) = N_0 \cdot \int_{v^*}^{V} w(v,t)dv \cong N_0 \cdot \int_{v^*}^{\infty} w(v,t)dv \qquad (3.22)$$

Nucleation rate, i.e. production rate of stable clusters, results from differentiation of the above integral, yielding

$$\dot{N} = N_0 \frac{d}{dt} \int_{v^*}^{\infty} w(v,t)dv = N_0 \left[\int_{v^*}^{\infty} \frac{\partial w}{\partial t}dv - w(v^*)\frac{dv^*}{dt} \right] \qquad (3.23)$$

Application of the continuity equation (9) and the Gauss-Ostrogradski theorem reduces first term on the right-hand side of Eq.(3.23) to flux in the critical point $v = v^*$; the total production rate of stable clusters results in the form

$$\dot{N} = \dot{N}_{th} + \dot{N}_{ath} = N_0 \left[j(v^*) - w(v^*) \cdot \frac{dv^*}{dt} \right] \qquad (3.24)$$

The first term - *thermal nucleation* - provides a flux of clusters from the region of embryos to the region of nuclei. A necessary condition is overcoming the potential barrier, ΔF^* . The second term - *athermal nucleation* - appears whenever external conditions (temperature, T , pressure, p , etc) change in time, making critical cluster size a function of time. Athermal nucleation does not require activation. Its rate is proportional to the rate of change of external conditions, and disappears when these are constant

$$\dot{N}_{ath} \propto \frac{dv^*}{dt} = \frac{\partial v^*}{\partial T}\frac{dT}{dt} + \frac{\partial v^*}{\partial p}\frac{dp}{dt} + \ldots \qquad (3.25)$$

A more general, multimensional version of nucleation rates will be discussed in Section 3.4.3.

3.3.2 Shape Effects

Crystals are, as a rule, anisotropic and their shape is rarely spherical. It is natural to expect that free energy of cluster formation would be a function not only of cluster *size* (volume) but also of its *shape*. Consider an anisotropic cluster with tetragonal symmetry and dimensions a (perpendicular to the fourfold symmetry axis) and c (along the symmetry axis) (Figure 3.2). Since surface energy on the base face, normal to c is different to that on the side face, normal to a, free energy of cluster formation becomes a function of the dimensions a and c taken separately

$$\Delta F(a,c) = (a^2 c - v_0)\Delta f + 2 \left[2\sigma_a(ac - v_0^{2/3}) + \sigma_c(a^2 - v_0^{2/3}) \right] \qquad (3.26)$$

σ_c and σ_a are, respectively, interface tensions on the faces normal to vectors c and a.

Minimization of free energy at constant cluster volume

$$\frac{\partial \Delta F}{\partial (c/a)}\bigg|_{v=\text{const}} = 0 \tag{3.27}$$

yields *optimum* (thermodynamically most probable) *shape* of the cluster

$$\left(\frac{c}{a}\right)_{opt} = \frac{\sigma_c}{\sigma_a} \tag{3.28}$$

and the corresponding family of free energies

$$\Delta F_{opt}(v) = (v - v_0)\Delta f + 6(v^{2/3} - v_0^{2/3})(\sigma_a^2 \sigma_c)^{1/3} \tag{3.29}$$

One can also calculate *critical dimensions* and *critical free energy*. Putting

$$\frac{\partial \Delta F}{\partial a} = \frac{\partial \Delta F}{\partial c} = 0 \tag{3.30}$$

one obtains

$$a^* = \frac{-4\sigma_a}{\Delta f} \tag{3.31}$$

$$c^* = \frac{-4\sigma_c}{\Delta f} \tag{3.32}$$

and

$$\Delta F^* = \Delta F_{max} = \frac{32\sigma_a^2 \sigma_c}{(\Delta f)^2} \tag{3.33}$$

The fact that optimum dimensions (a^*, c^*), are proportional to surface energies on the respective faces, corresponds with the crystallographic principle formulated by Wulf [13], according to which crystal growth rate in a

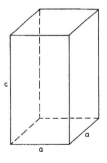

Fig. 3.2. Tetragonal cluster.

direction x , is inversely proportional to the area of face normal to x . Effects of cluster shape on free energy of nucleation were discussed for cylindrical clusters by Lauritzen and Hoffman [7], Mandelkern [8], and for tetragonal clusters by Ziabicki [14].

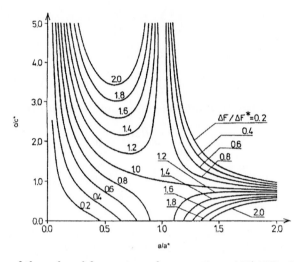

Fig. 3.3. Map of the reduced free energy of aggregation, $\Delta F/\Delta F^*$, for tetragonal clusters in reduced dimensions, a/a^* and c/c^* .

Figure 3.3 presents a map of dimensionless free energy of nucleation for tetragonal clusters. Transition from the region of unstable embryos to the region of stable nuclei (i.e. thermal nucleation) can be performed in a variety of paths. Different paths lead over different potential barriers, ΔF^* . The easiest path (one associated with the smallest barrier) leads through the saddle point ($a/a^* = c/c^* = 1$) corresponding to optimum shape of the cluster. Therefore, instead of solving tetragonal nucleation as a two-variable problem, some nucleation models assume optimum cluster shape ($a/c = a_{opt}/c_{opt} = \sigma_a/\sigma_c$) and consider *quasi- one-dimensional* process of cluster growth.

Ziabicki [15] proposed a more general model of anisotropic clusters - a convex polyhedron (Figure 3.4). Its shape is characterized by a pencil of n vectors, $\{r_1, r_2, \ldots , r_n\}$.

The polyhedron is constructed in the following way. First, a set of unit vectors $\{e_1, e_2, \ldots , e_n\}$ directed along the main crystal lattice directions is chosen. The set includes directions normal to most densely packed crystal planes. The directions $\{e_1, e_2, \ldots , e_n\}$ provide *material constants*, determined by symmetry of the crystal lattice and do not change on crystallization. What does change in the process of cluster growth, is the set of *scalar* variables - lengths of lattice vectors - $\{r_1, r_2, \ldots , r_n\}$. At the end of each vector

$r_i = e_i r_i$, a normal plane is placed. The polyhedron results from intersection of planes normal to all vectors, r_i . Note, that the scalar variables, r_i , are independent of each other and can be changed freely in the process of cluster growth. However, intersection of normal planes may produce *hidden faces* which are neglected in further considerations.

The areas of individual cluster (crystal) faces, s_i , result from the solution of the intersection problem and depend on the entire set of the variables, r_k

$$s_i = s_i(r_1, \dots, r_n) \quad \text{for} \quad i = 1, 2, \dots, n \tag{3.34}$$

Free energy of aggregation depends of cluster volume, v , and the sum of n surface terms. For a convex polyhedron it can be presented in the form

$$\Delta F(r) = \Delta f \cdot v(r) + \sum_{i=1}^{n} s_i \cdot \sigma_i = \sum_{i=1}^{n} s_i \cdot \left(\frac{1}{3} r_i \Delta f + \sigma_i \right) \tag{3.35}$$

where σ_i is surface energy density on the i-th face.

3.3.3 Concentration Effects

Crystallization in a solution may depend on concentration of the crystallizing molecules. The rate of addition in the reaction (Eq. (3.3)) depends on the probability of finding a reacting molecule, β_1 . When introduced to free energy of aggregation, concentration, c , contributes a logarithmic term [16]

$$\Delta f(c) = \Delta f_0 - \left(\frac{kT}{v_0} \right) \ln c \tag{3.36}$$

Concentration effect can be interpreted as resulting from the condition that the reacting molecules must meet in the final area, called *reaction cross-section*. A general treatment of the cross-section problem will be discussed

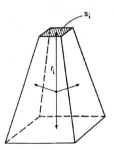

Fig. 3.4. Model of a convex polyhedral cluster. r_i - normal vector, s_i - area of the i-th face.

in Section 3.4.2. The term $(k \ln c)$ provides *entropy of mixing* of crystallizing molecules and molecules of the solvent. Because of dilution, magnitude of the free energy (driving force for the transition) is reduced. Using the condition (8) we obtain concentration-dependent melting (dissolution) temperature

$$\frac{T_m(c)}{T_{m0}} = \frac{1}{1 - \frac{kT_{m0}}{v_0 \Delta h_0} \ln c} \leq 1 \qquad (3.37)$$

lower than that for an undiluted system (T_{m0}). Figure 3.5 presents reduced melting temperature as a function of concentration. It is evident that extreme

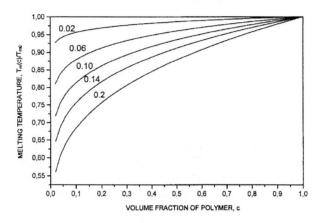

Fig. 3.5. Reduced melting temperature of an intermolecular crystal, $T_m(c)/T_{m0}$, v_s polymer concentration (volume fraction), c. Entropy ratio, $kT_{m0}/\Delta h_0 v_0$, indicated.

dilution practically excludes crystallization, requiring prohibitively deep undercooling. The parameter $(kT_{m0}/v_0 \Delta h_0)$ is inversely proportional to the entropy of melting in an undiluted system. The larger is this parameter, the stronger is effect of dilution.

Concentration dependent bulk free energy, Δf, appears also in the expressions for critical cluster size and shape (v^* in Eq. (3.13), a^*, c^* in Eqs. (3.31)-(3.32)), and therefore affects potential barrier for nucleation, and nucleation rate.

In solutions of low-molecular substances, c can be interpreted as *molar fraction* of crystallizing molecules. Eqs. (3.35-3.36) present non-linear form of the classical *Raoult law* - depression of freezing (melting) temperature due to dilution. Application of this treatment to intermolecular (bundle-like) crystallization of polymers requires that c is interpreted as *volume fraction* of the crystallizing polymer. Entropy of mixing long chain macromolecules with many internal degrees of freedom, is controlled by volume fraction of polymer in the solution, rather than molar fraction.

The situation is completely changed when nucleation (and crystal growth) is *intramolecular*, rather than intermolecular. This is the case of *folded-chain crystallization*. Since growth of the cluster proceeds by cooperative folding of subsequent segments of the same polymer chain, probability of finding next kinetic unit for addition is unity, and polymer concentration does not affect crystallization. This explains the fact (observed experimentally) that polymers can crystallize from dilute solutions without considerable depression of temperature [17, 18]. According to Eq.(3.37) crystallization temperature should fall to zero.

Concentration effects in a polymer crystal with *mixed morphology* (one containing intra- and intermolecular segments) will be discussed in Section 3.5.5.

3.3.4 Asymmetric Molecules. Reaction Cross-section in the Space of Orientations

When aggregating molecules are spherically symmetric (Figure 3.6a), the reaction of cluster growth is not subject to orientation restrictions. On the other hand, when reacting molecules are asymmetric (Figure 3.6b) *orientations* of the reacting molecules must be *consistent*. Out of all molecules approaching surface of the growing cluster, only those can be effectively attached, whose orientations fit into some tolerance angle, Δ_θ . This effect is well known in stereochemistry, in particular in biochemical reactions involving large asymmetric molecules. Effects of molecular shape on the entropy of crystal melting are also known [19]. The symmetry effect can be treated in the way similar to that of concentration [16], leading to bulk free energy of aggregation (crystallization) in the form

$$\Delta f(\Delta_\theta) = \Delta f_{sph} - \frac{kT}{v_0} \ln\left(\frac{\Delta_\theta}{\Omega}\right) \tag{3.38}$$

where Δ_θ is "tolerance angle", or *reaction cross-section in the space of orientations*, and Ω is full solid angle (measure of the space of orientations). The reference value, Δf_{sph} , is bulk free energy of aggregation for spherically symmetric (unorientable) molecules.

The logarithmic term in Eq.(3.38) describes *entropy of disorientation*, a component of the entropy of melting. It has been observed [19] that the more asymmetric are crystallizing molecules, the larger is entropy of melting and the smaller melting temperature, T_m. Application of the condition (8) to free energy density from Eq.(3.38) yields equation for melting temperature for asymmetric, orientable molecules

$$\frac{T_m(\Delta_\theta)}{T_{m,sph}} = \frac{1}{1 - \dfrac{kT_{m,sph}}{v_0 \Delta h} \ln\left(\dfrac{\Delta_\theta(T_m)}{\Omega}\right)} = \frac{1}{1 - \dfrac{k}{v_0 \Delta s_{sph}} \ln\left(\dfrac{\Delta_\theta(T_m)}{\Omega}\right)} \le 1 \tag{3.39}$$

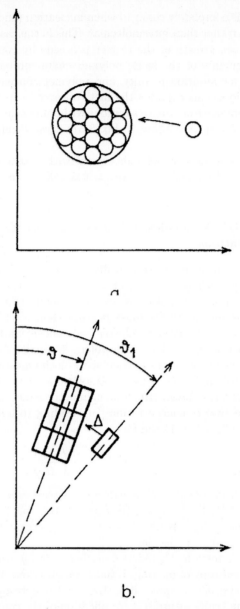

Fig. 3.6. Scheme of cluster growth. a) isotropic cluster resulting from aggregation of spherical particles, b) anisotropic cluster formed from elongated (rodlike) particles.

The same heat of melting, Δh , is assumed for spherical and asymmetric molecules. The relation between melting temperature and the reaction cross-section resembles concentration effect in intermolecular crystallization. Sensitivity to orientational cross-section depends on the ratio $(k/v_0\Delta s_{sph})$. Second term in the denominator of Eq.(3.39) represents ratio of the disorientation entropy, to entropy of melting for unorientable (spherical) molecules.

In our early paper [16] the "tolerance angle", Δ_θ , was introduced as a model parameter. More recently, together with Jarecki [20-22], we have derived Δ_θ from simple statistical mechanical considerations. Partition energy in a system of aggregating spherical particles was compared with a similar energy in the system of non-spherical particles. (Δ_θ/Ω) - reduced cross-section in the space of orientations - can be interpreted as the ratio of equilibrium aggregation constants for asymmetric and spherical particles. For ideal, non-interacting particles with uniaxial symmetry we have obtained [22]

$$\frac{\Delta_\theta}{\Omega} = \frac{K_{eq}}{K_{eq,sph}} \simeq \frac{kT}{4\pi^2\nu^2 I_1} = \frac{T}{T_0}\left(\frac{\Delta_\theta}{\Omega}\right)_0 \ll 1 \qquad (3.40)$$

where ν is vibration frequency of bonds in the cluster, and I_1 - moment of inertia of an aggregating (crystallizing) molecule. It is evident that Δ_θ/Ω is the smaller (i.e. orientation effects the stronger) the larger and more asymmetric are crystallizing kinetic units. The orientational cross-section is a function of temperature, and depends on intermolecular interactions, and non-uniform orientation distribution of crystallizing units [21-22]. Effects of symmetry are consistent with empirical observations discussed in the book by Ubbelohde [19]. Figure 3.7 presents reduced melting temperature as a function of the orientational cross-section $(\Delta_\theta/\Omega)_0$, calculated from Eqs.(3.39) and (3.40). The reference cross-section $(\Delta_\theta/\Omega)_0$ is related to melting temperature of spherical molecules, $T_{m,sph}$. It is evident that reduction of melting temperature (with respect to spherical molecules) can reach tens of degrees.

The concept of cross-section in the *space of orientations* is a natural extension of the cross-section for collisions in the *space of positions*. A more general treatment in the 6-dimensional configurational space will be discussed in Section 3.4.2.

For crystallization of polymers, orientation effects are especially important. The aggregating kinetic units - e.g. chain segments - are usually large and asymmetric what makes the ratio (Δ_θ/Ω) much smaller than unity. More than that, polymers often crystallize from pre-oriented systems, which introduces a new effect: *selective crystallization*.

3.3.5 Oriented Systems. Selective Crystallization

In a system of crystallizing particles with non-random orientation distribution, $w_1(\boldsymbol{\theta})$, the bulk free energy of aggregation includes an additional, orientation-dependent term [16]

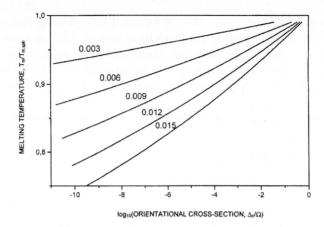

Fig. 3.7. Reduced melting temperature for rodlike molecules, $T_m/T_{m,sph}$, as a function of orientational cross-section $(\Delta_\theta/\Omega)_0$. Entropy ratio, $kT_{m,sph}/\Delta h_0 v_0$, indicated. The ratio $(\Delta_\theta/\Omega)_0$ corresponds to melting temperature for spherical molecules, $T_{m,sph}$.

$$\Delta f(\boldsymbol\theta) = \Delta f_{sph} - \frac{kT}{v_0}\ln\left[\int_{\Delta_\theta} w_1(\boldsymbol\theta)d^3\boldsymbol\theta\right] \tag{3.41}$$

$\boldsymbol\theta$ denotes vector of orientation represented, in general, by three Euler angles. For uniaxial orientation, $\boldsymbol\theta$ reduces to a single angle, ϑ .

The integral in Eq.(3.41) expresses probability of finding crystallizing molecule within the "rotational volume", Δ_θ , surrounding the growing cluster (Fig. 6b). For small Δ_θ , and not too sharp distributions, the integral can be replaced by the average value

$$\int_{\Delta_\theta} w_1(\boldsymbol\theta)d^3\boldsymbol\theta \cong \Delta_\theta \cdot w_1(\boldsymbol\theta) \tag{3.42}$$

which reduces Eq.(3.41) to

$$\Delta f(\boldsymbol\theta) \cong \Delta f_{sph} - \frac{kT}{v_0}\ln\left[\Delta_\theta w_1(\boldsymbol\theta)\right] = \Delta f_{un} - \frac{kT}{v_0}\ln\left[\Omega \cdot w_1(\boldsymbol\theta)\right] \tag{3.43}$$

The new reference value in Eq.(3.43) denotes bulk free energy density for a system of asymmetric, randomly oriented molecules, and Ω represents volume of the orientation space.

Physical nature of orientation effects is entropic. The logarithmic term in Eqs. (3.41) and (3.43) can be interpreted as entropy of mixing differently oriented molecules in the amorphous (uncrystallized) phase. Orientation dependence of the free energy of aggregation has far going physical consequences.

Following Δf , all kinetic and thermodynamic characteristics of nucleation become orientation dependent. Crystallization becomes *selective*: orientation-dependent driving force causes differently oriented clusters (crystals) to be created at different temperatures and with different speeds. In systems crystallized from preoriented molecules, the average degree of crystal orientation is usually much higher than orientation of the crystallizing molecules.

The condition (8) applied to orientation-dependent free energy density from Eq. (3.43)

$$\Delta f(T; \boldsymbol{\theta})|_{\theta=\text{const}} = 0 \; ; \quad T = T_m(\boldsymbol{\theta}) \tag{3.44}$$

provides an equation for critical temperature

$$\frac{T_m(\boldsymbol{\theta})}{T_{m,un}} = \frac{1}{1 - \dfrac{kT_{m,un}}{v_0 \Delta h} \ln \left[\Omega \cdot w_1(\boldsymbol{\theta}; T_m) \right]} \tag{3.45}$$

which, unlike temperatures from Eqs. (3.37) and (3.39), does not necessarily represent thermodynamic equilibrium. When the system does not allow for immediate reorientation of crystallizing molecules (and growing clusters), $T_m(\boldsymbol{\theta})$ represents *critical temperature for the onset of crystallization*, and changes in time with orientation distribution of crystallizing units. $T_m(\boldsymbol{\theta})$ turns into *equilibrium melting temperature*, whenever, due to an external potential field, equilibrium orientation distribution $w_1(\boldsymbol{\theta})$ is fixed. Such cases will be discussed in Sections 3.5.1. - 3.5.3.

The reference values of free energy and temperature are related to an equilibrum, randomly oriented system of asymmetric particles. In the limit of ideal (delta) distribution

$$w_1(\boldsymbol{\theta}) \to \delta(\boldsymbol{\theta}) \; ; \quad \int_{\Delta_\vartheta} w_1(\boldsymbol{\theta}) d^3\theta \to 1 \tag{3.46}$$

free energy density, and equilibrium melting temperature reduce to ones characteristic of spherical molecules

$$\Delta f(\boldsymbol{\theta}) \to \Delta f_{sph} \delta(\boldsymbol{\theta}) \tag{3.47}$$

$$T_m(\boldsymbol{\theta}) \to T_{m,sph} \delta(\boldsymbol{\theta}) \tag{3.48}$$

The selectivity with respect to orientation concerns not only crystallization temperature, but also nucleation rates. Clusters oriented along the main orientation axis are formed much faster than those oriented along perpendicular directions. Some experimental data consistent with this idea will be discussed in Section 3.5.1.

3.4 Multidimensional Theory of Nucleation

3.4.1 Configurational Space. Distribution of Clusters and Crystallizing Units.

Consideration of shape and orientation effects as extension of the classical, one-dimensional theory [15,16] led the author to formulate a more general, multidimensional model of nucleation [10]. Configuration of crystallizing units (atoms, molecules, chain segments) is considered in a 6-dimensional configurational space, $\mathcal{Z}$. Vector ζ includes characteristics describing orientation (three Euler angles: ϑ , φ , ψ) and position of the center of mass (x_o, y_o, z_o) of the crystallizing unit

$$\zeta = \left\{ \vartheta, \varphi, \psi, \frac{x_0}{v_0^{1/3}}, \frac{y_0}{v_0^{1/3}}, \frac{z_0}{v_0^{1/3}} \right\} \tag{3.49}$$

$$\zeta \in \mathcal{Z} \tag{3.50}$$

Molecular volume of the crystallizing unit, v_0 , is used for reduction of dimensions. Configuration distribution is defined as probability density in the space $\mathcal{Z}$

$$w_1(\zeta, t) = \frac{1}{N_{1,0}} \frac{dN_1}{d^6\zeta} \tag{3.51}$$

where $d^6\zeta$ is volume element, and $N_{1,0}$ - total number of crystallizing units. In a general case, w_1 should be determined from an appropriate continuity equation, to be solved together with the Fokker-Planck equation for cluster distribution. In special cases, the distribution of aggregating molecules, $w_1(\zeta, t)$, may be assumed constant in time, and/or uniform.

Configuration, ξ , of an *n-hedral, convex* cluster, includes the set of scalar variables, $\{r_1, r_2, \ldots, r_n\}$ - lengths of vectors normal to individual crystal faces (cf. Figure 3.4), orientational and positional characteristics included in ζ , and a set of k dimensionless variables $\{p_1, p_2, \ldots, p_k\}$ characterizing internal structure. All these variables create vector ξ in the N-dimensional configurational space $\mathcal{V}$

$$\xi = \left\{ \frac{r_1}{v_0^{1/3}}, \frac{r_2}{v_0^{1/3}}, \ldots, \frac{r_v}{v_0^{1/3}}, \vartheta, \varphi; \psi, \frac{x_0}{v_0^{1/3}}, \frac{y_0}{v_0^{1/3}}, \frac{z_0}{v_0^{1/3}}, p_1, p_1, \ldots, p_k \right\} \tag{3.52}$$

$$\xi \in \mathcal{V} \tag{3.53}$$

The 6-dimensional orientational and positional space for aggregating molecules, $\mathcal{Z}$, is a subspace in the N-dimensional configurational space for clusters, $\mathcal{V}$

$$\mathcal{Z} \subset \mathcal{V} \tag{3.54}$$

The requirement of consistent orientation and position of the growing cluster and of the molecule to be attached to it, makes orientation/position components of the vector ξ equal to the corresponding components of vector ζ . Thus, *configuration of the cluster implies unique configuration of the crystallizing units*

$$\xi \Rightarrow \zeta \tag{3.55}$$

The metric tensor of the space $\mathcal{V}$ was derived in the following form [23]

$$g_{ik} = \begin{bmatrix} 1 & 0 & \ldots & 0 & 0 & 0 & 0 & 0 & 0 & 0 & 0 & 0 & \ldots & 0 \\ 0 & 1 & \ldots & 0 & 0 & 0 & 0 & 0 & 0 & 0 & 0 & 0 & \ldots & 0 \\ \ldots & \ldots & \ldots & \ldots & \ldots & \ldots & & \ldots & \ldots & \ldots & \ldots & \ldots & & \ldots \\ 0 & 0 & \ldots & 1 & 0 & 0 & 0 & 0 & 0 & 0 & 0 & 0 & \ldots & 0 \\ 0 & 0 & \ldots & 0 & 1 & 0 & 0 & 0 & 0 & 0 & 0 & 0 & \ldots & 0 \\ 0 & 0 & \ldots & 0 & 0 & 1 & \cos\vartheta & 0 & 0 & 0 & 0 & 0 & \ldots & 0 \\ 0 & 0 & \ldots & 0 & 0 & \cos\vartheta & 1 & 0 & 0 & 0 & 0 & 0 & \ldots & 0 \\ 0 & 0 & \ldots & 0 & 0 & 0 & 0 & 1 & 0 & 0 & 0 & 0 & \ldots & 0 \\ 0 & 0 & \ldots & 0 & 0 & 0 & 0 & 0 & 1 & 0 & 0 & 0 & \ldots & 0 \\ 0 & 0 & \ldots & 0 & 0 & 0 & 0 & 0 & 0 & 1 & 0 & 0 & \ldots & 0 \\ 0 & 0 & \ldots & 0 & 0 & 0 & 0 & 0 & 0 & 0 & 1 & 0 & \ldots & 0 \\ 0 & 0 & \ldots & 0 & 0 & 0 & 0 & 0 & 0 & 0 & 0 & 1 & \ldots & 0 \\ \ldots & \ldots & \ldots & \ldots & \ldots & \ldots & & \ldots & \ldots & \ldots & \ldots & \ldots & & \ldots \\ 0 & 0 & \ldots & 0 & 0 & 0 & 0 & 0 & 0 & 0 & 0 & 0 & \ldots & 1 \end{bmatrix} \tag{3.56}$$

Note, that scalar variables describing cluster dimensions (r_i), are mutually independent. The non-diagonal components result from rotation around the Euler angles φ and ψ . In a general case, the number of the degrees of freedom (dimensions of the space, $\mathcal{V}$) is $N = n+k+6$. The first two groups of variables, those describing cluster dimensions $\{r_i\}$, and orientation $\{\vartheta, \varphi, \psi\}$, have been explained in Sections 3.3.2. and 3.3.5. The effect of cluster shape results from its anisotropy, and orientation appears in the bulk free energy (and, consequently, in flux j and other kinetic characteristics) whenever asymmetric molecules are subjected to orienting forces. Physical situations in which crystallization is affected by molecular orientation include crystallization in the presence of an external orienting field, flow, deformation and/or stress. The model has been completed with additional degrees of freedom: *position of the center of mass, x_0* , and parameters of *internal structure, p_k* . Consideration of the position of cluster (crystallizing molecule), x_0 , is justified by the presence of concentration gradients imposed by an external field. Potential energy of the aggregating unit (and that of the growing cluster) may depend on the position with respect to the field pole. Such a situation exists when ponderable particle is placed in a gravitational field or, when electrically charged molecule is subjected to an electrostatic field. Gravitational effects are common in phase transitions in the cosmic space and, in the laboratory scale, can be created in an ultracentrifuge. Concentration gradients may affect crystallization also when the field is removed. The last group of degrees of freedom is related to *structural defects*. The variables p_i may characterize concentration and distribution of vacancies in the crystalline aggregates, "*kinks*" of the polymer chain, "*cilia*" on the surface of polymer crystals, etc. The defects affect internal energy and entropy of clusters and contribute to free energy of aggregation. Cluster distribution density, $w(\boldsymbol{\xi}, t)$, is thus defined as

$$w_1(\boldsymbol{\xi}, t) = \frac{1}{N_0} \frac{dN}{d^N \boldsymbol{\xi}} \tag{3.57}$$

where $d^N \boldsymbol{\xi}$ is volume element in the configurational space, and N_0 is total number of particles in the system. $w(\boldsymbol{\xi}, t)$ can be obtained from the Fokker-Planck equation [10]

$$\frac{\partial w(\boldsymbol{\xi}, t)}{\partial t} + \mathrm{div}_\xi \boldsymbol{j} = 0 \tag{3.58}$$

N-dimensional flux in $\mathcal{V}$ includes, like in the classical theory, diffusional and potential contributions

$$\boldsymbol{j}(\boldsymbol{\xi}; t) = \boldsymbol{j}_{dif} + \boldsymbol{j}_{pot} = -\boldsymbol{D}_\xi \left[\nabla_\xi w(\boldsymbol{\xi}) + \frac{w}{kT} \nabla_\xi \Delta F(\boldsymbol{\xi}, t) \right] \tag{3.59}$$

and div_ξ and ∇_ξ denote, respectively, divergence and gradient operators in the space $\mathcal{V}$. Using the metric tensor from Eq.(3.56) one obtains [23]

$$\mathrm{div}_\xi \boldsymbol{j} = \sum_{i=1}^{n} \frac{\partial j^n}{\partial r_i} + \frac{1}{\sin \vartheta} \frac{\partial}{\partial \vartheta} (\sin \vartheta j^\vartheta)$$

$$+ \frac{\partial j^\varphi}{\partial \varphi} + \frac{\partial j^\psi}{\partial \psi} + \frac{\partial j^x}{\partial x_0} + \frac{\partial j^y}{\partial y_0} + \frac{\partial j^z}{\partial z_0} + \sum_{j=1}^{k} \frac{\partial j^{pj}}{\partial p_j} \tag{3.60}$$

$$\nabla_\xi \rho = \sum_{i=1}^{n} \boldsymbol{e}_{ri} \frac{\partial \rho}{\partial r_i} + \boldsymbol{e}_\vartheta \frac{\partial \rho}{\partial \vartheta} + \frac{1}{\sin^2 \vartheta} \boldsymbol{e}_\varphi \left[\frac{\partial \rho}{\partial \varphi} - \cos \vartheta \frac{\partial \rho}{\partial \psi} \right]$$

$$+ \frac{1}{\sin^2 \vartheta} \boldsymbol{e}_\psi \left[\frac{\partial \rho}{\partial \psi} - \cos \vartheta \frac{\partial \rho}{\partial \varphi} \right] + \boldsymbol{e}_x \frac{\partial \rho}{\partial x_0} + \boldsymbol{e}_y \frac{\partial \rho}{\partial y_0} + \boldsymbol{e}_z \frac{\partial \rho}{\partial z_0} + \sum_{j=1}^{k} \boldsymbol{e}_{pj} \frac{\partial \rho}{\partial p_j} \tag{3.61}$$

Diffusion tensor, $\boldsymbol{D}_\xi$, includes components related to growth (change of cluster dimensions, r_k), rotation (Euler angles), translation of the center of mass, x_0 , as well as variation of internal structure, p_i .

Introduction of configurational variables other than cluster dimensions, has far going physical consequences which will be discussed in Sections 3.4.2. and 3.4.3.

3.4.2 Thermodynamic Characteristics

Density of the bulk free energy of aggregation can be presented in the form

$$\Delta f(\xi, t) = \Delta f_0(\xi) - \left(\frac{kT}{v_0}\right) \ln \left[N_{1,0} \cdot \int_\Delta w_1(\zeta) d^6\zeta \right] \qquad (3.62)$$

The expression under logarithm describes probability that one of $N_{1,0}$ crystallizing units satisfies the necessary condition of aggregation - appearance in the positional (Δx) and orientational (Δ_θ) cross-section around the growing cluster

$$\Delta = \Delta_x \cdot \Delta_\theta \qquad (3.63)$$

Since Δ is usually small, the integral in Eq.(3.62) can be expressed by density, w_1, taken at the configuration ζ, matching configuration of the growing cluster, ξ

$$\int_\Delta w_1(\zeta) d^6\zeta \cong \Delta \cdot w_1[\zeta(\xi)] \qquad (3.64)$$

The approximate form of the bulk free energy density reduces to

$$\Delta f(\xi, t) \cong \Delta f_0(\xi) - \left(\frac{kT}{v_0}\right) \ln [N_{1,0} \cdot w_1(\zeta(\xi)) \cdot \Delta] \qquad (3.65)$$

Components of vector ζ, describing configuration of the *crystallizing unit*, are identical with corresponding components of *cluster* configuration, ξ. Cluster configuration, ξ, implies unique configuration of the crystallizing unit, ζ. Application of the condition Eq.(3.8) to free energy density from Eq. (3.65), yields critical temperature for the onset of crystallization which, in general, does not represent thermodynamic equilibrium

$$\Delta f(T; \xi)|_{\xi=\text{const}} = 0 ; \quad T = T_m(\xi) \qquad (3.66)$$

Thermodynamic stability of the cluster is defined by the set of simultaneous inequalities

$$\frac{\partial \Delta F(\xi)}{\partial r_i} \leq 0 \quad i = 1, 2, \ldots n ; \quad \xi \in V^* \qquad (3.67)$$

The cluster is stable when free energy of aggregation is a non-increasing function of all its n dimensions, r_i. Clusters which satisfy the inequalities

(67) form in the configurational space a *region of stability*, $\mathcal{V}^*$, separated from the remaining part of the space $\mathcal{V}$ by the critical hypersurface $\mathcal{S}^*$

$$\mathcal{V}^* \subset \mathcal{V} ; \quad \mathcal{S}^* \subset \mathcal{V} \tag{3.68}$$

Nucleation does not take place when any one of the inequalities (67) is not satisfied.

3.4.3 Nucleation Rates and Mechanisms

The number of stable clusters (nuclei) in the system is determined by integral of the distribution function taken over the stability region, $\mathcal{V}^*$

$$N(t) = N_0 \int\!\!\int\limits_{\mathcal{V}^*}\!\!\int w(\boldsymbol{\xi}, t) d^N \boldsymbol{\xi} \tag{3.69}$$

Differentiation with respect to time yields nucleation rate

$$\dot{N} = N_0 \frac{d}{dt} \int\!\!\int\limits_{\mathcal{V}^*}\!\!\int w(\boldsymbol{\xi}, t) d^N \boldsymbol{\xi} = N_0 \left[\int\!\!\int\limits_{\mathcal{V}^*}\!\!\int \frac{\partial w}{\partial t} d^N \boldsymbol{\xi} - \frac{d\boldsymbol{R}^*}{dt} \int\limits_{\mathcal{S}^*}\!\!\int w(\boldsymbol{\xi}) d\boldsymbol{S} \right] \tag{3.70}$$

$d^N \boldsymbol{\xi}$, and $d\boldsymbol{S} = d^{N-1} \boldsymbol{\xi}'$ are volume elements, respectively, of the space, $\mathcal{V}$, and on the surface, $\mathcal{S}^*$. Application of the Ostrogradski-Gauss theorem and use made of the Fokker-Planck equation (58) reduces nucleation rate to two integrals taken over the critical hypersurface $\mathcal{S}^*$

$$\dot{N} = \dot{N}_{th} + \dot{N}_{ath} = N_0 \left[\int\limits_{\mathcal{S}^*}\!\!\int \boldsymbol{j}(\boldsymbol{\xi}) \cdot d\boldsymbol{S} - \frac{d\boldsymbol{R}^*}{dt} \cdot \int\limits_{\mathcal{S}^*}\!\!\int w(\boldsymbol{\xi}) d\boldsymbol{S} \right] \tag{3.71}$$

$\mathcal{R}^*$ denotes radius vector of the critical hypersurface $\mathcal{S}^*$. Like in the modified classical theory (Section 3.3.1) there appear two nucleation mechanisms: thermal and athermal. The change of the radius vector results from variable external conditions (temperature, pressure, etc) and disappears when the process is strictly isothermal (isobaric, ..., etc).

$$\frac{d\boldsymbol{R}^*}{dt} = \frac{\partial \boldsymbol{R}^*}{\partial T} \frac{dT}{dt} + \frac{\partial \boldsymbol{R}^*}{\partial p} \frac{dp}{dt} + \dots ; \quad \boldsymbol{R}^* \subset \boldsymbol{V}^* \tag{3.72}$$

The multidimensional model implies new nucleation phenomena. Nucleation (i.e. production of thermodynamically stable clusters) does not necessarily consist in the *growth* of small clusters, as inferred by the classical theory. The flux vector in the space $\mathcal{V}$ has N components and each of them may give rise to thermal nucleation. The N-dimensional flux vector, $\boldsymbol{j}$, can be formally separated into several groups

$$j = j_{gr} + j_{rot} + j_{trans} + j_{heal} \tag{3.73}$$

The first group

$$j_{gr} = \{j_{r1}, j_{r2}, \dots, j_m\} \tag{3.74}$$

describes *growth* of clusters in n crystallographic directions, the only mechanism considered in the classical theory. The flux of *rotation*

$$j_{rot} = \{j_\vartheta, j_\varphi, j_\psi\} \tag{3.75}$$

and *translation*

$$j_{trans} = \{j_x, j_y, j_z\} \tag{3.76}$$

correspond to orientational and positional degrees of freedom and become important in external potential fields (cf. Section 3.4.4). Last, not least, *healing of internal defects*

$$j_{heal} = \{j_{p1}, j_{p2}, \dots, j_{pk}\} \tag{3.77}$$

may give rise to the change of free energy and produce a flux of thermodynamically stable clusters (nuclei). Individual mechanisms become active whenever cluster distribution is non-uniform and/or free energy of aggregation depends on configuration, ξ (cf. Eq. 3.59). Rotation of dipoles in an electric or magnetic field (without change of their dimensions) is sufficient for the change of the potential energy, and may provide the source of stable nuclei. The same is true with translational motion of charged particles in an electrostatic field, or removal of defects leading to the change of internal structure parameters, p_k .

In the multidimensional model, there appear also various athermal nucleation mechanisms. The shift of the critical hypersurface characterized by the rate $(d\mathcal{R}^*/dt)$ can be due to many different degrees of freedom, not only cluster size as in the classical theory (cf. Section 3.3.1).

Figure 3.8 presents schematically two fundamental groups of nucleation mechanisms, treated as motions in the configurational space.

3.4.4 Effects of Potential Fields and Configuration Distributions

Consider molecules crystallizing in an external potential field, E . Interaction with the field modifies bulk free energy of aggregation and affects thermodynamic and kinetic characteristics of nucleation. When the molecules exhibit dipole properties (permanent or induced) interaction with the field may induce *effects of orientation*. Orientation may also appear in potential flows of large, non-spherical molecules and clusters. *Positional effects* may be expected in systems where energy of a particle is controlled by its distance

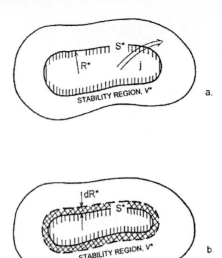

Fig. 3.8. Nucleation as motion in the configurational space, $\mathcal{V}$. a) thermal nucleation - flux of clusters ($\boldsymbol{j}$) through critical surface, $\boldsymbol{S}^*$. b) athermal nucleation - shifting of the critical surface $\boldsymbol{S}^*$ and expansion of the stability region, $\boldsymbol{V}^*$.

from the field pole. Charged molecules in an electrostatic field, and ponderable particles in a gravitational field provide natural examples.

Configuration of crystallizing units (and clusters) is described by the 6-dimensional vector, $\boldsymbol{\zeta}$, including orientation with respect to the field axes and position of the center of mass (Eq. 3.49). Potential field introduces non-uniform distribution of crystallizing units and affects distribution of the resulting clusters. When the field is removed, non-uniform distribution starts to relax, but configuration gradient may remain for a period of time and affect nucleation. Bulk free energy of aggregation

$$\Delta f(\zeta, \boldsymbol{E}) = \Delta f_{00} + \Delta u(\zeta, \boldsymbol{E}) - \frac{kT}{v_0} \ln \left[N_{1,0} \int_\Delta w_1(\zeta, t) d^6\zeta \right] \cong$$
$$\cong \Delta f_{00} + \Delta u(\zeta; \boldsymbol{E}) - \frac{kT}{v_0} \ln \left[N_{1,0} \Delta \cdot w_1(\zeta, t) \right] \qquad (3.78)$$

includes two configuration dependent terms. The first, potential term, Δu , is the difference between field induced potential energy density of a cluster and similar energy density of the crystallizing unit

$$\Delta u(\zeta; \boldsymbol{E}) = \frac{U_{cl}(\zeta; \boldsymbol{E})}{v} - \frac{U_1(\zeta; \boldsymbol{E})}{v_0} \qquad (3.79)$$

The second additional term in Eq. (3.78) describes entropy of mixing of crystallizing units subjects to configuration distribution $w_1(\zeta, t)$. In the absence of any fields ($\Delta u = 0$), in a system with uniform configuration distribution

$$w_1(\zeta) = w_{1,un} = (V\Omega)^{-1} \tag{3.80}$$

one obtains bulk free energy of aggregation in the form

$$\Delta f_{un} = \Delta f_{00} - \frac{kT}{v_0} \ln\left(N_{1,0}\Delta \cdot w_{1,un}\right) = \Delta f_{00} - \frac{kT}{v_0} \ln\left(\frac{c\Delta_x\Delta_\theta}{\Omega}\right) \tag{3.81}$$

Ω is measure of the orientation space, V - volume, and $N_{1,0}$ - total number of crystallizing units in the system. Since $c = N_{1,0}/V$ is number concentration of crystallizing units, Eq.(3.81) explains the source of concentration effects discussed in Section 3.3.3 and effects of molecular symmetry discussed in Section 3.3.4. Using Δf_{un} as a reference value, one can present bulk free energy of crystallization in the form

$$\Delta f(\zeta; E) \cong \Delta f_{un} + \Delta u(\zeta, E) - \frac{kT}{v_0} \ln\left[w_1(\zeta)V\Omega\right] =$$
$$= \Delta f_{un} + \Delta u(\zeta; E) - \frac{kT}{v_0} \ln\left[\frac{w_1(\zeta)}{w_{1,un}}\right] \tag{3.82}$$

In the vicinity of equilibrium, configuration distribution of aggregating molecules approaches the Boltzmann form

$$w_1(\zeta) \to w_{1,eq}(\zeta) = \frac{\exp\left[-U_1(\zeta; E)/kT\right]}{Z_1} \tag{3.83}$$

where

$$Z_1(E, T) = \int \exp\left[-U_1(\zeta; E)/kT\right]d^6\zeta \tag{3.84}$$

is configurational integral. Free energy of aggregation reduces to

$$\Delta f_{eq}(\zeta; E) = \Delta f_{00} + \frac{U_{cl}(v, \zeta; E)}{v} + \frac{kT}{v_0} \ln\left[\frac{Z_1(E)}{N_{1,0}\Delta}\right] \tag{3.85}$$

or

$$\Delta f_{eq}(\zeta; E) = \Delta f_{un} + \frac{U_{cl}(v, \zeta; E)}{v} + \frac{kT}{v_0} \ln\left[\frac{Z_1}{V\Omega}\right] \tag{3.86}$$

and determines equilibrium configuration distribution of a class of clusters characterized by volume v

$$w_{eq}(\zeta)\Big|_{v, E=\text{const}} = \text{const.} \exp\left[\frac{U_{cl}(v, \zeta; E)}{kT}\right] \tag{3.87}$$

Application of the condition (8) yields equation for critical temperature - a function of the configurational variable ζ

$$\Delta f(T;\zeta)\big|_{\zeta=\text{const}} = 0 ; \quad T = T_m(\zeta; E) \tag{3.88}$$

$$\frac{T_m(\zeta; E)}{T_{m,un}} = \frac{1 - \dfrac{U_{cl}(\zeta; T_m)}{v \Delta h}}{1 + \dfrac{k T_{m,un}}{v_0 \Delta h} \ln \left[\dfrac{Z_1(T_m)}{V \Omega} \right]} \tag{3.89}$$

The reference melting temperature in Eq.(3.89) corresponds to an equilibrium system free from potential fields, and having uniform configuration distribution. Configuration-dependent melting temperature, $T_m(\zeta)$, induces *selective crystallization and melting behavior*. In an external field, the system behaves as if it were divided into *classes* characterized with different values of the configurational variable, ζ . Individual classes are formed and melt in different temperatures. $T_m(\zeta)$ describes   *local equilibrium* in each class. Differentiation of critical cluster size, and critical free energy of aggregation, provides the source of configuration-dependent nucleation rates. Examples will be discussed in Sections 3.5.1. - 3.5.4.

Consider the situation, when field-dependent potential density of the crystallizing molecules, does not change upon aggregation, i.e. when

$$\Delta u = \frac{U_{cl}}{v} - \frac{U_1}{v_0} = 0 \tag{3.90}$$

The above condition is satisfied when the field is switched off. In some approximation, Δu can also be neglected when the field is active. In situation characterized by Eq.(3.90), ζ-dependent free energy of aggregation is controlled by instantaneous configuration distribution, $w_1(\zeta, t)$. The condition (8) leads to the equation for time-dependent critical temperature, $T_m(\zeta, t)$

$$\frac{T_m(\zeta; t)}{T_{m,un}} = \frac{1}{1 + \dfrac{k T_{m,un}}{v_0 \Delta h} \ln \left[V \Omega \cdot w_1(\zeta, t; T_m) \right]} \tag{3.91}$$

Eq.(3.91) describes the onset of crystallization in a transient system with configuration gradient.

The multidimensional theory of nucleation explains experimentally observed facts (selective crystallization, healing effects) and predicts new phenomena, not observed sofar. For the analysis of specific problems, one often considers *restricted models* based on limited number of variables. Such models are justified by external conditions, nature of the problem analyzed, as well as tractability considerations. A frequent simplification consists in replacing multidimensional clusters with clusters with *thermodynamically most probable shapes* which reduces the set of n cluster dimensions $\{r_i\}$ to one variable - volume. In many specific models, *uniform concentration* of crystallizing molecules is assumed. Effects of spatial and angular concentration distribution will be discussed in Section 3.5.6.

3.5 Some Applications

3.5.1 Selective Crystallization of Rigid Rods in Elongational Flow

Consider a system of rigid, rodlike molecules with volume v_0 which crystallize in a uniaxial, isochoric, steady flow with velocity gradient

$$\nabla V = q^* \begin{bmatrix} -\dfrac{1}{2} & 0 & 0 \\ 0 & -\dfrac{1}{2} & 0 \\ 0 & 0 & 1 \end{bmatrix} \tag{3.92}$$

It is assumed that the resulting clusters exhibit optimum shapes $(l/d)_{opt} = \sigma_l/\sigma_d$, thus reducing cluster configuration to two variables, $\boldsymbol{\xi} = \{v, \vartheta\}$. Configuration of the crystallizing molecules is described by a single orientation angle $\zeta = \{\vartheta\}$. Uniaxial (elongational) velocity field is typical for many industrial processes involving molten polymers, first of all, melt spinning of fibers. The flow field induces particle-field interactions

$$U_1(\vartheta) = \frac{-3kTq^*R_1}{4D_{rot,1}} \cos^2 \vartheta = kTH_1 \cos^2 \vartheta \tag{3.93}$$

for a crystallizing units, and

$$U_{cl}(\vartheta) = \frac{-3kTq^*R_{cl}}{4D_{rot,cl}} \cos^2 \vartheta = kTH_{cl} \cos^2 \vartheta \tag{3.94}$$

for the resulting cluster. The interaction potential is proportional to velocity gradient, q^* , and shape factor, R , a function of the aspect ratio, disappearing for spherical particles ($p = 1$). H is inversely proportional to rotatational diffusion coefficient, D_{rot}, which can be expressed through viscosity of the medium, η , particle volume, v , and another shape function, $f(p)$. Using Arrhenius form of the viscosity function, one obtains

$$D_{rot} = \frac{kT}{\eta v f(p)} = \text{const.} \frac{kT}{v f(p)} \exp(-E_\eta/kT) \tag{3.95}$$

where E_η is activation energy of viscous flow. Note, that the potentials H_1 and H_{cl} , are functions of temperature. For a viscous fluid, the product of velocity gradient and viscosity is equivalent to normal stress difference, $\Delta p = p_{zz} - p_{xx}$

$$(H_1, H_{cl}) \propto (q^*/D_{rot}) \propto (q^*\eta) \propto \Delta p \tag{3.96}$$

The difference of potential densities reduces to the difference of shape functions

$$\Delta u(\vartheta) = \frac{U_{cl}}{v} - \frac{U_1}{v_0} = -\text{const. } \Delta p \cos^2 \vartheta [R(p_{cl})f(p_{cl}) - R(p_1)f(p_1)] \quad (3.97)$$

p_{cl} and p_1 , denote aspect ratios of the cluster and of the crystallizing molecule. The energy difference Δu disappears when cluster growth does not not lead to a change of shape $(p_{cl} = p_1)$.

The bulk free energy of aggregation, $\Delta f(T, \vartheta)$ includes potential component, $\Delta u(T, \vartheta)$, and entropy term related to orientation distribution of the crystallizing molecules, $w_1(\vartheta)$

$$\Delta f(T, \vartheta) = \Delta f_0(T) + \Delta u(T, \vartheta) - \frac{kT}{v_0} \ln\left[w_1(\vartheta)\Delta_\theta\right] \quad (3.98)$$

Taking a system at rest with randomly distributed molecules as a reference, $(\Delta u = 0, w_1 = \Omega^{-1})$ we arrive at the bulk free energy of aggregation

$$\Delta f(T, \vartheta) = \Delta f_{un}(T) + \Delta u(T, \vartheta) - \frac{kT}{v_0} \ln\left[\Omega \cdot w_1(\vartheta)\right] \quad (3.99)$$

Ω is measure of the orientation space. Consider equilibrium distribution of the flowing units

$$w_1(\vartheta) = w_{1,eq}(\vartheta) = \frac{\exp\left[-H_1 \cos^2 \vartheta\right]}{Z_1} \quad (3.100)$$

where Z_1 is temperature-dependent statistical integral

$$Z_1[H_1(T), T] = \int e^{-H_1 \cos^2 \vartheta} \sin \vartheta d\vartheta \quad (3.101)$$

The orientation-dependent term is reduced to the cluster component, yielding

$$\Delta f(T, \vartheta) = \Delta f_{un}(T) + \frac{U_{cl}(v, \vartheta)}{v} + \frac{kT}{v_0} \ln\left[\frac{Z_1(T)}{\Omega}\right] =$$
$$= \Delta f_{un}(T) + \frac{kTH_{cl}(T) \cos^2 \vartheta}{v} + \frac{kT}{v_0} \ln\left[\frac{Z_1(T)}{\Omega}\right] \quad (3.102)$$

Application of the condition of equilibrium, yields equation for orientation-dependent transition (melting) temperature, T_m

$$\frac{T_m}{T_{m,un}} = \frac{1}{1 - \dfrac{kT_{m,un}H_{cl}(T_m) \cos^2 \vartheta}{v\Delta h} - \dfrac{kT_{m,un}}{v_0 \Delta h} \ln \dfrac{Z_1(T_m)}{\Omega}} \quad (3.103)$$

Note, that both, H_{cl} , and Z_1 depend on temperature. Figure 3.9 presents melting temperature as a function of orientation. It is evident, that $T_m(\vartheta)$

is elevated above the reference level, $T_{m,un}$ (unoriented system) for crystals nearly parallel to flow axis (small angles ϑ , maximum at $\vartheta = 0$). In the range of large angles $T_m(\vartheta)$ is suppressed and reaches minimum at $\vartheta = \pi/2$. The maximum is the higher, and the minimum the deeper, the higher is absolute value of the flow potential, H_{cl} .

For well oriented crystals, critical crystallization (melting) temperature, T_m , can be elevated by tens of degrees, compared with an unoriented system.

Orientation-dependent free energy of aggregation in the vicinity of thermodynamic equilibrium (Eq. 3.102) leads to orientation-dependent critical cluster volume, $v^*(\vartheta)$

$$\frac{v^*(\vartheta)}{v_0^*} = \left(\frac{\Delta f_{un}}{\Delta f(\vartheta)}\right)^3 = \left(\frac{\Delta f_{un}}{\Delta f_{un} + \dfrac{kTH_{cl}\cos^2\vartheta}{v} + \dfrac{kT}{v_0}\ln\left[\dfrac{Z_1}{\Omega}\right]}\right)^3 \qquad (3.104)$$

Selectivity of crystal nucleation, consisting in orientation-dependent crystallization temperature and crystallization rates, is a characteristic of *oriented systems*. Orientation-dependent critical cluster volume, $v^*(\vartheta)$, determines stability region in the configurational space and implies two different mechanisms of thermal nucleation (Figure 3.10). In elongational flow, cluster can reduce its free energy and achieve status of a stable nucleus either via *growth* at constant orientation, or via *rotation* towards flow axis at constant dimensions.

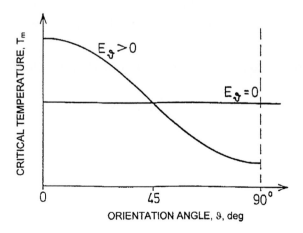

Fig. 3.9. Melting temperature, T_m , in a uniaxial potential field as a function of orientation angle ϑ (schematic).

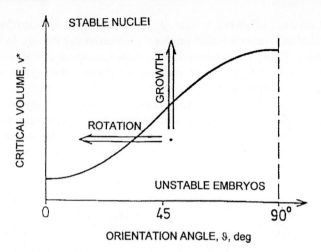

Fig. 3.10. Critical cluster volume, v^* , in a uniaxial potential field as a function of orientation angle ϑ (schematic). Two nucleation mechanisms indicated: growth at constant orientation, and rotation towards field axis without change of dimensions.

3.5.2 Crystallization of Flexible Chains. Constraints and Morphological Effects

When crystallization takes place in a system of flexible polymer chains, there appear elastic and orientation effects in the aggregation free energy. Bulk free energy density related to addition to the growing cluster of one kinetic unit from the chain composed of N segments and having end-to-end vector $\boldsymbol{h}$, reads

$$\Delta f(T, \boldsymbol{a}, \boldsymbol{h}, N) = \Delta f_0(T) + \Delta f_{el}(T, \boldsymbol{h}, N) + \Delta f_{or}(T, \boldsymbol{a}, \boldsymbol{h}, N)$$

$$= \Delta f_0(T) + \Delta f_{el}(T, \boldsymbol{h}, N) - \frac{kT}{v_0} \ln \left[w_1(\boldsymbol{a}, \boldsymbol{h}, N) \cdot \Delta_\theta \right]$$

$$(3.105)$$

The first additional term, Δf_{el} , denotes change in the *elastic free energy* of the system, the second one - *orientation effect*, i.e. the probability of finding the kinetic unit with appropriate orientation. Orientation is characterized by the segment vector $\boldsymbol{a}$, and $w_1(\boldsymbol{a}; \boldsymbol{h}, N)$ denotes orientation distribution of crystallizing units. The kinetic unit in the process of cluster growth (crystallization) is a group of n statistical chain segments (Kuhn segments). Attachment of one group of n segments to the growing cluster (crystal) is accompanied by change of the amorphous chain configuration from $(\boldsymbol{h}, N)$ to $(\boldsymbol{h}', N')$. The corresponding change of the elastic free energy density is thus

$$\Delta f_{el} = \frac{f_{el}(\boldsymbol{h}', N') - f_{el}(\boldsymbol{h}, N)}{v_0} \qquad (3.106)$$

where v_0 denotes volume of the crystallizing unit. Elastic free energy and segment orientation distribution can be calculated numerically in the entire range of chain deformations [24, 25]. In this example, we will confine our considerations to small deformations, describable by the Gaussian statistics, for which elastic free energy of a polymer chain reduces to

$$f_{el}(\boldsymbol{h}, N) \cong \frac{3kT h^2}{2N a^2} \tag{3.107}$$

and segment orientation distribution

$$w_1(\boldsymbol{a}, \boldsymbol{h}) \cong C \exp\left(\frac{3\boldsymbol{h} \cdot \boldsymbol{a}}{N a^2}\right) \tag{3.108}$$

Δf_{el} from Eq.(3.106) depends on conformation changes of the crystallizing macromolecule, and can be split into two effects: loss of entropy-controlled free energy of $n = N - N'$ crystallizing segments moved from the amorphous phase to the growing cluster (crystal), and change of elastic free energy of the N' uncrystallized segments in the residual chain. In the Gaussian approximation

$$\Delta f_{el} \cong \frac{(N - N') \, f_{el}(\boldsymbol{h}, N)}{v_0} \frac{}{N} + \frac{N'}{v_0} \left[\frac{f_{el}(\boldsymbol{h}', N')}{N'} - \frac{f_{el}(\boldsymbol{h}, N)}{N}\right] =$$
$$= -\frac{3kT(N - N')h^2}{2N^2 a^2 v_0} + \frac{3kT N'}{2a^2 v_0}\left[\frac{(h')^2}{(N')^2} - \frac{h^2}{N^2}\right] \tag{3.109}$$

The first term associated with crystallizing segments is always negative, the second one - positive or negative, dependently on whether the residual chain is relaxed, or additionally stressed upon crystallization. Both effects depend on the source of crystallizing segments (intermolecular or intramolecular), and external conditions (stress, constraints, etc). Two modes of cluster growth: intermolecular (extended-chain, bundle-like) and intramolecular (folded-chain) will be discussed separately. It will be shown that *selective crystallization* - the characteristic feature of rigid molecules in an orienting field (Section 3.5.1), is also expected for extended-chain or bundle-like growth, in which kinetic units (groups of chain segments) are provided by many macromolecules. On the other hand, in intramolecular, folded-chain crystallization, where all segments come from a single macromolecule, such effects are not predicted.

Intermolecular (Extended-chain, Bundle-like) Crystallization Consider an *intermolecular* cluster (Figure 3.11), composed of g stems, each containing n statistical chain segments. Attachment of $n = N - N'$ segments (one stem) originating from the chain $(\boldsymbol{h}, N)$ yields

$$\Delta f_{el}(T, \boldsymbol{h}, \boldsymbol{h}', N) = -\frac{3kT n h^2}{2N^2 a^2 v_0} + \frac{3kT(N - n)}{2a^2 v_0}\left[\frac{(h')^2}{(N - n)^2} - \frac{h^2}{N^2}\right] \tag{3.110}$$

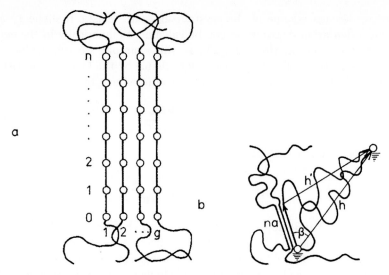

Fig. 3.11. Intermolecular crystallization. a) Scheme of a bundle-like cluster composed of g stems originating from different chains. Each stem contains n statistical chain segments; b) Conformation changes in the amorphous chain with fixed ends.

Since individual groups of segments are provided by different chains, what should appear in the driving force for nucleation is elastic free energy *averaged* over the appropriate distribution of chain configurations

$$\langle \Delta f_{el} \rangle = -\frac{3kTn\langle h^2 \rangle}{2N^2 a^2 v_0} + \frac{3kT(N-n)}{2a^2 v_0} \left[\frac{\langle (h')^2 \rangle}{(N-n)^2} - \frac{\langle h^2 \rangle}{N^2} \right] \qquad (3.111)$$

When the system is relaxed, the elastic free energy reduces to zero

$$\langle h^2 \rangle = Na^2 \; ; \quad \langle (h')^2 \rangle = N'a^2 \; ; \quad \langle \Delta f_{el} \rangle = 0 \qquad (3.112)$$

Therefore, elastic effects contribute to crystallization only in the presence of an external field and/or deformation. When elastic stress in the amorphous part of the crystallizing polymer does not change on crystallization, second term in Eq. (3.111) disappears, and nucleation is controlled by the first, negative term

$$\langle \Delta f_{el} \rangle = -\frac{3kTn\langle h^2 \rangle}{2N^2 a^2 v_0} \qquad (3.113)$$

Such a situation seems to be typical for crystallization in steady flow, or stress field. Alternatively, when ends of the crystallizing chain are fixed (as is the case in crosslinked, constrained systems, Figure 3.11b)

$$h' = h - na \; ; \quad \cos \beta = \frac{h \cdot a}{ha} \qquad (3.114)$$

the elastic free energy depends on dimensions of the cluster and its inclination to the end-to-end vector $\boldsymbol{h}$

$$
\begin{aligned}
\langle \Delta f_{el} \rangle &= \frac{3kTn}{2N(N-n)a^2 v_0}[\langle h^2 \rangle + N(na^2 - 2\langle \boldsymbol{h} \cdot \boldsymbol{a} \rangle)] = \\
&= -\frac{3kTn}{2N(N-n)a^2 v_0}[\langle h^2 \rangle + N(na^2 - 2\langle ha \cos \beta \rangle)]
\end{aligned} \tag{3.115}
$$

It is evident that the second term in Eq.(3.115) assumes positive values (increase of free energy of the system, reduction of crystallization power) when the growing cluster (crystal) is inclined to the end-to-end vector $\boldsymbol{h}$ at large angles β. On the other hand, cluster formed along the vector $\boldsymbol{h}$ ($\beta \cong 0$) may reduce tension in the amorphous part, stimulating crystallization. Stress relaxation accompanying crystallization of stretched polymer networks was observed and analyzed fifty years ago by Flory [26] and has often been discussed.

All n segments within each kinetic unit (stem) are parallel. Probability of finding first segment in the group is proportional to segment distribution function, $w_1(\boldsymbol{a})$. Once orientation of the first segment is defined, subsequent $n - 1$ segments originating *from the same chain* are provided as a matter of course, with probability $p = 1$. Attachment of the first segment of the second stem from another amorphous chain (segment $n + 1$) requires finding a segment with orientation parallel ($\boldsymbol{a}$) or antiparallel ($-\boldsymbol{a}$) to the preceding stem. Since *sense of the vector $\boldsymbol{a}$* is indistguishable, the original segment distribution, Eq.(3.108) should be *symmetrized* to yield

$$
\begin{aligned}
w_{1,sym}(\boldsymbol{a}, \boldsymbol{h}, N) &= \frac{1}{2}[w_1(\boldsymbol{a}) + w_1(-\boldsymbol{a})] \\
&= \frac{1}{2}C \left[\exp\left(\frac{3\boldsymbol{h} \cdot \boldsymbol{a}}{Na^2}\right) + \exp\left(\frac{-3\boldsymbol{h} \cdot \boldsymbol{a}}{Na^2}\right) \right] = C \, \mathrm{ch}\left(\frac{3\boldsymbol{h} \cdot \boldsymbol{a}}{Na^2}\right)
\end{aligned} \tag{3.116}
$$

Table 3.1 presents orientations and probabilities of individual chain segments in a bundle-like cluster composed of g stems ($g \cdot n$ segments).

Orientation-dependent free energy density, Δf_{or}, should be averaged over the appropriate distribution of chains providing individual groups of segments

$$
\begin{aligned}
\langle \Delta f_{or} \rangle(\boldsymbol{a}) &= -\frac{kTn}{v_0}\langle \ln[w_{1,sym}(\boldsymbol{a}, \boldsymbol{h}, N)\Delta_\theta] \rangle = -\frac{kTn}{v_0}\left\langle \ln\left[\mathrm{ch}\left(\frac{3\boldsymbol{h} \cdot \boldsymbol{a}}{Na^2}\right)\right]\right\rangle \\
&= -\frac{9kT}{2v_0}\left[\frac{\langle (\boldsymbol{h} \cdot \boldsymbol{a})^2 \rangle}{N^2 a^4} - \frac{15\langle (\boldsymbol{h} \cdot \boldsymbol{a})^4 \rangle}{4N^4 a^8} + \dots\right]
\end{aligned} \tag{3.117}
$$

Terms with N^{-4} and higher negative powers of chain length, result from expansion of the ln(ch x) function. For "affine" distribution of Gaussian chains subjected to isochoric, uniaxial deformation along the axis Z

Table 3.1. Table 1. Orientations of individual segments in a bundle-like cluster (cf. Figure 3.11a)

segment #	orientation	probability, p_i
1	a	$w_{1,sym}(a)$
2	a	1
...	...	...
n	a	1
$n+1$	a (or $-a$)	$w_{1,sym}(a)$
$n+2$	a (or $-a$)	1
...	...	...
$2n$	a (or $-a$)	1
...	...	...
$n(g-1)+1$	a (or $-a$)	$w_{1,sym}(a)$
...	...	...
ng	a (or $-a$)	1
joint probability		
$\prod_{i=1}^{g\cdot n} p_i$		$[w_{1,sym}(a)]^g$

$$\Lambda = \begin{bmatrix} \dfrac{1}{\sqrt{\lambda}} & 0 & 0 \\ 0 & \dfrac{1}{\sqrt{\lambda}} & 0 \\ 0 & 0 & \lambda \end{bmatrix} \tag{3.118}$$

conformation distribution of polymer chains assumes the form

$$W(h,N) = \text{const. } \exp\left[\frac{3(\Lambda^{-1}h)^2}{2Na^2}\right]$$

$$= \left(\frac{3}{2\pi Na^2}\right)^{3/2} \exp\left[-\frac{3}{2Na^2}\left(\lambda(x^2+y^2)+\frac{z^2}{\lambda^2}\right)\right] \tag{3.119}$$

Segment vector, a , in a fixed coordinate system can be expressed through polar angles

$$a = a \begin{bmatrix} \sin\vartheta\cos\varphi \\ -\sin\vartheta\cos\varphi \\ \cos\vartheta \end{bmatrix} \tag{3.120}$$

which yields global orientation distribution of chain segments

$$\langle w_1(\vartheta)\rangle = \langle w_{1,sym}(\vartheta)\rangle = C \exp\left\{\frac{3}{2N}\left[\cos^2\vartheta\left(\lambda^2-\frac{1}{\lambda}\right)+\frac{1}{\lambda}\right]\right\} \tag{3.121}$$

When averaging is performed with a uniaxial end-to-end distribution $W(h)$, the global orientation function is insensitive to prior symmetrization. Substituting elastic (Eq. 3.113) and orientation (Eq. 3.117) terms averaged with the end-to-end distribution (3.119) to Eq. (3.105), we obtain aggregation free energy density for bundlelike crystallization in a stressed (but not constrained) system in the form

$$\langle \Delta f(T, \vartheta) \rangle = \Delta f_0 - \frac{nkT}{2Nv_0} \left[\lambda^2 + \frac{2}{\lambda} \right]$$
$$- \frac{3kT}{2Nv_0} \left[\cos^2 \vartheta \left(\lambda^2 - \frac{1}{\lambda} \right) + \frac{1}{\lambda} \right] + \frac{1}{N^2}(\ldots) \qquad (3.122)$$

The present analysis slightly differs from our earlier treatment based on pre-averaged orientation distribution applied to uniaxially and biaxially oriented polymer systems [24, 25, 27].

Introducing reference free energy density for an unoriented and undeformed system ($\lambda = 1$)

$$\Delta f_{un} = \Delta f_0 - \frac{3nkT}{2Nv_0} - \frac{3kT}{2Nv_0} + \frac{1}{N^2}(\ldots) \qquad (3.123)$$

one obtains from Eq.(3.122)

$$\langle \Delta f(T, \vartheta) \rangle = \Delta f_{un} - \frac{nkT}{2Nv_0} \left[\lambda^2 + \frac{2}{\lambda} - 3 \right]$$
$$- \frac{3kT}{2Nv_0} \left[\cos^2 \vartheta \left(\lambda^2 - \frac{1}{\lambda} \right) + \frac{1}{\lambda} - 1 \right] + \frac{1}{N^2}(\ldots) \qquad (3.124)$$

It can be shown that similar results are obtained when, instead of uniaxial deformation, the system is subjected to steady elongational flow, Eq. (3.92). In both cases, orientation effects exhibit $\cos^2 \vartheta$ symmetry. The same symmetry has been discussed in Section 3.5.1. for orientation of rigid rods. From Eq. (3.124) one can calculate deformation- and orientation-dependent melting temperature, reduced with the reference value for an unoriented and unstressed system

$$\frac{T_m(\vartheta)}{T_{m,un}} = 1 \left/ \left[1 - \frac{kT_{m,un}}{2N\Delta h v_0} \left[3\cos^2 \vartheta \left(\lambda^2 - \frac{1}{\lambda} \right) + n\lambda^2 \right. \right. \right.$$
$$\left. \left. \left. + \frac{2n+3}{\lambda} - 3(n+1) + \frac{1}{N}(\ldots) \right] \right] \right. \qquad (3.125)$$

Critical cluster volume, which controls nucleation rate results in the form

$$\frac{v^*(\vartheta)}{v_{un}^*} = \Delta f_{un}^3 \left/ \left[\Delta f_{un} - \frac{kT}{2Nv_0} \left[3\cos^2 \vartheta \left(\lambda^2 - \frac{1}{\lambda} \right) + n\lambda^2 \right. \right. \right.$$
$$\left. \left. \left. + \frac{2n+3}{\lambda} - 3(n+1) + \frac{1}{N}(\ldots) \right] \right]^3 \right. \qquad (3.126)$$

Nucleation rates (controlled by v^*) are orientation-dependent, strongly enhanced in the range of small ϑ and suppressed for ϑ close to $\pi/2$. Similar effects have been predicted for crystallization of rigid, rodlike molecules in elongational flow (Section 3.5.1). Figures 3.9 and 3.10 apply also to the present case.

The notion of selective crystallization in oriented systems has been discussed in early sixties by Krigbaum and Roe [28]. Later, the effect has been derived from the nucleation theory with the condition of consistent orientations by Ziabicki and Jarecki [16, 29]. Krigbaum and Roe assumed that in an oriented system, formation of a critical cluster with orientation ϑ is controlled by the probability of finding simultaneously g^* segments with the same orientation. The resulting crystal orientation distribution was postulated in the form [28]

$$w_{cr}(\vartheta) = C[w_1, (\vartheta)]^{g^*} \tag{3.127}$$

where $w_1(\vartheta)$ is orientation distribution of segments in the crystallizing amorphous phase, and $g^* = v^*/v_0$ is number of kinetic elements in the critical cluster. On the other hand, Ziabicki and Jarecki [29] obtained *kinetically controlled* cluster (crystal) orientation distribution in the form

$$w_{cr}(\vartheta) \propto j(\vartheta) \propto \exp\left[\frac{-\Delta F^*(\vartheta)}{kT}\right]$$

$$= C \exp\left[\frac{-\Delta F_0^*}{kT}\left(\frac{\Delta f_{un}}{\Delta f_{un} - \dfrac{kT}{v_0}\ln[w_1(\vartheta)\cdot\Omega]}\right)^2\right] \tag{3.128}$$

Both results predict concentration of crystals in the orientation region rich in crystallizing segments (in the case of uniaxial orientation, near $\vartheta = 0$), and practical elimination in regions with infrequent amorphous segments. Critical cluster size (v^*, or g^*), a constant in the Krigbaum-Roe treatment, in our nucleation theory is a function of orientation. Therefore Krigbaum-Roe distribution is somewhat overestimated in the region of maximum ($\vartheta = 0$). Figure 3.12 presents orientation distributions in a polymer system crystallized in an oriented state. Weak amorphous orientation (amorphous orientation factor, $f_a = 0.144$) induces very sharp crystal distribution ($f_c = 0.967$ for the distribution 128 and $f_c = 0.984$ for Krigbaum- Roe distribution, Eq. (3.127)). In the example studied, constant critical cluster size in Eq. (3.127) was assumed, corresponding to an unoriented system ($g_{un}^* = 96.03$). The actual cluster size in the range of small angles was considerably reduced, in the the range of small angles ($\vartheta = 0$), amounting to 9.08. This explains overly sharp distribution predicted by Krigbaum and Roe. It is evident that intermolecular (bundle-like or extended-chain) crystallization, like crystallization of rigid rods, is sensitive to orientation. *Selectivity of crystallization* (preference for crystals oriented along the main orientation axis, discrimination of

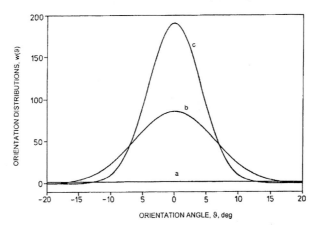

Fig. 3.12. Orientation distributions in an oriented system. a) - orientation of the crystallizing amorphous segments, $w_1(\vartheta)$, Eq. (3.100) with $H_1 = -1$; b) - orientation of critical clusters, $w_{cr}(\vartheta)$, from the Krigbaum-Roe model (Eq. 3.126, $g_0^* = 96.03$); c) - nucleation-controlled crystal orientation, $w_{cr}(\vartheta)$, Eq.(3.127). Reference melting temperature, $T_{m,un} = 553K$, crystallization temperature, $T = 473K$.

perpendicular ones), concerns thermodynamic and kinetic characteristics of crystallization.

Intramolecular (Folded-chain) Crystallization Consider a regularly chain-folded cluster (Figure 3.13a) containing g folds, each containing n statistical chain segments, all originating from the same macromolecule, (h, N). Unlike in the intermolecular (bundlelike) crystallization discussed above, we will analyze free energy of formation of the entire cluster, $\Delta F(g)$, rather than free energy density, Δf, related to attachment of a single kinetic unit (one fold). The total free energy is specific for configuration of the crystallizing chain (h, N), and should not be averaged over chain configurations

$$\Delta F(T, \boldsymbol{a}, g, \boldsymbol{h}, N) = \Delta F_0(T, g) + \Delta F_{el}(T, g, \boldsymbol{h}, N) + \Delta F_{or}(T, \boldsymbol{a}, g, \boldsymbol{h}, N) \tag{3.129}$$

The elastic contribution (except for averaging) resembles one for bundle-like crystallization

$$\Delta F_{el}(g, T, \boldsymbol{h}, \boldsymbol{h}', N) =$$
$$= -g \cdot n \frac{f_{el}(\boldsymbol{h}, N)}{N} + (N - g \cdot n) \left[\frac{f_{el}(\boldsymbol{h}', N - g \cdot n)}{N - g \cdot n} - \frac{f_{el}(\boldsymbol{h}, N)}{N} \right]$$
$$= -\frac{3kTg \cdot nh^2}{2N^2 a^2} + \frac{3kT(N - g \cdot n)}{2a^2} \left[\frac{(h')^2}{(N - g \cdot n)^2} - \frac{h^2}{N^2} \right] \tag{3.130}$$

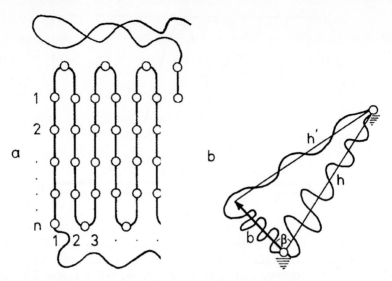

Fig. 3.13. Intramolecular crystallization. a) Scheme of a folded-chain cluster composed of g folds ($g \cdot n$ statistical segments) originating from a single chain; b) Conformation changes in the amorphous chain with fixed ends.

Here, like in Eq.(3.110), the second term can be positive or negative, depending on the configuration of the residual chain, h'. In the case of crystallization with unchanged free energy of the amorphous part, elastic effect reduces to

$$\Delta F_{el}(T, g, h, h', N) = -g \cdot n \frac{f_{el}(h, N)}{N} \cong -\frac{3kTg \cdot nh^2}{2N^2a^2} \qquad (3.131)$$

On the other hand, for folded-chain crystallization in a constrained (crosslinked) system with fixed chain ends (Figure 3.13b), characterized by chain conformation

$$h' = h - b = h - gb_0 \; ; \quad \cos \beta = \frac{h \cdot b_0}{hb_0} \qquad (3.132)$$

where b is cluster vector perpendicular to chain fold, we obtain

$$\Delta F_{el}(T, g, h, N) =$$

$$= \frac{3kTg}{2N(N - g \cdot n)a^2}[nh^2 + N(b_0^2 g \cdot n - 2h \cdot b_0)] = \qquad (3.133)$$

$$= \frac{3kTg}{2N(N - g \cdot n)a^2}[nh^2 + N(b_0^2 g - 2h \, b_0 \cos \beta)]$$

Morphological characteristics (cluster dimensions, number of segments per fold, inclination of cluster to chain vector) determine sign and magnitude of the free energy contribution. Variation of morphological details can strongly affect thermodynamics and kinetics of crystallization. An example of such an analysis can be found in the paper by Kosc and Ziabicki [30] concerned with crystallization of stretched model networks.

More dramatic difference between inter-, and intramolecular crystallization concerns *orientation effects*. Segments within each fold are parallel, and their orientation alternates from one fold to the next one. Probability of finding first segment in the first fold is proportional to segment orientation distribution $w_1(a;h)$. When first segment has been defined, all subsequent segments originating from the same chain are provided automatically with probability $p = 1$ (Table 3.2.).

Table 3.2. Orientations and probabilities of individual segments in a folded-chain cluster (cf. Figure 3.13a)

segment #	orientation	probability, p_i
1	a	$w_1(a)$
2	a	1
...	...	...
n	a	1
$n+1$	$-a$	1
$n+2$	$-a$	1
...	...	...
$2n$	$-a$	1
...	...	...
$n(g-1)+1$	a for odd g $-a$ for even g	1
...	...	...
ng	a for odd g $-a$ for even g	1
joint probability $\prod_{i=1}^{g \cdot n} p_i$		$w_1(a)$

The total orientation term for the g-fold (gn-segment) cluster is independent of cluster size and reduces to orientation of the first segment

$$\Delta F_{or}(T, a, g, h, N) = -kT \ln [w_1(a, h, N)\Delta_\theta] \cong -\frac{3kT(h \cdot a)}{Na^2} + \text{const}$$
(3.134)

Assuming unchanged elastic energy of the uncrystallized part of the chain (Eq. 3.131), we arrive at the free energy of cluster formation in the form

$$\Delta F(T, g, a, h, N) = \Delta F_0 - \frac{3n \cdot gkTh^2}{2N^2a^2} - \frac{3kT(h \cdot a)}{Na^2}$$
(3.135)

which, referred to an unoriented and unstressed system (ΔF_{un}), yields

$$\Delta F(T, g, \boldsymbol{a}, \boldsymbol{h}, N) = \Delta F_{un} - \frac{3n \cdot gkT}{2N} \left[\frac{h^2}{Na^2} - 1 \right] - \frac{3kT(\boldsymbol{h} \cdot \boldsymbol{a})}{Na^2} \quad (3.136)$$

The fact that orientation term is independent of cluster size, g, leads to far going thermodynamic and kinetic consequences. Melting temperature, T_m, calculated from free energy density in the thermodynamic limit

$$\lim_{g \to \infty} \frac{\Delta F_{un}(T, g) + \Delta F_{el}(T, g, \boldsymbol{h}, N) + \Delta F_{or}(T, \boldsymbol{a}, \boldsymbol{h}, N)}{g} = 0 \quad (3.137)$$

and elastic free energy from Eq. (3.131) is independent of cluster orientation

$$\frac{T_m(\boldsymbol{h}, N)}{T_{m,un}} = \frac{1}{1 - \dfrac{3nkT_{m,un}}{2N^2 a^2 v_0 \Delta h_0} (h^2 - Na^2)} \quad (3.138)$$

According to Eq.(3.138), T_m is elevated for chains with high end-to-end distance ($h^2 > Na^2$) and suppressed for coiled ones ($h^2 < Na^2$), what seems to suggest preference for crystallization of highly extended chains. Maximization of ΔF with respect to cluster size yields critical cluster volume, v^*

$$\frac{v^*(T, \boldsymbol{h}, N)}{v_{un}^*} = \left[\frac{\Delta f_{un}}{\Delta f_{un} - \dfrac{3nkT}{2N^2 a^2 v_0} (h^2 - Na^2)} \right]^3 \quad (3.139)$$

which also *does not depend on orientation*. On the other hand, maximum free energy of cluster formation results as a weak function of orientation, viz.

$$\Delta F^*(T, \boldsymbol{a}, \boldsymbol{h}, N) =$$

$$= \Delta F_{un}^*(T) \left[\frac{\Delta f_{un}}{\Delta f_{un} - \dfrac{3nkT}{2N^2 a^2 v_0} (h^2 - Na^2)} \right]^2 - kT \ln [w_1(\boldsymbol{a}, \boldsymbol{h}, N) \cdot \Omega]$$

$$= \Delta F_{un}^*(T) \left[\frac{\Delta f_{un}}{\Delta f_{un} - \dfrac{3nkT}{2N^2 a^2 v_0} (h^2 - Na^2)} \right]^2 - \frac{3kT(\boldsymbol{h} \cdot \boldsymbol{a})}{Na^2} + C \quad (3.140)$$

Kinetically controlled orientation distribution of *chain-folded crystals* resulting from Eq.(3.140) is identical with orientation distribution of *crystallizing amorphous segments*

$$w_{cr}(\vartheta) \propto \exp \left[\frac{-\Delta F^*(\vartheta)}{kT} \right] = w_1(\vartheta; \boldsymbol{h}, N) \quad (3.141)$$

The appearance of a folded-chain cluster (and, consequently, crystal) with orientation ϑ is controlled by the probability of finding first segment of the first fold. This is determined by the amorphous distribution, $w_1(\vartheta)$. Compared with intermolecular crystals, orientation distribution for chain-folded crystals is very broad. In Figure 3.12, w_1 (identical with w_{cr} from Eq. 3.141) is very broad.

The fact that selective effects in crystallization thermodynamics and kinetics are confined to intermolecular (extended-chain or bundle-like) crystallization, throws some light on observations made on flexible chain polymers crystallizing in the conditions of orientation. X-Ray and optical studies of the crystallization of molten polypropylene show crystallization temperatures in elongational flow elevated by up to 22 degs compared with quiescent conditions [31]. Time lag of crystallization and crystallization rates are also strongly affected by the flow [32]. Another evidence of selective crystallization is provided by the work of Peters on injection molding of polypropylene melt. The degree of crystal orientation was found to be evidently higher than that of amorphous segments [33], which is an evidence for selective crystallization. All these observations seem to suggest orientation-sensitive intermolecular aggregation. Selective crystallization cannot be attributed to regular chain folding, although it does not exclude some amount of folded-chain crystals superimposed on bundle-like structures (mixed morphology, cf. Section 3.5.7. below). More experimental facts speaking in favor of selective crystallization will be discussed in Sections 3.5.3 and 3.5.4.

3.5.3 High-speed Melt Spinning

An example of industrial realization of oriented polymer crystallization is provided by high-speed melt spinning. Formation of synthetic fibers with take-up speeds above 4000 m/min leads to uniaxial orientation of chain segments, and orientation-controlled crystallization. Results of studies on polyesters, polyamides, polypropylene and other polymers, show that crystals obtained in high-speed spinning are highly oriented along fiber axis, and that, nearly independently of spinning speed [34]. An example of such a behavior for polyethylene terephthalate (PET) fibers is shown in Figure 3.14 [27]. Orientation factors - f_c and f_a - moments of the respective orientation distributions $w_{cr}(\vartheta)$ and $w_1(\vartheta)$ - characterize average orientation of crystals and amorphous segments in the system. High crystal orientation in Figure 3.14 ($f_c = 0.90 - 0.98$) provides experimental evidence for selective crystallization described in Section 3.5.2. In a system oriented in elongational flow, preferentially formed are crystals parallel to fiber axis ($\vartheta \cong 0$), while formation of perpendicular crystals is thermodynamically and kinetically suppressed. Relatively low degrees of orientation of the amorphous phase ($f_a = 0.2 - 0.4$) are due to other mechanisms, like rotation in the flow field, and "consumption" of best oriented segments by growing crystals [27].

3.5.4 Melting of Oriented Crystals Under Tension

The model discussed in Section 3.5.2. predicts selective crystallization, as well as, *selective melting* of oriented crystals in the presence of flow or stress. Sajkiewicz and Wasiak [35] studied crystal orientation in high-molecular-weight polyethylene ($M_w > 10^6$) crystallized and melted under tension. The tension is associated with extension of polymer chains and orientation of chain segments. According to Eq. (3.125), gradual heating of an oriented, crystalline polymer under tension should lead to *selective melting*. The effect can be explained using Figure 3.9. Below the minimum melting temperature, $T_{m,min} = T_m(\vartheta = \pi/2)$, all crystals are stable. When temperature is raised to $T_{m,min}$, the first crystals that melt are perpendicular to fiber axis. With increasing temperature, melting includes crystals with smaller and smaller orientation angles. This leads to narrowing of crystal orientation distribution, and an increase of crystal orientation factor, f_c , with increasing temperature of heating. The data presented in Figure 3.15 [35] are consistent with the above predictions, and provide another piece of evidence for selective crystallization.

3.5.5 Crystallization in an Electric Field

Electric field provides another example of a potential field affecting polymer crystallization. Poly-(vinylidene difluoride) (PVDF) crystallizes in various structural modifications. The polar γ-form exhibits permanent polarization, P_c , and gives rise to dipole effects; two apolar forms (α and β) exhibit only anisotropic polarizability responsible for a quadrupole effect. We have considered the restricted nucleation model described in Section 3.5.1. Cylindrical clusters with thermodynamically optimum shape are characterized by volume, v , and angle ϑ formed by cylinder axis with axis of the electric field

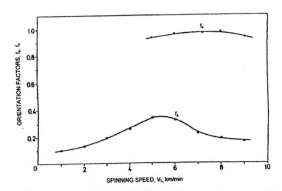

Fig. 3.14. Orientation factors for crystals (f_c) and amorphous segments (f_a) in poly(ethylene terephthalate) fibers as functions of spinning speed, V_L [27].

[36, 37]. E is field intensity. Application of a constant field yields cluster interaction potential in the form

$$U_{cl}(v, \vartheta; E) = U_{di} + U_{quad}$$
$$= - vP_c E \cos \vartheta - \frac{1}{2} v E^2 \varepsilon_0 \left[\varepsilon_c - 1 + \Delta\varepsilon \left(\cos^2 \vartheta - \frac{1}{3} \right) \right] \quad (3.142)$$

where P_c is permanent polarization, ε_0 and ε_c - dielectric constants of the amorphous and crystalline phases, and $\Delta\varepsilon$ - anisotropy of dielectric constants. In the case of the polar form, the interaction potential is dominated by dipole ($\cos \varepsilon$) effect, while in the apolar forms, crystal-field interactions are limited to the quadrupole ($\cos^2 \varepsilon$) effect. The analysis performed in refs. [36, 37] leads to the conclusion, that in the presence of the field, formation of polar (γ) crystals oriented along the field is preferred. Formation of perpendicularly oriented γ-crystals and apolar (α and β) crystals is suppressed. Formation of apolar crystals is not very sensitive to orientation, either. Field effects are the stronger, the higher is crystallization temperature. Figure 3.16 presents calculated critical crystallization temperature as a function of field intensity, E [37]. Experiments of Marand, Stein and Stack [38], and Sajkiewicz [39] confirm these expectations. Experimentally observed fraction of γ-PVDF crystals as a function of crystallization temperature is shown in Figure 3.17.

3.5.6 Crystallization in a Gravitational Field. Ultracentrifuge

Consider a solution crystallizing in an ultracentrifuge (Figure 3.18).

Spherically symmetric molecules with mass, m_1, and volume v_0 (density $d_1 = m_1/v_0$) are dissolved in a liquid of density d_0. It is assumed that clusters are spherically symmetric and their density is nearly equal to that of

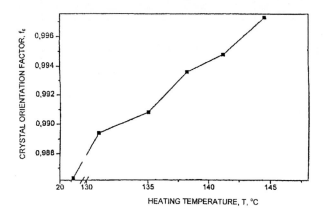

Fig. 3.15. Orientation factor for polyethylene crystals, f_c, heated under constant tension as a function of heating temperature [35].

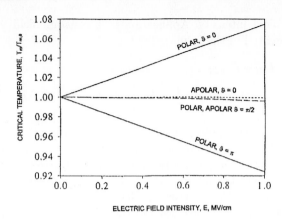

Fig. 3.16. Reduced critical crystallization (melting) temperature, $T_m/T_{m,0}$ for polar (γ) and apolar ($\alpha + \beta$) crystals of polyvinylidene difluoride (PVDF) as a function of field intensity, E [37]. The reference temperature corresponds to $E = 0$.

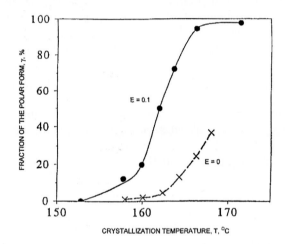

Fig. 3.17. Fraction of polar (γ) crystals of poly(vinylidene difluoride) as a function of crystallization temperature, T [38]. Field intensity constant.

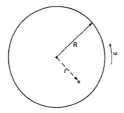

Fig. 3.18. Scheme of crystallization in an ultracentrifuge.

crystallizing molecules, $d_{cl} \cong d_1$. Configuration of the crystallizing molecule is reduced to one variable - distance from the rotor axis, $\zeta = \{r\}$, and configuration of the cluster is described by its volume and position, $\xi = \{v, r\}$. Molecule placed at the distance x from rotor axis, interacts with gravitational field produced in the ultracentrifuge with the potential

$$U_1(r) \cong -m_1 r^2 \omega \left(1 - \frac{d_0}{d_1}\right) = E_{g,1} x^2 \qquad (3.143)$$

where ω is angular velocity, and $x = r/R$, relative radial position in the rotor. With unchanged density, cluster potential is obtained in the form

$$U_{cl}(v, r) = -mr^2 \omega \left(1 - \frac{d_0}{d_1}\right) = E_{g,cl} x^2 \cong \frac{m}{m_1} E_{g,1} x^2 \cong \frac{v}{v_0} E_{g,1} x^2 \quad (3.144)$$

If different thermal expansion coefficients are neglected, the potentials $E_{g,1}$, and $E_{g,cl}$ are independent of temperature. In the steady state, crystallizing molecules assume equilibrium concentration distribution

$$w_{1,eq}(x; E_{g,1}) = \frac{\exp[-E_{g,1} x^2 / kT]}{Z_1} \qquad (3.145)$$

which modifies free energy and kinetic characteristics of aggregation. In the conditions of equilibrium, position-dependent melting temperature is obtained from the equation

$$\frac{T_m(x; E_{g,1})}{T_{m,0}} = \frac{1 - \dfrac{E_{g,1} x^2}{v_0 \Delta h}}{1 + \dfrac{kT_{m,0}}{v_0 \Delta h} \ln Z_1(T_m)} \qquad (3.146)$$

Position-dependent critical cluster volume

$$\frac{v^*(x; E_g)}{v_0^*} = \left(\frac{\Delta f_0}{\Delta f_0 + \dfrac{1}{v_0}(E_{g,1} x^2 + kT \ln Z_1)}\right)^3 \qquad (3.147)$$

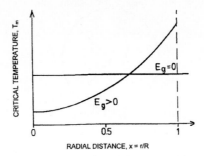

Fig. 3.19. Melting temperature, T_m , as a function of radial position in the ultra-centrifuge, $x = r/R$ (schematic).

implies *position-dependent* crystallization. Figure 3.19 presents melting temperature, and Figure 3.20 - critical cluster volume, as functions of relative distance from rotor axis, x . It is evident that critical conditions of crystallization are first satisfied in the peripheral layer ($r = R$, $x = 1$); nucleation rate in this layer reaches maximum. The dependence of critical cluster size on radial position implies two thermal mechanisms of nucleation: *growth* of subcritical clusters, and their *translation* from the region of higher potential (close to rotor axis), to the peripheral region, without change of dimensions.

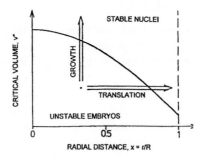

Fig. 3.20. Critical cluster volume, v^* , as a function of radial position in the ul-tracentrifuge, $x = r/R$ (schematic). Two nucleation mechanisms indicated: growth at constant position, and translation towards rotor periphery without change of dimensions.

3.5.7 Polymer Crystallization with Mixed Morphology

In polymer systems, specific assumptions about molecular configuration and the mode of crystal growth strongly affect thermodynamics and kinetics of crystal nucleation. An important role is played by polymer concentration [16, 40, 41]. In the *intramolecular*, folded- chain crystallization, the probability of finding a kinetic element (chain segment) for cluster growth is, in first approximation, independent of concentration. In contrast to that, the probability of *intermolecular*, bundlelike growth is directly proportional to concentration of the crystallizing segments, c .

Consider a mixed cluster composed of $m = v/v_0$ kinetic units (stems) (Figure 3.21). n stems are attached on the cooperative (regular folding) prin-

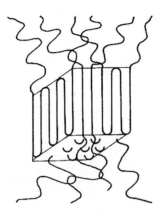

Fig. 3.21. Mixed cluster composed of folded-chain and bundle-like stems.

ciple, and the remaining $(m-n)$ stems originate from various polymer chains. The bulk free energy density of regular folding is taken as a reference

$$\Delta f_{fold} = \Delta f_0 \qquad (3.148)$$

Free energy density associated with intermolecular, bundlelike attachment differs from Δf_{fold} by two additional terms

$$\Delta f_{bundle} = \Delta f_0 + \frac{U_{bundle} - U_{fold}}{v_0} - \frac{kT}{v_0} \ln c = \Delta f_0 + \frac{\delta U_{bundle}}{v_0} - \frac{kT}{v_0} \ln c$$
$$(3.149)$$

The potential energy, U_{fold} , includes energy of chain bending, and U_{bundle} takes into account energy of packing free chain ends on crystal surface. δU can be either positive or negative. Negative δU can be expected in dilute solutions, where bending energy plays the decisive role. In condensed systems (concentrated solutions and melts) δU may assume positive values because of packing requirements. Packing effects have often been used as an argument against bundlelike crystallization.

In the free energy of mixed clusters, there appears an additional entropy term related to possible arrangements of two different kinds of stems

$$\Delta F(m,n) = \Delta F_{fold} + (m-n)\delta U - (m-n)kT \ln c$$
$$+ kT\left[n\ln\frac{n}{m} + (m-n)\ln\frac{n-m}{m}\right] \quad (3.150)$$

The reference value, ΔF_{fold} , denotes free energy of aggregation for a cluster composed of regularly folded segments, and c is molar fraction of crystallizing units. Basic surface energy in ΔF_{fold} is assumed constant, since morphology-affected contributions (bending potential, etc.) have already been included in the extra potential, δU . Combination of energetical and entropic effects in the free energy of aggregation, ΔF , offer different picture than that based on potential energy alone. E.g. in very dilute systems, folded-chain aggregation is favored in spite of higher potential energy. Minimization of the free energy with respect to the number of folded segments

$$\left.\frac{\partial \Delta F}{\partial n}\right|_{m=const.} = 0 \quad (3.151)$$

yields thermodynamically most probable fraction of folded segments

$$\left(\frac{n}{m}\right)_{opt} = \frac{1}{1 + ce^{-\delta U/kT}} \quad (3.152)$$

The corresponding optimum free energy density for a mixed cluster results in the form

$$\Delta f_{opt} = \left(\frac{n}{m}\right)_{opt}\Delta f_{fold} + \left(\frac{m-n}{m}\right)_{opt}\Delta f_{bundle}$$
$$+ \frac{kT}{mv_0}\left[n\ln\frac{n}{m} + (m-n)\ln\frac{m-n}{m}\right]_{opt} =$$
$$= \Delta f_{fold} - \frac{kT}{v_0}\ln[1 + ce^{-\delta U/kT}] \quad (3.153)$$

The condition of thermodynamic equilibrium, Eq.(3.8), leads to the equation for concentration- dependent melting temperature

$$T_m(c) = T_{m,fold}\frac{1}{1 - \dfrac{kT_{m,fold}}{\Delta h_0 v_0}\ln[1 + ce^{-\delta U/kT_m(c)}]} \quad (3.154)$$

The reference heat of melting, and the reference melting temperature, correspond to an ideal, folded-chain crystal. Let us discuss the behavior of "optimized" mixed crystals in asymptotic conditions. At infinitely small concentration $(c \to 0)$ and /or infinite positive energy difference $(\delta U \to +\infty)$ ideal, folded-chain crystal is obtained

$$\left(\frac{n}{m}\right)_{opt} = 1 \tag{3.155}$$

with melting temperature

$$T_m(c) = T_{m,fold} \tag{3.156}$$

In an undiluted polymer ($c \rightarrow 1$) the fraction of folded segments depends on the difference of potential energies

$$\left(\frac{n}{m}\right)_{opt} = \frac{1}{1 + e^{-\delta U/kT}} \tag{3.157}$$

In the case of very large, negative energy difference ($-\delta U \gg 1$), i.e. in the conditions of strong thermodynamic discrimination of folded-chain growth, pure bundlelike crystals are obtained

$$\left(\frac{n}{m}\right)_{opt} = 0 \tag{3.158}$$

and the solution of Eq.(3.154) yields equilibrium melting temperature much higher than the reference temperature for folded-chain crystals

$$\frac{T_m(c)}{T_{m,fold}} = 1 - \frac{\delta U}{\Delta h_0 v_0} \gg 1 \tag{3.159}$$

(It should be considered, though, that large, negative potential difference δU implies very low reference melting temperature, $T_{m,fold}$). When potential energy difference is zero ($\delta U = 0$) the fraction of chain-folded crystals is a function of concentration alone

$$\left(\frac{n}{m}\right)_{opt} = \frac{1}{1 + c} \tag{3.160}$$

and oscillates between $\frac{1}{2}$ (undiluted polymer) and 1 (infinitely dilute solution). In intermediate situations, thermodynamically most probable structure and melting temperature of mixed crystals depend on concentration and potential difference, δU. Figure 3.22 presents optimum fraction of folded stems, and Figure 3.23 - melting temperature of an optimized mixed cluster as a function of polymer concentration. It should be mentioned that mixed crystals, composed of segments supplied by intra- and intermolecular sources, would be sensitive to segment orientation.

3.5.8 Nucleation Controlled by Translational and Rotational Diffusion

The distribution of crystallizing molecules in the 6-dimensional (translational and rotational) space depends on external conditions and aggregation rate.

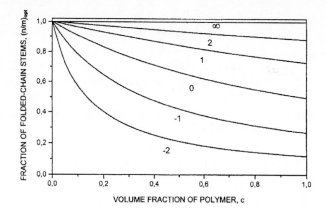

Fig. 3.22. Optimum fraction of folded-chain stems, $(n/m)_{opt}$ as a function of polymer concentration (volume fraction), c . Reduced potential difference, $\delta U/kT$, indicated.

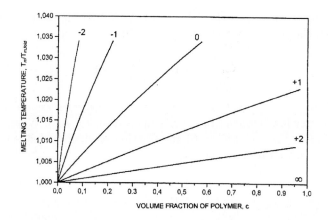

Fig. 3.23. Reduced melting temperature for mixed crystals, $T_m/T_{m,fold}$, as a function of polymer concentration (volume fraction), c . Reduced potential difference, $\delta U/kT_{m,fold}$, indicated. Entropy ratio assumed, $kT_{m,fold}/\Delta h_0 v_0 = 0,0723$.

When aggregation is infinitely slow, unperturbed distribution of crystallizing units is maintained. In the absence of any force fields the distribution is uniform. In equilibrium with an external potential field - a Boltzmann-like distribution is obtained. When aggregation rate is comparable with that of molecular diffusion, concentration of aggregating molecules on the surface of the growing cluster depends on the rate of attachment (consumption) and the rate of transport from remote parts of the system. At very high reaction rates, kinetics of aggregation are completely controlled by diffusion, leading to the situation known as *diffusion limited aggregation* (DLA). It should be stressed that due to the requirement of consistent configurations we are concerned about *position* and *orientation* of each crystallizing kinetic unit.

Effects of translational and rotational diffusion in the kinetics of chemical reactions were discussed by many authors, starting with Smoluchowski [42] and Collins [43], followed by Šolc and Stockmayer [44], Schurr and Schmitz [45, 46] and others.

Consider axially symmetric molecules in the 5-dimensional configurational space, ζ , consisting of three coordinates of the center of mass, $x_0 = \{x_0, y_0, z_0\}$, and two polar angles - $\{\varphi, \vartheta\}$. The distribution of crystallizing units, $w_1(\zeta, t)$ is determined by the equation of diffusion with the appropriate sink ("consumption" of some kinetic units by growing clusters)

$$\frac{\partial w_1}{\partial t} - \text{div}\left[\boldsymbol{D}^{trans}\left(\nabla w_1 + \frac{w_1}{kT}\nabla U_1\right)\right] - D^{rot}\boldsymbol{R}\cdot\left(\boldsymbol{R}w_1 + \frac{w_1}{kT}\boldsymbol{R}U_1\right) =$$
$$= J\{w_1(\zeta), w[\boldsymbol{\xi}(\zeta)]\}$$

(3.161)

where

$$\boldsymbol{R} = \boldsymbol{e} \times \frac{\partial}{\partial \boldsymbol{e}} \; ; \quad R_i = \varepsilon_{ijk}e_j\frac{\partial}{\partial e_k}$$

(3.162)

is rotation operator, ε_{ijk} is Levi-Civita permutation symbol, and J - a functional describing consumption of kinetic units by attachment to growing clusters. The position- and orientation-dependent potential $U_1(\zeta)$ can accomodate effects of steady-state irrotational flow.

Jarecki [47-49] analyzed effects of translational and rotational diffusion in the theory of nucleation using the model presented in Figure 3.24. An anisotropic (axially symmetric) sphere with radius R_1 is reacting with an anisotropic spherical cluster of radius R . Orientation is assumed uniaxial, which reduces the configurational space to four dimensions $\{x_0, y_0, z_0, \vartheta\}$. Jarecki analyzed steady-state concentration of crystallizing molecules on the surface of a selected cluster. In such a treatment, effects of "consumption" can be described by the boundary condition

$$\boldsymbol{n} \cdot (\boldsymbol{D}^{trans}\nabla w_1|_{surf}) = v_0[k^+w_1 - k^-]$$

(3.163)

According to Eq.(3.163), diffusional flux normal to the surface of the growing cluster is balanced by net flux of spheres involved in the reactions of association (k^+) and dissociation (k^-). The rate constants depend on molecular mobility and thermodynamic driving force, δF . The kinetics of aggregation take into account the condition of consistent positions and orientations of the reacting species. Concentration of crystallizing spheres on cluster surface ($r = R + R_1$) may be presented in the form [48]

$$\frac{w_1(r = R + R_1)}{w_{1,eq}} = 1 + [1 - e^{\delta F/kT}] \cdot F(\vartheta, R, D^{trans}, D^{rot}, k^+)$$

(3.164)

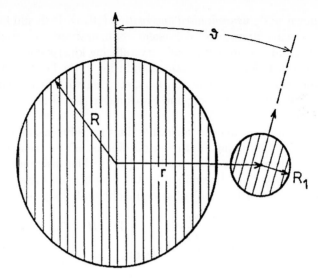

Fig. 3.24. Scheme of reaction between an anisotropic spherical cluster (radius R) and an anisotropic sphere (radius R_1).

where $\delta F = v_0(\partial \Delta F/\partial v)$ is thermodynamic driving force for aggregation. The correction function, can be expressed through two dimensionless groups and the orientation angle ϑ

$$F = F \left(\frac{D^{rot} R^2}{D^{trans}}, \frac{k^+ R}{D^{trans}}, \vartheta \right) \tag{3.165}$$

Concentration of molecules far apart from the reacting cluster is assumed uniform. Figure 3.25 presents deviation of the concentration on cluster surface from the uniform level [48]. At aggregation rates small compared to translational diffusion ($k^+ R/D^{trans} = 0.01$), reduction of surface concentration is relatively small, except for unphysical situation when rotational diffusion is infinitely slow and translational diffusion very fast. For the model considered, translational and rotational diffusion rates are comparable. When aggregation rate is large compared to translational diffusion ($k^+ R/D^{trans} = 100$) surface concentration of crystallizing molecules is reduced and the more so, the smaller is orientation angle, ϑ , and the smaller is the ratio of rotational to translational diffusion. In the limiting case ($D^{rot} R^2/D^{trans} = 0$), all molecules with orientations within the "tolerance angle" Δ_θ (rotational cross-section for aggregation), are captured, their concentration approaches zero and units with "improper" orientations (ϑ outside Δ_θ) remain on the surface and their concentration does not differ from equilibrium. Jarecki analyzed also effect of an external orienting field. The field modifies orientation distribution of the crystallizing molecules and speeds up aggregation when

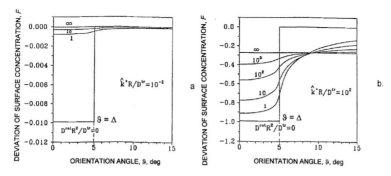

Fig. 3.25. Deviation of concentration of aggregating spheres on the surface of the growing cluster, F, as a function of orientation angle, ϑ [48]. a) $k^+R/D^{trans} = 0.01$; b) $k^+R/D^{trans} = 100$.

"tolerance" angle, Δ , is small and rotational dffusion not too large compared with translational diffusion [47].

3.5.9 Crystallization Memory as a Transient Nucleation Problem

"Memory" of previous structures is a common phenomenon in polymer crystallization. Crystallization rates and resulting structures depend on the original structure and thermal history of the sample. Ziabicki and Alfonso [50] developed a simple model based on transient, one-dimensional nucleation theory, to describe effects of prolonged melting on subsequent crystallization. The thermal history is shown schematically in Figure 3.26. Crystalline sample is heated at the temperature $T_1 > T_m$, and then rapidly cooled down to the temperature $T_{cr} < T_m$ in which isothermal crystallization takes place. What changes at $T = T_1$, is cluster size distribution, $w(g, t)$, which determines concentration of athermal nuclei at the start of crystallization. When the sample is heated, the original distribution, $w_{00}(g)$, tends to equilibrium, $w = w_{eq}(g)$. Approximate solution of the transient one-dimensional nucleation equation, Eq.(3.10) yields cluster size distribution after heating period t_1

$$w(g, t_1; T_1) \cong w_{eq}(g, T_1) + (w_{00} - w_{eq}) \exp\left[-\frac{t_1}{\tau(T_1)}\right] \cong w_{00} \exp\left[-\frac{t_1}{\tau(T_1)}\right]$$
$$(3.166)$$

Integration yields the total number of large clusters (potential nuclei) left in the system after time t_1 spent at T_1

$$N_0(t_1; T_1) = N_{ath}(t_1; T_1) + N_{het} \cong N_{00} \exp\left[-\frac{t_1}{\tau(T_1)}\right] + N_{het} \qquad (3.167)$$

N_{00} denotes the initial number of potential *homogeneous* nuclei, and N_{het} - the number of *heterogeneous* nuclei (foreign particles, surface defects, etc.),

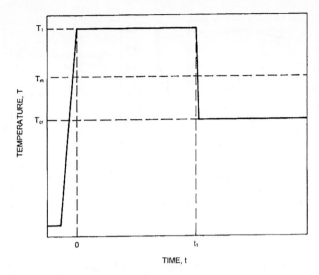

Fig. 3.26. Thermal history in the memory experiment

insensitive to melting. Interpreting isothermal crystallization at $T = T_{cr}$ as a result of 3-dimensional growth of N_0 predetermined nuclei, and measuring crystallization time, t_p , corresponding to maximum crystallization rate as a function of melting period t_1 , we obtain the total number of predetermined nuclei

$$N_0(t_1; T_1) = \text{const. } t_p^{-3}(t_1; T_1) \tag{3.168}$$

Heterogeneous contribution is obtained by extrapolation to infinite melting time

$$N_{het}(T_1) = \lim_{t_1 \to \infty} N_0(t_1; T_1) \tag{3.169}$$

and subtracted from the total number of acting nuclei. Relaxation time, τ , is obtained from the plot

$$\frac{1}{\tau(T_1)} = -\frac{\partial \ln[t_p^{-3}(t_1, T_1) - t_p^{-3}(\infty, T_1)]}{\partial t_1} \tag{3.170}$$

Experiments on several isotactictic polypropylenes [51] revealed relaxation times of the order of many minutes, and activation energy 77 - 95 kJ/mole, which lies in the range of activation energy of autodiffusion. Different relaxation times were obtained for different molecular weights.

3.5.10 Memory Effects Related to Flow Orientation

The model of transient nucleation has also been applied to crystallization of flow-oriented and partially relaxed polymer melts. Flow of the melt produces

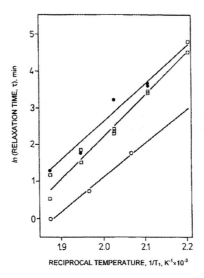

Fig. 3.27. "Relaxation time", τ , for crystallization memory in several polypropylenes as a function of reciprocal heating temperature, T_1 [52].

orientation of crystallizing units and clusters. Flow oriented system is relaxed for time t_R in the temperature $T_R > T_m$, cooled down and crystallized in a temperature $T_c < T_m$. Crystallization from melt-oriented polymer is controlled by athermal nucleation and results in characteristic microscopic structures [52-54].

To describe orientation effects, nucleation model developed in Section 3.5.1. has been used [55]. Uniaxial, rodlike aggregating molecules and cylindrical clusters with optimum length-to-radius ratio were considered. Steady elongational flow with intensity H produces equilibrium orientation distribution of the aggregating units

$$w_1(\vartheta, t_R = 0) = w_{1,eq}(\vartheta, H, T_p) = \frac{1}{Z_1(H, T_p)} \exp[-H(T_p)\cos^2\vartheta] \quad (3.171)$$

where Z_1 is configurational integral. The above distribution provides initial condition for the relaxation process $(t_R = 0)$. Cluster distribution, w , is considered in the space of cluster sizes (dimensionless volume), g , and orientations characterized with a single orientation angle ϑ (angle between cluster axis and the direction of flow). Equilibrium distribution of clusters produced by flow assumes the form

$$w(g, \vartheta; t_R = 0) = w_{eq}(g, \vartheta, H, T_p) =$$

$$= \text{const. } \exp\left[\frac{-\Delta F_0(g, T_p)}{kT_p} + (g-1)\left(H(T_p)\cos^2\vartheta + \ln\frac{Z}{4\pi}\right)\right] \quad (3.172)$$

where ΔF_0 is free energy of formation of a g-size cluster at T_p in the absence of flow.

Relaxation of the distributions $w_1(\vartheta)$ and $w(g, \vartheta)$ takes place in the temperature $T_R > T_m$. The target orientation distribution of aggregating molecules is random

$$w_1(\vartheta; t_R = \infty) = w_{1,eq}(\vartheta, H = 0) = \frac{1}{4\pi} \quad (3.173)$$

and so is target distribution of clusters, controlled by free energy of aggregation in quiescent conditions

$$w(g, \vartheta; t_R = \infty) = w_{eq}(g, \vartheta; H = 0, T_R) = \text{const. } \exp\left[\frac{-\Delta F_0(g; T_R)}{kT_R}\right] \quad (3.174)$$

We will describe relaxation in a simplified form by assuming linear relaxation equations. For aggregating rodlike molecules we assume

$$\frac{\partial w_1}{\partial t} = \frac{w_{1,eq} - w_1}{\tau_{rot,0}} \quad (3.175)$$

which yields transient orientation distribution in the form

$$w_1(\vartheta, t_R) = \frac{1}{4\pi} + \left(w_{1,eq}(\vartheta, H) - \frac{1}{4\pi}\right)\exp\left[\frac{-t_R}{\tau_{rot,0}}\right] \quad (3.176)$$

The solution, Eq.(3.176), satisfies initial (Eq. 3.171) and asymptotic conditions (Eq. 3.173). The relaxation time involved, $\tau_{rot,0}$, is related to rotation of the aggregating molecule in a viscous medium. Similarly, for clusters we assume

$$\frac{\partial w(g, \vartheta)}{\partial t} = (w_{eq} - w)\left[\frac{1}{\tau_{rot,cl}} + \frac{1}{\tau_{gr}}\right] \quad (3.177)$$

Relaxation of clusters involves two different mechanisms: one related to cluster rotation ($\tau_{rot,cl}$) and another one, associated with cluster growth (τ_{gr}). For large clusters the "growth relaxation time" is much shorter than the rotational one

$$\tau_{gr} \ll \tau_{rot,cl} \quad (3.178)$$

and rotation can be neglected. Omitting the $\tau_{rot,cl}$ term in Eq. (3.177) and using the initial condition, Eq. (3.172), we arrive at the transient cluster distribution

$$w(g, \vartheta, t_R, T_R) \cong w_{eq}(g; H = 0, T_R)$$

$$+[w_{eq}(g, \vartheta, H, T_p) - w_{eq}(g; H = 0, T_R)] \exp\left[\frac{-t_R}{\tau_{gr}}\right] \quad (3.179)$$

Eq. (3.179) satisfies also the asymptotic condition (3.174).

There are three ways in which the residual structure formed in flow and relaxation affects concentration of athermal nuclei and, consequently, crystallization of the supercooled system. The first is critical transition temperature, T_m, controlled by partially relaxed aggregating units, $w_1(\vartheta, t_R)$

$$T_m(\vartheta, t_R, T_R) = \frac{T_{m,0}}{1 - \dfrac{kT_c}{\Delta h v_0} \ln[4\pi w_1(\vartheta, t_R, T_R)]} \quad (3.180)$$

where Δh is heat of melting, and T_{m0} - equilibrium melting temperature in an unoriented (completely relaxed) system. Maximum T_m corresponds to orientation parallel to flow axis ($\vartheta = 0$). If crystallization temperature, T_c, fits into the range

$$T_m(\vartheta = 0) > T_c > T_m(\vartheta = \frac{\pi}{2}) \quad (3.181)$$

and crystal orientation is truncated at the upper limit, ϑ_{max}, determined by the condition

$$T_m(\vartheta_{max}) = T_c \quad (3.182)$$

Crystals with orientations $\vartheta > \vartheta_{max}$ cannot be formed at, or above, T_c.

The second effect is related to critical cluster size, g^*, which appears as a lower integration limit in Eqs. (3.184) and (185), below. From the model described in Sections 3.4.4. and 3.5.1, g^* is a function of crystallization temperature and orientation angle, ϑ

$$g^*(\vartheta, t_R, T_R, T_c) = [g_0^*(T_c)]\left(\frac{1}{1 - \dfrac{kT_c}{\Delta f_{un}(T_c)v_0} \ln[4\pi w_1(\vartheta, t_R, T_R)]}\right)^3$$

$$(3.183)$$

g_0^* denotes critical cluster size, and Δf_{un} - bulk free energy of aggregation, in an unoriented (completely relaxed) system. In the course of relaxation, the values of g^* for well oriented crystals increase, reducing the respective nucleation rates. g^* values for perpendicular clusters are very small and do not contribute much to crystallization, also if their values (according to Eq. 3.183) are reduced in the course of relaxation.

The third effect is related to transient cluster distribution, $w(g, \vartheta, t)$. In the course of relaxation, concentration of highly oriented clusters decreases, and critical cluster size (providing lower size limit for stable nuclei) increases, reducing concentration of oriented athermal nuclei effective in the crystallization temperature, T_c . The angular density of athermal nuclei active in the system after relaxation for period t_R at $T = T_R > T_m$ and cooled down to crystallization temperature $T = T_c$, reads

$$\frac{dN_{ath}(\vartheta, t_R, T_R)}{\sin \vartheta d\vartheta} = \text{const.} \int_{g^*(\vartheta)}^{\infty} w(g, \vartheta, t_R) dg \qquad (3.184)$$

and concentration of athermal nuclei in the orientation range $(\vartheta_1, \vartheta_2)$ is given by

$$N_{ath}(\vartheta_1, \vartheta_2, t_R) = \text{const.} \int_{\vartheta_1}^{\vartheta_2} \sin \vartheta d\vartheta \int_{g^*(\vartheta)}^{\infty} w(g, \vartheta, t_R) dg \qquad (3.185)$$

In an oriented system, maximum density corresponds to clusters oriented parallel to flow axis ($\vartheta = 0$), and it is well oriented clusters which control crystallization. Concentration of stable athermal nuclei in the orientation range $\vartheta \in (0, 10^\circ)$, is plotted vs. reduced time in Figure 3.28. The number of

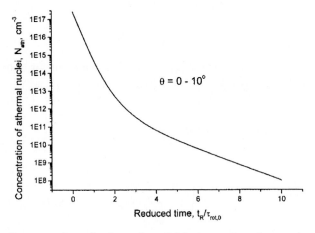

Fig. 3.28. Concentration of athermal nuclei in isotactic polypropylene subjected to flow orientation ($H = -0.25$, $T_p = 510K$), and partial relaxation at the temperature $T_R = 510K$. Concentration of athermal nuclei in the orientation range $\vartheta \in (0, 10^\circ)$ calculated from Eq.(3.185) as a function of time at T_R, crystallization temperature, $T_c = 427K$ [55].

oriented nuclei, and the expected number of oriented crystals decreases with increasing temperature, T_R , and duration, t_R , of relaxation. Experiments made by Alfonso and Azzurri [56] confirm these expectations.

3.6 Closing Remarks

The nucleation model presented in this chapter does not describe all aspects of polymer crystallization. Multidimensional theory of nucleation is a natural generalization of the classical concepts of nucleation of phase transitions, and is not specific for crystallization or polymers. Nucleation is considered as a motion of molecular (atomic) clusters in a configurational space covering various degrees of freedom. The theory explains orientation and positional effects in external potential fields, and suggests new mechanisms of nucleation, absent in the classical, one-dimensional theory. Molecular orientation of polymers subjected to flow or deformation, leads to selective crystallization. What the present paper does not describe in detail is specific modes in which polymer crystals are formed from long chain molecules (crystallization morphology). In polymer literature, many details of crystallization morphology have been discussed [7, 8, 57-62]. This problem has only been touched in Sections 3.5.2 and 3.5.7. Comparison of a simple model of bundle-like vs. folded-chain crystallization (Section 3.5.2) reveals fundamental difference in the concentration and orientation behavior. The model of mixed nucleation analyzed in Section 3.5.7. comprising two kinds of chain segments, intramolecular (folded chains) and intermolecular (bundle-like) illustrates rational way in which crystallization morphology can be treated. Clusters (crystals) with composite morphology may result from superposition of various possible structures. Thermodynamic and kinetic characteristics of individual structural components, completed with entropy of mixing yield predictions about the resulting structures without *a priori* assumptions about the result. In the example described in Section 3.5.7, thermodynamic optimization, taking into account external conditions (polymer concentration, stress) yields prediction of the most probable morphological structure. A similar approach has been applied by Lauritzen, DiMarzio and Passaglia [63, 64] to the problem of "cilia" - free chain ends appearing on the surface of plate-like polymer crystals Last, not least, it has been demonstrated that memory effects in unoriented (Section 3.5.9) and oriented (Section 3.5.10) polymers can be effectively explained using simple nucleation models.

References

[1] Volmer M, Weber A (1926), Z. Phys. Chem., 119 : 277

[2] Kaischew R, Stransky I (1934), Z. Phys. Chem., B26 : 317

[3] Becker R, Döring W (1935, 1938), Ann. Phys., 24 : 719; ibid., 32 : 128

[4] Zeldovich JB (1943), Acta Physicochim. URSS, 18 : 1

[5] Frenkel J (1946), Kinetic Theory of Liquids, Oxford University Press, London

[6] Turnbull D, Fisher JC (1949), J. Chem. Phys., 17 : 71

[7] Lauritzen JI, Hoffman JD (1960), J. Research NBS, 64A : 73

[8] Mandelkern L, Crystallization of Polymers (1964), McGraw and Hill, New York

[9] Frank FC, Tosi M (1961), Proc. Royal Soc. A263 : 323

[10] Ziabicki A (1986), J. Chem. Phys. 85 : 3042

[11] Fisher JC, Hollomon JH, Turnbull D (1948), J. Appl. Phys., 19 : 775

[12] Ziabicki A (1968), J. Chem. Phys., 48 : 4374

[13] Wulf GV (1923, 1924), Z. Krist, 59 : 135, 335; ibid., 60 : 1

[14] Ziabicki A (1967), Faserforsch. Textiltech., 18 : 142

[15] Ziabicki A (1968), J. Chem. Phys., 48 : 4368

[16] Ziabicki A (1976), J. Chem. Phys., 66 : 1638

[17] Wunderlich B, Arakawa T (1964), J. Polymer Sci. (Phys), A2 : 3697

[18] Keller A (1957), Phil. Mag. 2 : 1171

[19] Ubbelohde AR (1978), The Molten State of Matter, Wiley, Chichester

[20] Ziabicki A, Jarecki L (1982), IFTR Report #1/1982

[21] Ziabicki A, Jarecki L (1994), J. Chem. Phys., 101 : 2267

[22] Ziabicki A, Jarecki L (1995), Macromol. Symposia, 90 : 31

[23] Ziabicki A (1990), Arch. Mech. 42 : 703

[24] Ziabicki A, Jarecki L (1999), Crystal Nucleation in Uniaxially and Biaxially Deformed Polymer Systems, unpublished report

[25] Ziabicki A, Jarecki L (1998), Proceedings of the SPE-RETEC Conference "Orientation of Polymers, September 23-25, 1998, Boucherville, Canada, p.335.

[26] Flory P (1947), J. Chem. Phys. 15 : 397

[27] Ziabicki A, Jarecki L (1985), in ref. [34], p. 225.

[28] Krigbaum WR, Roe RJ (1964), J. Polymer Sci., 2A : 4391

[29] Ziabicki A, Jarecki L (1978), Colloid & Polymer Sci., 256 : 332

[30] Kość M, Ziabicki A (1982), Macromolecules, 15 : 1507

[31] Göschel U, Swartjes FHM, Meijer HEH, Eyere P (2000), Crystallisation in isotactic polypropylene: effects of memory, chain orientation and nucleants, EUPOC 2000, Gargnano, Italy, May 28-June 2, 2000.

[32] Swartjes FHM, Peters GWM, Göschel U, Meijer HEH (1999), A Novel Flow Cell for the Investigation of Elongational Flow Induced Nucleation, PPS-15 Conference on Polymer Processing, s'Hertogenbosch, Holland, May 31-June 4, 1999.

[33] Peters G (1999), Modelling of flow-induced crystallization, COST P1 Meeting, Vught, Holland, December 4-6, 1999

[34] Ziabicki A,. Kawai H (Eds., 1985), High Speed Fiber Spinning, Interscience, New York

[35] Sajkiewicz P (1989), IFTR Report #12/1989; Sajkiewicz P, Wasiak A (1999), Colloid & Polymer Sci., 277 : 646

[36] Jarecki L, Kość M, Ziabicki A (1990), IFTR Report # 22/1990

[37] Ziabicki A, Jarecki L (1996), Macromol. Symposia, 104 : 65

[38] Marand HL, Stein RS,. Stack GM (1988), J. Polymer Sci. (Phys) 26 : 1361

[39] Sajkiewicz P (1994), J. Polymer Sci. (Phys), 32 : 313

[40] Ziabicki A (1977), IFTR Report # 60/1977

[41] Ziabicki A (1977) in: Phase Transitions in Bulk Polymers, Europhysics Conference Abstracts, 2E : 33-43

[42] Smoluchowski MW (1917), Z. Physik, 29 : 129

[43] Collins FC, Kimball GE (1949), J. Colloid Sci., 4 : 425

[44] Šolc K, Stockmayer WH (1971), J. Chem. Phys., 54 : 2982

[45] Schurr JM (1970), Biophys. J., 10 : 700

[46] Schmitz KS, Schurr JM (1972, 1976), J. Phys. Chem., 76 : 534; ibid., 80 : 1934

[47] Jarecki L (1991), Colloid & Polymer Sci., 269 : 11-27

[48] Jarecki L (1996), Colloid & Polymer Sci., 272 : 784-796

[49] Jarecki L (1995), Colloid & Polymer Sci., 273 : 138-149

[50] Ziabicki A, Alfonso GC (1994), Colloid & Polymer Sci. 272 : 1027

[51] Alfonso GC, Ziabicki A (1995), Colloid & Polymer Sci. 273 : 317

[52] Monasse B (1992), J. Materials Sci. 27 : 6047

[53] Alfonso GC, Scardigli P (1997), Macromol. Symposia, 118 : 323

[54] Sajkiewicz P, Wasiak A, Kukla D, Boguszewski M (2000), J. Materials Sci. Lett. 19 : 847

[55] Ziabicki A, Alfonso GC, to be published

[56] Alfonso GC, Azzurri F, to be published

[57] Pennings AJ, van der Mark JMAA, Kiel AM (1970), Kolloid Z. u. Z. für Polymere, 237 : 336.

[58] Allegra G (1977), J. Chem. Phys. 66 : 5453

[59] Calvert PD, Uhlmann DR (1972), J. Appl. Phys. 43 : 944

[60] DiMarzio EA, Guttman CM (1982), J. Appl. Phys. 53 : 6581

[61] Sadler DM (1985), J. Polymer Sci. (Phys) 23 : 1533

[62] Sadler DM (1986), Polymer Comm. 27 : 140

[63] Lauritzen JI, DiMarzio EA, Passaglia E (1966), J. Chem. Phys. 45 : 4444

[64] DiMarzio EA (1967), J. Chem. Phys. 47 : 3451

4 Kinetic Theory of Nucleation in Polymers

D. Reguera[1], J.M. Rubí[1], and L.L. Bonilla[2]

[1] Departament de Física Fondamental, Universitat de Barcelona, Diagonal 647, 08028 Barcelona, Spain
[2] Escuela Politécnica Superior, Universidad Carlos III de Madrid, Avenida de la Universidad 30; 28911 Leganés, Spain

4.1 Introduction

Crystallization of polymers is a fascinating branch of polymer physics which has a significant relevance in industrial applications. Its importance arises from the fact that mechanical properties of any crystal polymer are determined by its morphology and internal structure, which in turn is dictated by the crystallization kinetics. That is the reason why, mathematical modeling aiming to describe and control the kinetics of polymer crystallization has achieved great interest.

In spite of the striking differences between crystallization of polymers and that of simple substances, polymer crystallization theories originate largely from theories developed earlier for simple substances.

One of these theories is the general Avrami-Kolmogoroff formalism, which provides a stochastic geometrical model for the development and space-filling of the new phase. The underlying physical ingredients of this theory are the nucleation and growth rates, i.e. the rate of appearance and growth of the new crystalline phase.

Nucleation process strongly determines the subsequent evolution of the crystallization of the sample thus constituting a key mechanism in the crystallization process. Due to this fact, nucleation has been subject of thorough investigations and has motivated the development of many theories. The most extended one, the classical nucleation theory, was originally formulated for isothermal homogeneous conditions. However, crystallization in polymers involves some peculiarities that make both assumptions not realistic and thus invalid. Since most mathematical approaches, as for example the Avrami-Kolmogoroff theory, requires the previous knowledge of nucleation and growth rates, it becomes important to develop a theory for nucleation and growth as realistic as possible.

Our purpose in this paper is precisely to establish the basis for a kinetic description of crystallization in polymers by proposing a mesoscopic formalism able to derive kinetic equations for the probability distribution functions. These functions depend on parameters specifying the state of a cluster, i.e. geometrical parameters as the size or the geometry, and phase space variables as the velocity, orientation, spatial position, etc., and their evolution is in general coupled to that of the thermal bath in which the emerging clusters are embedded.

The nucleation theory we propose goes beyond the classical formalism and can account systematically for the real conditions in which crystallization takes places. In particular, we will discuss some specific examples to describe nucleation and growth in spatially inhomogeneous systems, in the presence of external fields or temperature gradients. Moreover, we will analyze a less studied and specially interesting case: nucleation under the influence of an external flow.

This chapter has been organized in the following manner. In sections 4.1.1 and 4.1.2, we will review the main features of the crystallization process pointing out its similarities with an activated process. The classical picture of nucleation will be developed in section 4.2. Sections 4.3 and 4.4 are devoted to introduce our formalism, illustrated with its application to the case of isothermal homogeneous nucleation. We will then study the case of inhomogeneous non-isothermal nucleation, showing how the process is drastically altered by the presence of cross effects. In addition, we will discuss the influence of external flows in the nucleation mechanism. Finally, in order to describe the advanced stages of the crystallization, we will use our results in the classical Avrami-Kolmogoroff model, to analyze the behavior of the crystallinity. Further extensions and improvements will be outlined in section 4.9, whereas the last section is intended as a summary of the main characteristics of our theory.

4.1.1 The Polymer Crystallization Process

Crystallization of polymers is the process of structural reorganization of a melted polymer leading to the appearance of an ordered structure. In practice, crystallization proceeds by the melting and subsequent cooling down of the sample. In general this process strongly depends on temperature. For temperatures higher than the melting temperature T_m, thermal agitation breaks up any ordered structure and the polymer remains melted in a "liquid-like" phase. Contrarily, for temperatures lower than the melting temperature, the thermodynamic stable phase is the crystalline one. The transition from melted to crystalline polymer is what is known as crystallization. This process involves diffusion of monomers units towards the crystal front. However, the diffusion mechanisms are prevented at temperatures below the glass transition temperature T_g. Consequently, crystallization takes place for temperatures in the range $T_g < T < T_m$.

We have stated that for temperatures lower than T_m, the thermodynamic stable phase is the crystalline polymer. The question is how crystals are formed from the melted phase.

In the liquid phase, the density is subjected to fluctuations due to thermal agitation. These fluctuations may eventually create small aggregates (*clusters*) of polymers having the same properties than the crystalline phase. The small crystals are continuously being created and destroyed by fluctuations. The reason for the break up of the crystals is that, although below T_m the crystal phase is the thermodynamic stable one, the formation of a crystal involves the creation of an interface between the liquid and the crystal, with the corresponding energetic cost due to the surface tension. Therefore, the appearance of a little crystal involves the competition between two effects. One is the decrease of energy due to the fact that the chemical potential of the crystal phase is lower than that of the melt, which competes with the energetic cost due to the surface tension associated to the creation of an interface.

For clusters of small size surface effects are dominant. As a consequence, their growth is energetically unfavorable and the small crystals tend to spontaneously dissolve. However, there exist a typical size beyond which, the volume effects dominate over the surface ones, and the growth of the cluster is favored by a global reduction of the energy. This size which determines the stability of the clusters is called *critical size* and the process of formation of crystals of size bigger than or equal to the critical size is known as *nucleation*.

Basically, one can distinguish between two different types of nucleation: *homogeneous nucleation* which occurs on the bulk of a pure substance; and *heterogeneous nucleation*, which takes place in the presence of boundaries, surfaces, impurities or pre-existing crystals.

Nucleation is the initial step in the crystallization process, for it determines the appearance of the first crystal nuclei which are the germ of the second stage of crystallization process: the *growth*. At this stage, nuclei larger than the critical size, tend to grow through the addition of monomer units or alternatively acting as sites of heterogeneous nucleation (nucleation on the surface of the growing crystal).

Crystals grow freely until they progressively begin to compete to fill the whole space. They may eventually hit each other, and therefore the growth is stopped at the contact surface. This phenomenon, called *impingement,* is relevant at later stages of the process and determines the final morphology of the system.

In the final stage of crystallization, the impingement impedes the subsequent growth of the crystals. Therefore, amorphous non-crystallized matter remains trapped among clusters. Moreover, clusters themselves are not fully crystalline, but may contain some amorphous inclusions. This trapped material can eventually self-organize, joining to the main crystal structure and therefore increasing the crystallinity of the sample. This process of reorganization is what it is normally referred to as *secondary crystallization* or *perfection.*

In summary, we can schematically distinguish between three stages in the entire crystallization process:

– Nucleation, which is the process of formation of the initial crystalline embryos.

- Growth, which describes the space-filling of this crystals which is limited by impingement.
- Secondary crystallization, accounting for the reorganization of the amorphous inclusions to achieve a more crystalline material.

What we call a "crystallized polymer" is in fact a crystal aggregate with amorphous inclusions.

4.1.2 Nucleation and Growth as Activated Processes

As we have seen, the first step in the kinetics of many phase transitions is the formation of small embryos of the new phase within the bulk metastable substance. This is an activated process: a free energy barrier must be overcome in order to form embryos of a critical size, beyond which the new phase grows spontaneously.

Many nonequilibrium processes in nature may be described in terms of the crossing of a free energy barrier which separate two accessible stable states of the system, corresponding to the local minima at each side of the barrier. When the system localized at the left well acquires energy, it may surmount the barrier thus manifesting the characteristics of the state at the right well. Processes as thermal emission in semiconductors [1], chemical reactions [2], nucleation [3], adsorption [4], etc. share these common features and are commonly referred to as activated processes.

It is interesting to emphasize the essential difference between activated processes and the second category of processes usually occurring in nonequilibrium situations, namely transport processes. The latter constitute the response to the application of an external force (external currents or gradients) and may emerge even at very low values of the applied force. Contrarily, as commented previously, activated processes may only be induced whenever the applied perturbation exceeds a threshold value.

This difference has an important consequence. Whereas transport processes may exhibit a wide linear regime in which the external force and the established current are proportional, the regime in which activated processes develop is highly nonlinear. In this context, we can compare the linear Fourier, Fick or Ohm laws, in which the corresponding currents are proportional to the conjugated thermodynamic forces or gradients, with the exponential laws usually appearing in activated processes, as the nonlinear expression of the law of mass action giving the reaction rate of a chemical reaction in terms of its affinity.

A mesoscopic treatment of the process requires the formulation of a kinetic equation describing the evolution in time of the probability density of finding the system in a determined configuration. From this equation we can derive the evolution equations for the different moments providing expressions for the relaxation times, nucleation rates, etc., that can be contrasted with experiments.

In order to provide a general and realistic scheme one has to consider that the system undergoing an activated process may be in contact with a heat bath whose state may be not of equilibrium. An example illustrating this situation will be discussed in detail in Section 4.5: nucleation in the presence of temperature gradients.

In this particular example, the length scale over which the relevant variables of the heat bath vary significantly is long enough to justify an hydrodynamic description. We then may consider that the dynamics of the bath is governed by Navier-Stokes equations consistent with nonequilibrium thermodynamics. Contrarily, changes in the nuclei occur at mesoscopic scales and the dynamics has to be described in terms of probability distributions.

An unified formalism, based on nonequilibrium thermodynamics, can be adopted in which the entropy of the whole system may depend explicitly on thermodynamic variables, as assumed in its classical version, or fields and functionally on the probability density defined at mesoscopic scales. By application of the rules dictated by nonequilibrium thermodynamics, we will be able to derive kinetic equations in the general case in which the dynamics of the systems and the heat bath are coupled.

4.2 Nucleation Theory

Before starting our analysis of nucleation and growth in polymers we will proceed by very briefly reviewing the main aspects of nucleation processes (see for example Refs. [5]-[9]).

The aim of a self-consistent theory of nucleation is the description of the evolution of the population of clusters of the new phase. In a supercooled substance small embryos of the new stable phase (clusters) are constantly being created and destroyed from density fluctuations. These clusters are assumed to shrink and grow in size by gaining or losing single monomers. The net rate at which clusters of size n become clusters of size $n + 1$ at time t is then given by

$$J(n) = k^+(n)f(n,t) - k^-(n+1)f(n+1,t) \tag{4.1}$$

where $f(n,t)$ is the number density of clusters constituted by n monomers at time t; $k^+(n)$ is the rate at which a n-cluster gains monomers; and $k^-(n)$ is the rate at which it loses particles.

Consequently, the variation of the population of clusters of size n at each instant t is given by the following master equation

$$
\begin{aligned}
\frac{\partial f(n,t)}{\partial t} &= k^+(n-1)f(n-1,t) + k^-(n+1)f(n+1,t) \\
&\quad - k^+(n)f(n,t) - k^-(n)f(n,t)
\end{aligned} \tag{4.2}
$$

which, in view of eq. (4.1), can be rewritten as

$$\frac{\partial f(n,t)}{\partial t} = J(n-1,t) - J(n,t) \tag{4.3}$$

The basic variables in the master equation (4.2) are $k^+(n)$ and $k^-(n)$, the rates of attachment and detachment of single molecules, which may adopt different forms depending on the theory one applies.

Several models have been proposed to obtain the rate $k^+(n)$. For instance, an expression for this rate was obtained by Turnbull and Fisher [10] using reaction rate theory [5]

$$k^+(n) = 4n^{2/3}\Gamma \exp\left(-\frac{\delta\Delta G}{2k_BT}\right) \tag{4.4}$$

Here $\delta\Delta G = \Delta G(n+1) - \Delta G(n)$, where $\Delta G(n)$ is the free energy of formation of a cluster of n monomers, the factor $4n^{2/3}$ arises from consideration of the number of available attachment sites on the surface of a spherical cluster and Γ is the unbiased molecular jump frequency at the cluster interface, that can alternatively be expressed as

$$\Gamma = \frac{k_BT}{h} \exp\left(-\frac{\epsilon}{k_BT}\right) \tag{4.5}$$

where ϵ is the activation energy defined as the difference between the energy of the activated state and the average energies of the initial and final states, and h is the Planck constant. This jump frequency Γ is often identified with the jump frequency for bulk diffusion, through the atomic jump distance λ and the spatial diffusion coefficient D, as

$$\Gamma = \frac{6D}{\lambda^2} \tag{4.6}$$

Consequently, the attachment rate $k^+(n)$ is proportional to the bulk diffusion coefficient D. As we will illustrate in the following sections, the presence of flows, thermal gradients and hydrodynamic interactions may alter significantly this diffusion coefficient and thus the kinetic of the nucleation process.

4.2.1 Phenomenological Theories: Classical Nucleation Theory

Although the rate $k^+(n)$ can be obtained from first principles (using for instance kinetic theory), the value of the detachment rate $k^-(n)$ is more difficult to obtain in an independent way. To avoid this difficulty, phenomenological theories, as the classical nucleation theory, resort to a relationship between the known quantity $k^+(n)$ and $k^-(n)$. The peculiar characteristics of the equilibrium state provide this desired relationship. In fact, at equilibrium, the flux J must vanish. As a consequence, from equation (4.1) one obtains the expression

$$k^+(n)z(n) - k^-(n+1)z(n+1) = 0; \; k^-(n+1) = \frac{z(n)}{z(n+1)}k^+(n) \quad (4.7)$$

where $z(n)$ represents the equilibrium distribution of clusters. Therefore, the kinetic problem of calculating the nucleation rate becomes the thermodynamic problem of evaluating the equilibrium cluster distribution. From the theory of thermodynamic fluctuations [11], the equilibrium cluster size distribution is given by

$$z(n) = z(1) \exp - \frac{\Delta G(n)}{k_B T} \quad (4.8)$$

where $\exp - \frac{\Delta G(n)}{k_B T}$ is related with the probability of obtaining a cluster of the new phase containing n monomers, and $\Delta G(n)$ is the minimum reversible free energy of formation of an n-cluster. Therefore, the problem is formulated now in terms of the free energy of formation of clusters.

Several proposals exist giving an expression for the free energy $\Delta G(n)$, but the most extended one is that of the classical nucleation theory. In this theory, the free energy of a n-cluster is calculated by assuming that clusters have the same properties as the macroscopic bulk phase with a sharp interface. Making this capillarity approximation, $\Delta G(n)$ can be expressed, following Gibbs, as the sum of volume and interfacial energy contributions

$$\Delta G(n) = v(n)\Delta g + \sum \sigma_i A_i \quad (4.9)$$

Here $v(n)$ is the volume of the cluster, Δg is the bulk free energy per unit volume, and σ_i is the interface tension of the surface i with area A_i. For undercooled substances, Δg depends on the undercooling $\Delta T = T_m - T$, where T_m is the equilibrium melting temperature and T is the temperature. In fact, neglecting the temperature dependence of the heat of fusion Δh, as a first approximation Δg is given by

$$\Delta g = \frac{\Delta h \Delta T}{T_m} \quad (4.10)$$

Consequently, Δg is positive above T_m and is negative below that temperature.

For the simplest case of spherical clusters, the change of crystallization free energy splits up into bulk and surface contributions (see Fig. 4.1) and is given by

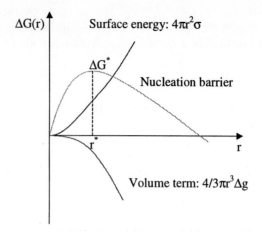

Fig. 4.1. Free energy of formation of a crystal embryo in the classical nucleation picture, and its surface and bulk contributions.

$$\Delta G = \frac{4\pi r^3}{3} \Delta g + 4\pi r^2 \sigma \qquad (4.11)$$

with r being the radius of the spherical crystal. This equation can alternatively be written in terms of the number of monomers as

$$\Delta G = an + bn^{2/3} \sigma \qquad (4.12)$$

where a is the free energy per monomer and $b = (36\pi)^{1/3} \, \bar{v}^{2/3}$, with $\bar{v}$ the molecular volume.

The *critical size* n^* may be defined as the size at which the free energy has a maximum

$$\frac{\partial \Delta G(n)}{\partial n} \bigg|_{n^*} = 0 \qquad (4.13)$$

Clusters smaller than the critical size tend to shrink, while those larger than n^* can be considered as nuclei of the new phase as they will on average grow. For the case of spherical clusters, the critical size and the critical radius are

$$n^* = \frac{32\pi}{3\bar{v}} \frac{\sigma^3}{\Delta g}, \qquad (4.14)$$

$$r^* = \frac{2\sigma}{\Delta g} \tag{4.15}$$

which correspond to the maximum value of the free energy

$$\Delta G^* = \frac{16\pi}{3} \frac{\sigma}{\Delta g^2} \tag{4.16}$$

This maximum is referred to as the *nucleation barrier,* and the rate $J(n^*, t)$ at which critical-sized embryos are formed is the *nucleation rate.*

4.2.2 Steady-State Nucleation Rate

Classical nucleation theory assumes that a steady state, in which the populations of different cluster sizes no longer depend on time, is rapidly established. In this situation, following eq. (4.3), the flux has the same value irrespective of the size i.e. $J(n, t) = J$. This stationary value of the flux is often referred to as the *nucleation rate,* and by recurrence from equation (4.1) it is given by

$$J = N_{tot} \left[\sum_{n_1}^{n_2} \frac{1}{k^+(n)z(n)} \right]^{-1} \tag{4.17}$$

where N_{tot} is the total number of clusters, and cluster sizes n_1 and n_2 are such that for $n < n_1$, $f(n) = z(n)$ and for $n > n_2$, $f(n) = 0$. Consequently, the value n_1 represents the smaller cluster distinguishable from equilibrium fluctuations in the metastable phase, and n_2 is a postcritical large enough stable cluster. It has been shown that the nucleation rate does not strongly depend on the values at the boundaries n_1 and n_2 [12].

Replacing the summation in equation (4.17) by an integral, introducing the "equilibrium" distribution $z(n)$, and evaluating the resulting integral by the steepest descent method (see Ref. [9]), one obtains the classical expression for the nucleation rate

$$J_{CL} = k^+(n^*)ZN_{tot} \exp\left[-\frac{\Delta G^*}{k_B T}\right] \tag{4.18}$$

where the asterisk denotes quantities evaluated at the critical point and

$$Z = \sqrt{\frac{-\frac{\partial^2 \Delta G}{\partial n^2}\big|_{n*}}{2\pi k_B T}} \tag{4.19}$$

is frequently referred to as the Zeldovich factor [13].

This result is obtained under the assumption that the steady-state is instantaneously achieved which implies that nucleation proceeds at a constant rate. While frequently true, this assumption becomes wrong in some specific cases. A finite time is required in order for the concentration of clusters and the fluxes to attain their stationary values. This time lag can be of the same order than the measure time, specially in the case of polymers and glasses, and transient kinetics may thus be important in the description of crystallization. It would then be desirable to describe the time-dependent nucleation under given experimental conditions. To this end, it is convenient to introduce a continuous version of the master equation (4.3).

4.2.3 Continuous Approaches

In the classical nucleation theory treatment, the kinetics of the nucleation process is described through a master equation for the evolution of the population of clusters. However, owing to the discrete nature of the growing process, resulting essentially from the progressive addition of atoms to the emerging cluster, the master equation provides a hierarchy of kinetic equations for the evolution of clusters of different sizes which is very difficult to solve. That is the main reason to abandon the master equation approach and to formulate a continuous diffusion equation to describe nucleation processes. The continuous diffusion equation corresponding to the master equation (4.2) adopts the general form

$$\frac{\partial f}{\partial t} = \frac{\partial}{\partial n} \left\{ D_n(n) \frac{\partial f}{\partial n} + b(n)f \right\} \tag{4.20}$$

where $b(n)$ and $D_n(n)$ are the drift and diffusion coefficients, respectively, which are related to the attachment rate $k^+(n)$ and the energy barrier $\Delta G(n)$. The diffusion coefficient is frequently identified with the forward rate

$$D_n(n) \cong k^+(n) \tag{4.21}$$

whereas the drift coefficient is usually obtained through the requirement that detailed balance is fulfilled

$$b(n) = \frac{D_n(n)}{k_B T} \frac{\partial \Delta G}{\partial n} \tag{4.22}$$

The proper choice of these coefficients constitute a controversial issue to the extent that several different expressions have been derived and proposed in the literature. For a review, see for instance Ref. [3] and references quoted therein. Most of them result from continuous approaches to the master equation (4.2). However, they present some drawbacks as the fact of not being able to reproduce the same results obtained from the original master equation they approximate or not satisfying detailed balance.

To circumvent these difficulties, we are going to introduce a new method based on non-equilibrium thermodynamics to obtain kinetic equations suitable to study nucleation processes. The advantages of introducing a non-equilibrium thermodynamics description are manifold. On one hand, it provides a systematic and clear procedure to derive kinetic equations suitable for any kind of situation, following a simple set of rules. As a consequence, the problem can be studied within a more general framework. On the other hand, we can obtain the coupled mesoscopic evolution of the nucleation process with the macroscopic evolution of the hydrodynamic fields temperature and velocity.

4.3 Kinetic Equations from Non-Equilibrium Thermodynamics

In this section, we shall review the basic concepts of the non-equilibrium thermodynamics approach to the kinetics of homogeneous nucleation and growth. Non-equilibrium thermodynamics [15] provides a powerful tool to set up systematically the equations characterizing irreversible processes. The approach we are going to follow is based on the formalism of internal degrees of freedom [14], [15]. It yields kinetic equations of the Fokker-Planck type for a distribution function, coupled to balance equations for mass, momentum, energy, etc. densities. The basic point of the formalism is the assumption that the thermodynamic state variables describing the system are in local equilibrium.

The governing equations for the variables will be set up by following systematically a small set of rules which can be summarized as follows:

1. Choose the variables describing the system. We assume that the state of a system depends, apart from the usual thermodynamic state variables (internal energy, mass density, temperature...), on some additional quantities $\{\gamma\}$ that we will call internal coordinates or internal degrees of freedom. For polymer crystallization, these variables can be for instance the size, orientation or velocity characterizing clusters. We will conceive changes in the configuration of this system as a diffusion process in the internal space and consequently we will perform a nonequilibrium thermodynamic description of this system in this internal space. The objective is to derive an equation for the evolution of $f(\{\gamma\}, t)$, the probability density of finding $\{\gamma\} \in (\{\gamma\}, \{\gamma\} + d\{\gamma\})$ at time t, by using the principles of non-equilibrium thermodynamics.

2. Write the conservation laws for the local conserved fields which follows from the continuum hypothesis. This hypothesis tacitly assumes that densities are smooth functions of space and time, and vary over scales which are much larger than any microscopic scale such as the mean free path. The set of conservation laws is not able by itself to specify the

evolution of the fields since in their formulation one introduces currents (of mass, momentum, energy, etc.) which must be determined. To close these equations, we need *constitutive equations* relating variables and their fluxes. This is in agreement with the fact that conservation laws are not enough to dictate the evolution of systems composed of many particles (thermodynamic systems). It becomes then obvious the need of complementing the former scheme with a thermodynamic description.

3. The problem of formulating a thermodynamic theory is solved by means of an additional hypothesis: the local equilibrium hypothesis, which assumes equilibrium at small length scales (but much larger than any microscopic scale). A local Gibbs equation, giving the entropy variations related to the evolution of the system in terms of the variations of the corresponding specific quantities, can then be proposed.

4. From the Gibbs equation, one obtains the entropy balance equation from which one may identify the rate of entropy change due to nonequilibrium processes occurring in the system. That quantity, referred to as entropy production, plays a central role in the development of the theory. It contains all possible sources of dissipation and consists of a sum of products between fluxes (effect) and conjugated forces (cause, usually proportional to gradients of system variables).

5. Nonequilibrium Thermodynamics postulates a linear relationship between fluxes and forces. The proportionality coefficients are called kinetic coefficients and obey symmetry relations called Onsager-Casimir reciprocal relations. The physical meaning of the kinetic coefficients needs to be ascertained from the physics of the problem at hand. It is clear that the linear relations between fluxes and forces are the sought constitutive equations. Putting together balance laws and constitutive relations, we obtain a complete set of differential equations for the variables characterizing the evolution of our system.

The application of the former scheme is best introduced with a specific example. Consequently, we shall first discuss the simplest case of isothermal homogeneous nucleation, to subsequently deal with extensions of our formalism to more general situations as the case of inhomogeneous nucleation in the presence of external gradients or flows.

4.4 Isothermal Homogeneous Nucleation

If nucleation occurs under homogeneous isothermal conditions, the description of the system becomes simpler because we do not have to worry about spatial dependencies. The only relevant variables are the ones providing a proper geometrical description of the growing cluster. When clusters are spherical, owing to isotropy conditions, they can univocally be described by their volume, their radii r or their number of single elements n. This is the simplest situation because it involves a single internal coordinate and it will be thus used to illustrate our theory in the following subsection. The extension to other geometries will be treated later on.

4.4.1 Crystallization of Spherical Clusters

When crystallization proceeds by the formation of spherical clusters, or spherulites, the nucleation process can be completely characterized by a single coordinate γ. This coordinate may represent for instance the number of monomers in a cluster, the cluster size or even a global order parameter indicating the degree of crystallization. The process of nucleation and growth of the new phase can be viewed as a diffusion process through the free energy barrier $\Phi(\gamma)$ that separates the metastable melted phase corresponding to γ_1 from the new stable crystal phase characterized by γ_2 (see Fig. 4.2).

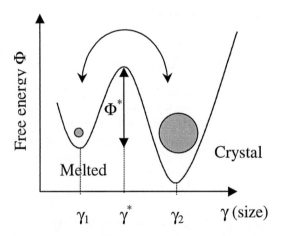

Fig. 4.2. Nucleation as a diffusion over the free energy barrier $\Phi(\gamma)$.

Keeping in mind our objective of deriving a kinetic equation for the evolution of $f(\gamma, t)$, our starting point is the formulation of the Gibbs equation for the entropy variations resulting from the diffusion process

$$\delta S = -\frac{1}{T} \int_{\gamma_1}^{\gamma_2} \mu(\gamma, t) \delta f(\gamma, t) \, d\gamma \qquad (4.23)$$

where $\mu(\gamma, t)$ is the chemical potential associated to the distribution function $f(\gamma, t)$ and T is the constant temperature of the system. In the absence of interactions between clusters, the chemical potential is given through the expression

$$\mu(\gamma, t) = k_B T \ln f(\gamma, t) + \Phi(\gamma) \qquad (4.24)$$

valid for ideal systems. Here $\Phi(\gamma)$ is the free energy barrier to overcome and k_B is Boltzmann's constant. The chemical potential can also be obtained from comparison of eq. (4.23) with the Gibbs entropy postulate [16]

$$S = -k_B \int f(\gamma, t) \ln \frac{f(\gamma, t)}{f_{eq}(\gamma)} \, d\gamma + S_0 \qquad (4.25)$$

where S_0 is the value of the entropy at equilibrium and $f_{eq}(\gamma)$ is the equilibrium distribution given through the corresponding Boltzmann factor

$$f_{eq}(\gamma) = \exp \frac{\mu_{eq} - \Phi(\gamma)}{k_B T} \qquad (4.26)$$

being μ_{eq} the equilibrium chemical potential.

The evolution of the probability density is governed by the continuity equation:

$$\frac{\partial f}{\partial t} = -\frac{\partial J}{\partial \gamma} \qquad (4.27)$$

where $J(\gamma, t)$ is a current or density flux in the internal space which has to be specified. To obtain its value, one proceeds to derive the expression of the entropy production $\sigma = dS/dt$ which follows from the continuity equation (4.27) and the Gibbs equation (4.23). Provided that the definition domain of the parameter γ is bounded, the current $J(\gamma, t)$ vanishes at the initial γ_1 and final γ_2 states of the process. After performing a partial integration one then arrives at

$$\sigma = -\frac{1}{T} \int_{\gamma_1}^{\gamma_2} J(\gamma, t) \frac{\partial \mu}{\partial \gamma}(\gamma, t) \, d\gamma \qquad (4.28)$$

This quantity has the usual form of a sum of flux-force pairs. Nonequilibrium thermodynamics assumes linear dependency between fluxes and forces and establishes linear phenomenological relations between them. In our case, following this scheme, we obtain the desired expression for the current

$$J(\gamma, t) = -\frac{L(\gamma)}{T} \frac{\partial \mu}{\partial \gamma}(\gamma, t) \, , \qquad (4.29)$$

where $L(\gamma)$ is a phenomenological coefficient which may in general depend on the internal coordinate. To derive this expression, locality in the internal space has also been taken into account, for which only fluxes, $J(\gamma, t)$, and forces, $T^{-1} \partial \mu(\gamma, t)/\partial \gamma$, with the same value of γ become coupled. We now insert (4.29) in the continuity equation (4.27), yielding the following Fokker-Planck equation

$$\frac{\partial f}{\partial t} = \frac{\partial}{\partial \gamma} \left\{ D_\gamma(\gamma, t) \frac{\partial f}{\partial \gamma} + b(\gamma, t) f \frac{\partial \Phi}{\partial \gamma} \right\} \tag{4.30}$$

where we have identified

$$b(\gamma, t) = \frac{L(\gamma)}{T f(\gamma, t)} \tag{4.31}$$

as a mobility in the internal space, and

$$D_\gamma(\gamma, t) = k_B T \, b(\gamma, t) \tag{4.32}$$

is the corresponding diffusion coefficient. Notice that the equilibrium distribution (4.26) is a stationary solution of equation (4.30) satisfying the condition $J = 0$.

The previous analysis is based upon the validity of the local equilibrium hypothesis in the internal space. In fact, the relationship (4.24) linking the chemical potential with the probability distribution is just a reformulation of the macrocanonic probability density of equilibrium fluctuations. Therefore, the obtained Fokker-Planck equation describes the decay of fluctuations towards equilibrium.

However, this equation is employed to describe the non-equilibrium evolution. The underlying reason is that one implicitly assumes that any non-equilibrium initial state $f(\gamma, 0)$ decays to equilibrium in the same way as equilibrium fluctuations. In addition, one also assumes that forces and currents are related through linear laws. Both assumptions are valid for systems not too far from equilibrium. Since deviations from equilibrium are determined by the magnitude of the gradients, small gradients are required. However, we have seen in eq. (4.25) that forces involve essentially the logarithm $\ln \frac{f}{f_{eq}}$. Therefore extremely large deviations from equilibrium are needed in order to obtain important forces. To sum up, under normal situations, while gradients remain moderate, the assumptions we have made are fully valid and in particular, the Fokker-Planck equation we have derived, constitute an accurate description of non-equilibrium processes.

Finally, to render the Fokker-Planck dynamics complete, we need to specify the value of the diffusion coefficient in the internal space, $D_\gamma(\gamma, t)$, as well as to determine the form of the barrier, $\Phi(\gamma)$. Both quantities can obviously be borrowed from well-established theories. In fact, in section 4.2 of this chapter we have provided the classical expression for the free energy barrier, eq. (4.12), and for the diffusion coefficient, eq. (4.21). However, one can alternatively obtain these quantities from simulations.

The value of the free energy barrier is related to the equilibrium distribution (4.26) which can be evaluated from Monte Carlo or molecular dynamics simulations as proposed by van Duijneveldt and Frenkel in Ref. [17]-[19] for the specific case of homogeneous nucleation.

To determine the diffusion coefficient $D_\gamma(\gamma, t)$ one can use the Landau-Lifshitz fluctuating hydrodynamics [11],[20]. This theory, originally formulated for transport processes and subsequently extended to activated processes [4], proposes the decomposition of the total current into a systematic and random contributions. In our interpretation of the nucleation process as a diffusion along an internal coordinate, we can apply that formalism to incorporate fluctuations to the former analysis. The total diffusion current splits up into systematic and random contributions in the form

$$J(\gamma, t) = J^s(\gamma, t) + J^r(\gamma, t) \qquad (4.33)$$

The systematic part $J^s(\gamma, t)$ matches expression (4.27) whereas the random part $J^r(\gamma, t)$ constitutes a Gaussian white noise stochastic process with zero mean and fluctuation-dissipation theorem given by

$$\langle J^r(\gamma, t) J^r(\gamma', t') \rangle = 2k_B L \delta(\gamma - \gamma') \delta(t - t') = 2D_\gamma f \delta(\gamma - \gamma') \delta(t - t') \qquad (4.34)$$

Integration of eq. (4.34) with respect to time and to the internal coordinate then leads to an expression of the diffusion coefficient in terms of the correlations of the random current

$$D_\gamma(\gamma, t) = \frac{1}{f(\gamma, t)} \int_{\gamma_1}^{\gamma_2} \int_0^\infty \langle J^r(\gamma, 0) J^r(\gamma', t) \rangle \, d\gamma' \, dt \qquad (4.35)$$

This equation constitutes a Green-Kubo formula and can be used to determine the diffusion coefficient in the internal space from realizations of the underlying stochastic process.

The incorporation of fluctuations in the internal space indicated previously implies the formulation of a more general kinetic equation for the nucleation process. Taking into account the random diffusion current in eq. (4.27) this expression transforms into the kinetic equation

$$\frac{\partial f}{\partial t} = \frac{\partial}{\partial \gamma} \left\{ D_\gamma(\gamma, t) \frac{\partial f}{\partial \gamma} + b(\gamma, t) f \frac{\partial \Phi}{\partial \gamma} \right\} + F^r \qquad (4.36)$$

containing a random source $F^r(\gamma, t) = -\frac{\partial}{\partial \gamma} J^r(\gamma, t)$. This equation, of the Fokker-Planck-Langevin type, expresses the fact that the distribution function may fluctuate around a mean value solution of the kinetics equation.

The discussion of the stochastic kinetic equations ends our analysis of the isotropic isothermal homogeneous nucleation.

4.4.2 Nucleation of Polyhedrical Clusters

In the general situation of anisotropic clusters, additional geometrical parameters have to be considered in order to perform an accurate description of the clusters. Several ways exist to describe the size and shape of the clusters. Here we will employ the model of a polyhedrical cluster of volume v with N faces of areas $s_1, s_2, ..., s_N$, following Ref. [21]. The complete description of the cluster is thus provided by the 3N-dimensional vector $\underline{\mathbf{R}}$

$$\underline{\mathbf{R}} = \{\vec{r}_1, \vec{r}_2, ..., \vec{r}_N\} \tag{4.37}$$

whose components are the N vectors $\vec{r_i}$ normal to the individual faces s_i.

According to equation (4.9) the minimum work of formation of the polyhedral cluster is

$$\Delta G(\underline{\mathbf{R}}) = v\Delta g + \sum \sigma_i s_i \tag{4.38}$$

In order to obtain the kinetic equation following our scheme, we must consider $\underline{\mathbf{R}}$ as the internal coordinate. Notice that in this case, the state of the system (the cluster) is not determined by a single coordinate as in the isotropic case, but by the set of internal coordinates $\underline{\mathbf{R}}$. The extension of our method set up previously to deal with this multidimensional case is straightforward. Nucleation can be viewed now as a diffusion in the multidimensional space of cluster shapes. The Gibbs equation accounting for entropy variations related to changes in configuration in $\underline{\mathbf{R}}$-space is

$$\delta S = -\frac{1}{T} \int \mu(\underline{\mathbf{R}}, t)\delta f(\underline{\mathbf{R}}, t)\, d\underline{\mathbf{R}} \tag{4.39}$$

where $d\underline{\mathbf{R}} \equiv d\vec{r}_1...d\vec{r}_N$ .. The corresponding entropy production is then given by

$$\sigma = -\frac{1}{T} \int \mathbf{J}(\underline{\mathbf{R}}, t) \cdot \frac{\partial\mu(\underline{\mathbf{R}}, t)}{\partial\underline{\mathbf{R}}}\, d\underline{\mathbf{R}} \tag{4.40}$$

where the current in $\underline{\mathbf{R}}$-space $\mathbf{J}(\underline{\mathbf{R}}, t)$ is defined through the corresponding conservation law of the probability density

$$\frac{\partial f(\underline{\mathbf{R}}, t)}{\partial t} = -\frac{\partial}{\partial\underline{\mathbf{R}}} \cdot \mathbf{J}(\underline{\mathbf{R}}, t) \tag{4.41}$$

and $\mu(\underline{\mathbf{R}}, t)$ is a generalized chemical potential in $\underline{\mathbf{R}}$-space similar in form to the one introduced in the one-dimensional case in equation (4.24). The linear laws for the currents obtained from the entropy production (4.40) are

$$\underline{\mathbf{J}}(\underline{\mathbf{R}}, t) = -\frac{1}{T}\underline{\mathbf{L}} \cdot \frac{\partial\mu}{\partial\underline{\mathbf{R}}} \tag{4.42}$$

which inserted in the conservation law (4.41) give the kinetic equation

$$\frac{\partial f}{\partial t} = \frac{\partial}{\partial \mathbf{R}} \cdot \left\{ \underline{\underline{\mathbf{D}}}_R(\mathbf{R}) \cdot \frac{\partial f}{\partial \mathbf{R}} + \frac{1}{k_B T} \underline{\underline{\mathbf{D}}}_R(\mathbf{R}) \cdot \frac{\partial \Delta G}{\partial \mathbf{R}} f \right\} \qquad (4.43)$$

describing the evolution of the probability density $f(\mathbf{R}, t)$. In this expression $\underline{\underline{\mathbf{D}}}_R = \frac{k_B \underline{\underline{\mathbf{L}}}}{f}$ is a diffusion tensor, which accounts for the rate of directional growth, and may in general depend on the configuration in $\mathbf{R}$-space. The existence of non-diagonal components of $\underline{\underline{\mathbf{D}}}_R$ may account for the simultaneous coupled growth in different directions.

If we neglect these couplings and identify the diagonal components of the diffusion-growth coefficient $\underline{\underline{\mathbf{D}}}_R$ with the forward rate k^+ given in expression (4.4), one can recover the kinetic equation for the nucleation of a polyhedral cluster proposed by A. Ziabicki [21][22].

The former analysis then illustrates how to apply our formalism to the case in which the characterization of the state of the system must be carried out by certain number of coordinates or degrees of freedom.

4.5 Nucleation in Spatially Inhomogeneous Non-Isothermal Conditions

Contrarily to the case of solidification of simple substances, crystallization of polymers occurs in a wide range of temperatures [23]. Moreover, in realistic situations the external conditions, specially those concerning the temperature of the system, may change in time thus altering the process significantly. These characteristics makes it necessary the study of the process under non-isothermal and inhomogeneous conditions. Our aim in this section is precisely to develop a more realistic model of solidification of polymers accounting for spatial inhomogeneities and temperature variations.

In the case of homogeneous isotropic nucleation, we can leave spatial dependencies aside as the process occurs in a similar way at any point of the system. We can thus focus our description on the evolution of the cluster size as a function of the external conditions (pressure, temperature, density). The homogeneity thus justifies the use of global thermodynamic approach.

However, when the system is inhomogeneous the conditions controlling crystallization vary from point to point of the material and therefore a local description of the process must be considered.

To this purpose, we will divide the whole sample into volume elements small enough in order that temperature, pressure, density... can be considered homogeneous, but large enough to guarantee that thermodynamic can still be applied. In terms of length scales, the previous statement means that the typical length of the volume element must be much smaller than the gradients present at the system and much bigger than any microscopic length.

In order to guarantee the consistentness of a thermodynamic analysis at the local level, we must introduce the local equilibrium hypothesis. This hypothesis assumes that, although the system is not globally in equilibrium, thermodynamics holds locally. That is, the local variables density, temperature, etc.. are related through the same thermodynamic relationships than those for homogeneous macroscopic systems.

It is important to remark, that crystallization involves two clearly differentiated length scales of macroscopic and mesoscopic natures. On one hand, the scale in which thermodynamic quantities as pressure, temperature, density, etc. vary. These quantities can be considered as uniform within each elementary cell, but may change along the sample. On the other hand, nucleation occurs on a mesoscopic scale.

Let us now consider a more general situation in which the clusters are embedded and evolve in the melted phase which acts as a heat bath. To account for spatial inhomogeneities, we will perform a local description in terms of $f(\gamma, \mathbf{x}, \mathbf{u}, t)/N$, the probability density of finding a cluster of size $\gamma \in (\gamma, \gamma + d\gamma)$ at $\mathbf{x} \in (\mathbf{x}, \mathbf{x} + d\mathbf{x})$, with velocity $\mathbf{u} \in (\mathbf{u}, \mathbf{u} + d\mathbf{u})$, at time t. N is the total number of clusters in the system. The whole system is then thermodynamically characterized by the local energy density *per cluster* $e(\mathbf{x}, t)$, the entropy density *per cluster* $s(\mathbf{x}, t)$, and the number density

$$\rho(\mathbf{x}, t) = \rho_m + f_c \equiv \rho_m + \int f(\gamma, \mathbf{x}, \mathbf{u}, t) d\mathbf{u} d\gamma \qquad (4.44)$$

where ρ_m is the density of the melted phase, assumed constant.

The Gibbs equation for the entropy variations of this system is now formulated for the corresponding local thermodynamic quantities *per unit of volume*

$$\delta \rho s = \frac{\delta \rho e}{T} - \frac{1}{T} \int \mu(\gamma, \mathbf{x}, \mathbf{u}, t) \, \delta f(\gamma, \mathbf{x}, \mathbf{u}, t) \, d\mathbf{u} d\gamma \qquad (4.45)$$

where $\mu(\gamma, \mathbf{x}, \mathbf{u}, t)$ is the generalized chemical potential. By neglecting interactions between clusters, the expression of the chemical potential corresponds to that of an ideal system,

$$\mu(\gamma, \mathbf{x}, \mathbf{u}, t) = k_B T \ln f(\gamma, \mathbf{x}, \mathbf{u}, t) + C(\gamma, \mathbf{u}) \qquad (4.46)$$

Here

$$C(\gamma, \mathbf{u}) = \frac{1}{2} m(\gamma) \mathbf{u}^2 + \Phi(\gamma) \qquad (4.47)$$

is the energy barrier of formation of a cluster of size γ, velocity $\mathbf{u}$ and mass $m(\gamma) = m_1 \gamma$, where m_1 is the mass of a single polymer and $\Phi(\gamma)$ is the energy of formation of a crystal with γ monomers *at rest*.

Our next step is to formulate the balance equations governing the evolution of the relevant quantities of the system: the probability density f, the internal energy $e(\mathbf{x}, t)$, and the entropy $s(\mathbf{x}, t)$. In absence of external forces, the continuity equation for f may in general be written as

$$\frac{\partial f}{\partial t} = -\mathbf{u} \cdot \nabla f - \frac{\partial}{\partial \mathbf{u}} \cdot \mathbf{J}_u - \frac{\partial J_\gamma}{\partial \gamma} \tag{4.48}$$

where $\mathbf{J}_u$ is a new current resulting from the interaction of the clusters with the heat bath.

The balance of local energy density can be formulated [15, 24] as follows

$$\frac{\partial \rho e}{\partial t} = -\nabla \cdot \mathbf{J}_q \tag{4.49}$$

which states that the *total* internal energy of the fluid element at $\mathbf{x}$ can only be altered by the presence of a heat flux $\mathbf{J}_q$. This equation follows from the conservation of the global energy, provided that the integral of the energy fluctuations over the whole space vanishes.

We can now calculate the rate of change of entropy per unit of volume by differentiating the Gibbs equation (4.45) with respect to time

$$\frac{\partial \rho s}{\partial t} = \frac{1}{T} \frac{\partial \rho e}{\partial t} - \frac{1}{T} \int \mu \frac{\partial f}{\partial t} d\gamma d\mathbf{u} \tag{4.50}$$

Introducing the energy balance (4.49) and the continuity equation (4.48), we obtain

$$\frac{\partial \rho s}{\partial t} = -\frac{1}{T} \nabla \cdot \mathbf{J}_q$$
$$+ \frac{1}{T} \int \left(\mathbf{u} \cdot \nabla f + \frac{\partial}{\partial \mathbf{u}} \cdot \mathbf{J}_u + \frac{\partial J_\gamma}{\partial \gamma} \right) d\gamma d\mathbf{u}. \tag{4.51}$$

Integrating by parts, assuming that the fluxes vanish at the boundaries, and introducing eq. (4.46) for the chemical potential, we can write down expression above in the form of a balance equation

$$\frac{\partial \rho s}{\partial t} = -\nabla \cdot \mathbf{J}_s + \sigma \tag{4.52}$$

where the entropy flux $\mathbf{J}_s$ is given by

$$\mathbf{J}_s = \frac{1}{T} \mathbf{J}'_q - k_B \int \mathbf{u} f (\ln f - 1) d\gamma d\mathbf{u} \tag{4.53}$$

and the entropy production σ, which must be positive semidefinite according to the second law of thermodynamics, is

$$\sigma = -\frac{1}{T^2} \mathbf{J}'_q \cdot \nabla T - \frac{1}{T} \int \mathbf{J}_u \cdot \frac{\partial \mu}{\partial \mathbf{u}} d\gamma d\mathbf{u} - \frac{1}{T} \int J_\gamma \frac{\partial \mu}{\partial \gamma} d\gamma d\mathbf{u} \tag{4.54}$$

In previous equations, $\mathbf{J}'_q$ is the actual heat flux

$$\mathbf{J}'_q = \mathbf{J}_q - \int \mathbf{u} C(\gamma, \mathbf{u}) \, f d\gamma d\mathbf{u} \tag{4.55}$$

in which the integral represents the contribution due to the convective flux of enthalpy of the clusters. In order to obtain the entropy flux we have additionally employed the identity

$$\int \mathbf{u} \cdot \nabla f \ln f \, d\gamma d\mathbf{u} = \nabla \cdot \int \mathbf{u} f (\ln f - 1) \, d\gamma d\mathbf{u} \tag{4.56}$$

By using the expression

$$\mu(\gamma, \mathbf{u}, \mathbf{x}, t) = k_B T \ln \frac{f}{f_{eq}} + \mu_c(\mathbf{x}, t) , \tag{4.57}$$

where $\mu_c(\mathbf{x}, t)$ is the local equilibrium chemical potential and the local equilibrium distribution f_{eq} is given by

$$f_{eq}(\gamma, \mathbf{u}, \mathbf{x}, t) = \exp\left(\frac{\mu_c - C(\gamma, \mathbf{u})}{k_B T}\right) , \tag{4.58}$$

we may identify from equation (4.54) the generalized forces

$$\frac{1}{T^2} \nabla T , \quad k_B \frac{\partial}{\partial \mathbf{u}} \ln\left(\frac{f}{f_{eq}}\right) , \quad k_B \frac{\partial}{\partial \gamma} \ln\left(\frac{f}{f_{eq}}\right) ,$$

conjugated to the heat, velocity and γ fluxes, respectively. It becomes then clear that the origin of the thermodynamic forces are temperature gradients or variations of the distribution function with respect to its local equilibrium value. Now, according to the tenets of nonequilibrium thermodynamics, we may postulate linear phenomenological relations between thermodynamic forces and fluxes. Assuming isotropy, implying that currents and forces of different tensorial nature are not coupled, and locality in the internal space since physical currents at each point γ are only determined by the local properties, one obtains

$$\mathbf{J}'_q = -\frac{L_{TT}}{T^2} \nabla T - k_B \int L_{Tu} \frac{\partial}{\partial \mathbf{u}} \ln\left(\frac{f}{f_{eq}}\right) d\gamma d\mathbf{u} \tag{4.59}$$

$$\mathbf{J}_u = -\frac{L_{uT}}{T^2} \nabla T - k_B L_{uu} \frac{\partial}{\partial \mathbf{u}} \ln\left(\frac{f}{f_{eq}}\right) \tag{4.60}$$

$$J_\gamma = -L_{\gamma\gamma} k_B \frac{\partial}{\partial \gamma} \ln \left(\frac{f}{f_{eq}} \right) \tag{4.61}$$

where the phenomenological coefficients satisfy Onsager reciprocal relations

$$L_{Tu} = -L_{uT} \tag{4.62}$$

We may identify in the previous equations $\lambda = \frac{L_{TT}}{T^2}$ as the thermal conductivity, $D_\gamma = \frac{k_B L_{\gamma\gamma}}{f}$ as the diffusion coefficient in γ-space, and $\beta = \frac{m L_{uu}}{fT}$ and $\xi = \frac{L_{uT}}{fT}$ as friction coefficients, the currents can be written as follows

$$\mathbf{J}'_q = -\lambda \nabla T + \int \xi m \left(f\mathbf{u} + \frac{k_B T}{m} \frac{\partial f}{\partial \mathbf{u}} \right) d\gamma d\mathbf{u}, \tag{4.63}$$

$$\mathbf{J}_u = -\xi \frac{f}{T} \nabla T - \beta \left(f\mathbf{u} + \frac{k_B T}{m} \frac{\partial f}{\partial \mathbf{u}} \right), \tag{4.64}$$

$$J_\gamma = -D_\gamma \left(\frac{\partial f}{\partial \gamma} + \frac{1}{k_B T} f \frac{\partial C}{\partial \gamma} \right) \tag{4.65}$$

Introducing now these expressions into the continuity equation (4.48), we finally obtain the Fokker-Planck equation describing the evolution of the inhomogeneous density distribution of clusters in a bath with a non-uniform temperature distribution

$$\frac{\partial f}{\partial t} = -\mathbf{u} \cdot \nabla f + \frac{\partial}{\partial \gamma} \left[D_\gamma \left(\frac{\partial f}{\partial \gamma} + \frac{1}{k_B T} f \frac{\partial C}{\partial \gamma} \right) \right] +$$
$$+ \frac{\partial}{\partial \mathbf{u}} \cdot \left[\beta \left(f\mathbf{u} + \frac{k_B T}{m} \frac{\partial f}{\partial \mathbf{u}} \right) \right] + \frac{\partial}{\partial \mathbf{u}} \cdot \left(\frac{\xi}{T} f \nabla T \right) \tag{4.66}$$

In addition, substitution of the expressions for the currents above, leads to the balance equation of the total internal energy

$$\frac{\partial \rho e}{\partial t} = -\nabla \cdot \left(\mathbf{J}'_q + \int \mathbf{u} C(\gamma, \mathbf{u}) \, f d\gamma d\mathbf{u} \right) =$$
$$= \lambda \nabla^2 T - \nabla \cdot \int \mathbf{u} C(\gamma, \mathbf{u}) \, f d\gamma d\mathbf{u} - \nabla \cdot \int \xi m f \mathbf{u} d\gamma d\mathbf{u} \tag{4.67}$$

This expression is coupled to the Fokker-Planck equation and describes the evolution of the temperature field. The local internal energy has two contributions: the internal energy of the clusters and that of the heat bath. The heat bath determines the local properties in each cell. In particular, the internal energy of the melted phase $\rho_m e_m$ determines the local temperature. If we neglect thermal expansion, variations of the internal energy of the bath can be related with temperature variations through the thermodynamic relation $\delta\left(\rho_m e_m\right) = c_V \rho_m \delta T$, where c_V is the specific heat at constant volume. The internal energy of the clusters may in turn vary due to the release of latent heat associated with the phase transformation. The heat released in the formation of a crystalline cluster of size γ is $cm(\gamma) f_c$, where c is the latent heat per unit mass. Consequently, variations of the total internal energy are given by

$$\delta\left(\rho e\right) = \delta\left(\rho_m e_m\right) - c\delta\rho_t \tag{4.68}$$

where $\rho_t(\mathbf{x}, t) = \int m(\gamma) f_c d\gamma$ represents the total density of crystallized material. Therefore, from the balance of *total* internal energy (4.67) we obtain the evolution of the temperature field

$$c_V \rho_H \frac{\partial T}{\partial t} = \lambda\nabla^2 T - \nabla \cdot \int \mathbf{u} C(\gamma, \mathbf{u}) \, f d\gamma d\mathbf{u} - \nabla \cdot \int \xi m f \mathbf{u} d\gamma d\mathbf{u} + c\frac{\partial\rho_t}{\partial t} \tag{4.69}$$

The former expression enables one to identify the basic mechanisms responsible for temperature variations: heat conduction, convection, thermal diffusion effects, and the release of latent heat in the crystallization process, corresponding to the different terms on the right hand side of that equation, respectively. Notice that, by neglecting the contributions arising from the diffusion of clusters, we can recover the usual equation for the evolution of temperature field

$$\rho_H c_v \frac{\partial T}{\partial t} = \lambda\nabla^2 T + c\frac{\partial\rho_t}{\partial t} \tag{4.70}$$

4.6 Homogeneous Nucleation in Spatially Inhomogeneous Systems. Diffusion Regime

The previous equations (4.66) and (4.69) completely specify the coupled evolutions of the temperature field and the probability distribution of the clusters. In particular, they can even describe inertial regimes at the earlier stages of the crystallization process. However, the process of equilibration in velocity space is usually much faster than the remaining ones. After times much larger than the characteristic time β^{-1} for the relaxation of the velocity, the system enters the diffusion and thermal diffusion regime for which the evolution is governed by a simpler set of equations. In this section we will discuss the main features of the diffusion regime.

Our starting point will be the Fokker-Planck equation (4.66) for the evolution of the probability distribution. After equilibration in velocity space, for which its distribution achieves its Maxwellian equilibrium form, the velocity does not longer constitute a relevant variable in the kinetic description of the system. Instead, we can describe the system by the reduced probability density, defined as

$$f_c(\gamma, \mathbf{x}, t) = \int f(\mathbf{x}, \mathbf{u}, t, \gamma) d\mathbf{u} \qquad (4.71)$$

and by the velocity in $(\gamma, \mathbf{x})$-space

$$\mathbf{v}_c(\gamma, \mathbf{x}, t) = f_c^{-1} \int \mathbf{u} f d\mathbf{u} \qquad (4.72)$$

The corresponding balance equation for the density follows after integration of eq. (4.48)

$$\frac{\partial f_c}{\partial t} = -\nabla \cdot f_c \mathbf{v}_c - \frac{\partial}{\partial \gamma} \int J_\gamma d\mathbf{u} \qquad (4.73)$$

where the integral of the current in γ-space with respect to velocities is

$$\int J_\gamma d\mathbf{u} = \int \left(f \frac{D_\gamma}{k_B T} \frac{\partial C(\gamma)}{\partial \gamma} + D_\gamma \frac{\partial f}{\partial \gamma} \right) d\mathbf{u} \qquad (4.74)$$

We can reasonably assume that the diffusion coefficient in γ-space, D_γ, is independent of the velocity, but in general the drift term $\frac{\partial C}{\partial \gamma}$ does depend on $\mathbf{u}$. In fact, from the energy barrier (4.47) this drift is

$$\frac{\partial C}{\partial \gamma} = \frac{1}{2} m_1 u^2 + \frac{\partial \Phi(\gamma)}{\partial \gamma} \qquad (4.75)$$

However, as it is well known, in the diffusion regime i.e. for times long enough compared to the characteristic time β^{-1} of Brownian relaxation, the system has achieved equilibration in the velocity space. Therefore the relation

$$\int \frac{1}{2} m u^2 f d\mathbf{u} = \frac{3}{2} k_B T f_c , \qquad (4.76)$$

reminiscent of the energy equipartition law, holds as in equilibrium. The integral of the current in γ-space can then be written as

$$\int J_\gamma d\mathbf{u} = D_\gamma \left[\frac{\partial f_c}{\partial \gamma} + f_c \frac{1}{k_B T} \left(\frac{\partial \Phi}{\partial \gamma} + \frac{3}{2} \frac{k_B T}{\gamma} \right) \right]$$

$$= D_\gamma \left[f_c \frac{1}{k_B T} \frac{\partial \tilde{\Phi}(\gamma)}{\partial \gamma} + \frac{\partial f_c}{\partial \gamma} \right] \qquad (4.77)$$

where $\widetilde{\Phi}(\gamma)$ is the modified nucleation barrier defined as

$$\widetilde{\Phi}(\gamma) = \Phi(\gamma) + \frac{3}{2}k_B T \ln \gamma \tag{4.78}$$

which now includes in an averaged way the effects of the Brownian diffusion of clusters. It is important to point out that diffusion of clusters increase nucleation barrier and consequently reduce nucleation rate as previous equation states. The reason for the increase of the barrier is that the thermal energy $k_B T$ available to surmount the activation barrier must be shared between increasing the size of the cluster and performing Brownian motion [25].

Taking this considerations into account, the density balance equation can finally be expressed as

$$\frac{\partial f_c}{\partial t} = -\nabla \cdot f_c \mathbf{v}_c + \frac{\partial}{\partial \gamma}\left[D_\gamma \left(f_c \frac{1}{k_B T}\frac{\partial \widetilde{\Phi}(\gamma)}{\partial \gamma} + \frac{\partial f_c}{\partial \gamma}\right)\right] \tag{4.79}$$

Moreover, the balance equation for the velocity is

$$\frac{\partial f_c \mathbf{v}_c}{\partial t} = -\nabla \cdot \int f \mathbf{u}\mathbf{u}d\mathbf{u} - \int \mathbf{u}\frac{\partial}{\partial \mathbf{u}}\cdot \mathbf{J}_u d\mathbf{u} - \frac{\partial}{\partial \gamma}\int \mathbf{J}_\gamma \mathbf{u}d\mathbf{u} =$$

$$= -\nabla \cdot \int f \mathbf{u}\mathbf{u}d\mathbf{u} - \beta f_c \mathbf{v}_c - \frac{\xi}{T} f_c \nabla T$$

$$+ \frac{\partial}{\partial \gamma}\left[D_\gamma \int \mathbf{u}\left(f \frac{1}{k_B T}\frac{\partial C}{\partial \gamma} + \frac{\partial f}{\partial \gamma}\right)d\mathbf{u}\right] \tag{4.80}$$

The term on the right hand side can be related with the kinetic expression of the pressure tensor of the "gas" of clusters, defined through [15]

$$\overrightarrow{P} = \int f \mathbf{u}\mathbf{u}d\mathbf{u} - f_c \mathbf{v}_c \mathbf{v}_c \tag{4.81}$$

which owing to the ideality of the system [15, 24] is diagonal and given by

$$\overrightarrow{P} = \frac{k_B T}{m}f_c \overrightarrow{U} = D\beta f_c \overrightarrow{U} \tag{4.82}$$

where $D = \frac{k_B T}{m\beta}$ is the *spatial* diffusion coefficient and $\overrightarrow{U}$ the unit tensor. Therefore, velocity evolution is determined by the kinetic coefficients β, ξ, and D_γ. However, velocity relaxation is usually faster than relaxation in the size-space. In fact, Stokes law provides a good estimate for the the friction constant, $\beta = \frac{6\pi\eta a}{m}$, indicating that its magnitude is very large for small clusters. Consequently, the inequality $\beta \gg D_\gamma$ is expected to hold, and therefore, we can neglect the contribution arising from the current J_γ. The resulting equation can be written as

$$f_c\frac{d\mathbf{v}_c}{dt} = -\nabla \cdot \left(D\beta f_c \overrightarrow{U}\right) - \beta\frac{\xi\beta^{-1}}{T}f_c\nabla T - \beta f_c\mathbf{v}_c \qquad (4.83)$$

where the total derivative is defined as

$$\frac{d}{dt} = \frac{\partial}{\partial t} + \mathbf{v}_c \cdot \nabla \qquad (4.84)$$

The diffusion regime is achieved for times $t \geq \beta^{-1}$. In this situation, we can neglect the time derivative, yielding

$$\mathbf{J}_D = f_c\mathbf{v}_c = -\nabla Df_c - \frac{\xi\beta^{-1}}{T}f_c\nabla T \qquad (4.85)$$

Inserting previous expression in the kinetic equation for the reduced probability eq. (4.79)

$$\frac{\partial f_c}{\partial t} = \nabla \cdot \left(\nabla Df_c + f_c\xi\beta^{-1}\nabla \ln T\right) + \frac{\partial}{\partial\gamma}\left[D_\gamma\left(f_c\frac{1}{k_BT}\frac{\partial\widetilde{\Phi}}{\partial\gamma} + \frac{\partial f_c}{\partial\gamma}\right)\right] \qquad (4.86)$$

This is the expression for the evolution of the cluster distribution function in spatially inhomogeneous systems in the *diffusion regime*.

It is convenient to rewrite the spatial flux of clusters as follows

$$\mathbf{J}_D = -D\nabla f_c - \frac{k_B}{m\beta}f_c\nabla T - \frac{\xi\beta^{-1}}{T}f_c\nabla T = -D\nabla f_c - D_{th}\frac{\nabla T}{T} \qquad (4.87)$$

where it becomes evident that J_D has two contributions: normal diffusion described by Fick law and a drift term that can be identified with thermal diffusion. The quantity

$$D_{th} = \left(D + \frac{\xi}{\beta}\right)f_c = Df_c\left(1 + \frac{\xi m}{k_BT}\right) \qquad (4.88)$$

is the *thermal diffusion coefficient*. A measure of the importance of thermal diffusion effects follows from the ratio between thermal and diffusion coefficients, called Soret coefficient

$$S = \frac{D_{th}}{D} \qquad (4.89)$$

Typical values of this coefficient for gaseous and liquid mixtures range from 10^{-3} to 10^{-5} K^{-1} in orders of magnitude, which indicates that mass diffusion is dominant. For polymer solutions, however, the Soret coefficient increases with the molecular weight to the extent of becoming significant at high values of the molecular weight [26]. Consequently, thermal diffusion effects may play a relevant role in the process.

We have seen that the coupling between thermal and diffusion effects modifies the diffusion current. In addition, it also modifies the heat flux. To show this feature it is more convenient to work with the unmodified heat current because it is the quantity usually measured in experiments

$$\mathbf{J}_q = \mathbf{J}'_q + \int \mathbf{u}\frac{1}{2}mu^2 f d\gamma d\mathbf{u} \tag{4.90}$$

Inserting this expression in equation (4.63) and using the result

$$\int \mathbf{u}\frac{1}{2}mu^2 f d\gamma d\mathbf{u} = \int k_B T \mathbf{J}_D d\gamma \tag{4.91}$$

we obtain

$$\mathbf{J}_q = -\lambda\nabla T + \int (k_B T + \xi m)\,\mathbf{J}_D d\gamma \tag{4.92}$$

Now, employing eq. (4.87) for the diffusion current $\mathbf{J}_D$, we finally arrive at

$$\mathbf{J}_q = -\tilde{\lambda}\nabla T - \int D_{th} k_B T \frac{\nabla f_c}{f_c} d\gamma \tag{4.93}$$

where

$$\tilde{\lambda} = \lambda + \int \frac{D_{th}^2 k_B T}{D f_c} d\gamma \tag{4.94}$$

is the modified heat conductivity.

In summary, the coupling of diffusion and thermal effects results in modifications of the fluxes and the transport coefficients leading to drastic changes in the crystallization process. The influence of this coupled effects in the process can be measured in terms of the ratio between thermal and spatial diffusion, or Soret coefficient, and the magnitude of the gradients. Precisely, polymers use to have low thermal conductivities and high values of the Soret coefficient. This is a signature that thermal-diffusion effects may become relevant in real crystallization.

In fact, in the particular case of condensation of gases under inhomogeneous conditions and presence of thermal gradients, it has been shown, that nucleation rate may be significantly altered by spatial diffusion and thermal-diffusion effects [27]. Although polymer crystallization and condensation are processes of different nature, they share some common features at their earlier stages. Consequently, these results may constitute a hint to assess the relevance of diffusion/thermal-diffusion effects in crystallization.

4.7 Nucleation in a Shear Flow

Crystallization of polymers often involves mechanical processing of the melt, such as extrusion, shearing or injection. This feature may have drastic consequences in the crystallization process. Therefore it would be desirable to build up a model that can take into account these processing in a consistent and proper way. Several models have been proposed to capture some features of the crystallization under external influences [28] but up to now, there is no theory providing a complete description of the crystallization process in that conditions.

It is worth pointing out that the presence of a shear flow breaks down the isotropy of the system, distinguishing between diffusion in different directions. Moreover, it introduces spatial inhomogeneities, which induce spatial, velocity and temperature fluxes. Cross effects between these currents may be in general very important, and consequently the evolution of the probability density, velocity and temperature fields will be governed by a highly coupled set of kinetic differential equations. These equations can be derived within our general framework, following the steps developed in the previous sections, but the resulting equations could be quite complex. For illustrative purposes, it is better to restrict our analysis to effects purely originated from the presence of the flow. Consequently, for the sake of simplicity, we will assume isothermal conditions.

The system we consider is a melted polymer in which the emerging crystallites are embedded. We assume that this melted substance acts as a heath bath imposing a constant temperature. The system is then subjected to conditions creating a shear flow characterized by a velocity profile $\mathbf{v}_0(\mathbf{x})$.

Following the scheme developed in previous sections, we will first formulate the Gibbs equation for the entropy variations of this system with respect to the reference state. In this particular situation, our reference state is the stationary state characterized by the steady velocity profile of the shear flow $\mathbf{v}_0(\mathbf{x})$, and the Gibbs equation must be formulated for a mass element followed along its center of gravity motion [15]

$$T\frac{ds}{dt} = \frac{de}{dt} + p\frac{d(1/\rho)}{dt} - \int \mu(\gamma, \mathbf{x}, \mathbf{u}, t) \frac{dc(\gamma, \mathbf{x}, \mathbf{u}, t)}{dt} \, du d\gamma \qquad (4.95)$$

Here $c(\gamma, \mathbf{x}, \mathbf{u}, t) = \frac{f(\gamma, \mathbf{x}, \mathbf{u}, t)}{\rho}$ is the number fraction, the chemical potential is defined through (4.46) and we have defined the total derivative as $\frac{d}{dt} \equiv \frac{\partial}{\partial t} + \mathbf{v}_c \cdot \nabla$. The quantity $C(\mathbf{u}, \gamma)$ in the chemical potential denotes the energy cost of formation of a state described by the internal variables $(\mathbf{u}, \gamma)$, given by

$$C(\gamma, \mathbf{u}) = \frac{1}{2} m(\gamma)(\mathbf{u} - \mathbf{v}_0)^2 + \Phi(\gamma) \tag{4.96}$$

where $\Phi(\gamma)$ represents the energy of formation of a cluster of size γ *at rest* and the first term is its kinetic energy with respect to the steady state velocity profile $\mathbf{v}_0$. By using the local equilibrium distribution f_{eq} given by

$$f_{eq}(\gamma, \mathbf{u}, \mathbf{x}, t) = \exp\left(\frac{\mu_c - C(\gamma, \mathbf{u})}{k_B T}\right), \tag{4.97}$$

the chemical potential can be written as

$$\mu(\gamma, \mathbf{u}, \mathbf{x}, t) = k_B T \ln \frac{f}{f_{eq}} + \mu_c(\mathbf{x}, t) \tag{4.98}$$

where $\mu_c(\mathbf{x}, t)$ denotes the local equilibrium chemical potential which is independent of the internal variables. This expression is analogous to the chemical potential one would obtain by comparison of the Gibbs equation (4.95) with the Gibbs' entropy postulate.

Our aim now is to formulate the equation governing the evolution of the probability density. In the absence of external force fields, the continuity equation for f may in general be written as

$$\frac{\partial f}{\partial t} = -\mathbf{u} \cdot \nabla f - \frac{\partial}{\partial \mathbf{u}} \cdot \mathbf{J}_u - \frac{\partial J_\gamma}{\partial \gamma} \tag{4.99}$$

where $\mathbf{J}_u$ is a current arising from the interaction of the clusters with the heat bath.

In order to derive the entropy balance equation, we also need the expression for the variations of the energy with respect to its value at the stationary state. Since we are assuming isothermal conditions, and neglecting viscous heating, we obtain the energy conservation law

$$\frac{d\rho e}{dt} = 0 \tag{4.100}$$

which states that the internal energy of the fluid element following its center of mass motion is constant.

Introducing the energy balance (4.100) and the continuity equation (4.99) into the Gibbs equation (4.95) , we obtain after some straightforward calculations the entropy balance equation

$$\rho\frac{ds}{dt} = -\nabla \cdot \mathbf{J}_s + \sigma \tag{4.101}$$

where the entropy flux $\mathbf{J}_s$ is given by

$$\mathbf{J}_s = -k_B \int (\mathbf{u} - \mathbf{v}_0) f(\ln f - 1) d\gamma d\mathbf{u} - \frac{1}{T} \int C(\gamma, \mathbf{u}) f(\mathbf{u} - \mathbf{v}_0) d\gamma d\mathbf{u} \tag{4.102}$$

and the entropy production σ, which must be positive according to the second law of thermodynamics, is

$$\sigma = -\frac{1}{T} \int \mathbf{J}_u \cdot \frac{\partial \mu}{\partial \mathbf{u}} d\gamma d\mathbf{u} - \frac{1}{T} \int J_\gamma \frac{\partial \mu}{\partial \gamma} d\gamma d\mathbf{u}$$
$$-\frac{1}{T} \int f(\mathbf{u} - \mathbf{v}_0) \cdot \nabla \frac{1}{2} m(\gamma)(\mathbf{u} - \mathbf{v}_0)^2 d\gamma d\mathbf{u} \tag{4.103}$$

The entropy production can be interpreted as a sum of products between forces-fluxes pairs. In this particular situation, the forces are $-\frac{1}{T}\frac{\partial \mu}{\partial \mathbf{u}}$, $-\frac{1}{T}\frac{\partial \mu}{\partial \gamma}$, and $-\frac{1}{T}\nabla \frac{1}{2}m(\gamma)(\mathbf{u} - \mathbf{v}_0)^2$ and their conjugated fluxes the velocity flux $\mathbf{J}_u$, the flux in γ-space J_γ, and the spatial relative current $\mathbf{J}_x \equiv f(\mathbf{u} - \mathbf{v}_0)$, respectively.

The next step is the formulation of linear phenomenological equations relating those fluxes and forces. The situation we are considering present some peculiarities. Spatial and velocity currents have both the same tensorial character (they are both vectors) which implies that in general they must be coupled. Moreover, the presence of the shear flow breaks down the isotropy of the system by introducing a given direction. Consequently, phenomenological coefficients are expected to be in general tensors. In addition, locality in the internal space will be assumed again. Taking all these considerations into account, the expressions for the currents are

$$\mathbf{J}_x = +\frac{m}{T} \overrightarrow{L}_{xx} \cdot \nabla \mathbf{v}_0 \cdot (\mathbf{u} - \mathbf{v}_0) + \frac{1}{T} \overrightarrow{L}_{ux} \cdot \frac{\partial \mu}{\partial \mathbf{u}} \tag{4.104}$$

$$\mathbf{J}_u = -\frac{1}{T} \overrightarrow{L}_{uu} \cdot \frac{\partial \mu}{\partial \mathbf{u}} + \frac{m}{T} \overrightarrow{L}_{ux} \cdot \nabla \mathbf{v}_0 \cdot (\mathbf{u} - \mathbf{v}_0) \tag{4.105}$$

$$J_\gamma = -\frac{1}{T} L_{\gamma\gamma} \frac{\partial \mu}{\partial \gamma} \tag{4.106}$$

where we have used the Onsager reciprocal relation

$$\overrightarrow{L}_{ux} = -\overrightarrow{L}_{xu} \tag{4.107}$$

and the result

$$\nabla \frac{1}{2} m(\gamma)(\mathbf{u} - \mathbf{v}_0)^2 = -m\nabla \mathbf{v}_0 \cdot (\mathbf{u} - \mathbf{v}_0)$$

It is useful to redefine the phenomenological coefficients, $\overrightarrow{L}_{ij}$, in a more convenient way. By identifying $D_\gamma = \frac{k_B L_{\gamma\gamma}}{f}$ as the diffusion coefficient in γ-space, which in turn can be identified with the rate of attachment of monomers to a cluster by equation (4.21); $\overrightarrow{\alpha} = \frac{m\overrightarrow{L}_{uu}}{fT}$ and $\overrightarrow{\zeta} = \frac{m\overrightarrow{L}_{ux}}{fT}$ as *friction tensors,* and introducing the explicit form of the chemical potential, equation (4.46), the currents $\mathbf{J}_u$ and J_γ can be written as

$$\mathbf{J}_u = -f \left[\overrightarrow{\alpha} + \overrightarrow{\zeta} \nabla \mathbf{v}_0 \right] \cdot (\mathbf{u} - \mathbf{v}_0) - \overrightarrow{\alpha} \frac{k_B T}{m} \cdot \frac{\partial f}{\partial \mathbf{u}}, \tag{4.108}$$

$$J_\gamma = -D_\gamma \left(\frac{\partial f}{\partial \gamma} + \frac{1}{k_B T} f \frac{\partial C}{\partial \gamma} \right) \tag{4.109}$$

Introduction of these expressions for the currents into the continuity equation yields

$$\frac{\partial f}{\partial t} = -\mathbf{u} \cdot \nabla f + \frac{\partial}{\partial \gamma} \left[D_\gamma \left(\frac{\partial f}{\partial \gamma} + \frac{1}{k_B T} f \frac{\partial C}{\partial \gamma} \right) \right] +$$
$$+ \frac{\partial}{\partial \mathbf{u}} \cdot \left(\left[\overrightarrow{\alpha} + \overrightarrow{\zeta} \nabla \mathbf{v}_0 \right] \cdot f (\mathbf{u} - \mathbf{v}_0) + \overrightarrow{\alpha} \frac{k_B T}{m} \cdot \frac{\partial f}{\partial \mathbf{u}} \right) \tag{4.110}$$

which corresponds to the Fokker-Planck equation describing the evolution of the inhomogeneous density distribution of clusters in the presence of a steady state shear flow. Notice that the phenomenological coefficients $\overrightarrow{\alpha}$ and $\overrightarrow{\zeta}$ still remain unspecified. Their identification must be carried out through a proper interpretation of the relaxation equations derived from the Fokker-Planck equation. This point will be discussed in the next section.

4.7.1 Balance Equations in the Diffusion Regime

The Fokker-Planck equation (4.110) we have derived retains information about the evolution of the cluster velocity distribution. Although the movement of clusters may play a significant role in the crystallization process, velocity dependencies of the process can hardly be measured because velocity distribution relaxes to equilibrium very fast. Therefore, as discussed in Section 4.6, we can perform a simplified description of the process. The relevant quantities will then be the distribution function for the density of clusters together with the velocity in γ-space defined in equations (4.71) and (4.72).

Integration of eq. (4.110) over velocities leads to the corresponding balance equation for the density of clusters in γ-space

$$\frac{\partial f_c}{\partial t} = -\nabla \cdot f_c \mathbf{v}_c + \frac{\partial}{\partial \gamma} \left(f_c \frac{D_\gamma}{k_B T} \frac{\partial \widetilde{\Phi}(\gamma)}{\partial \gamma} + D_\gamma \frac{\partial f_c}{\partial \gamma} \right) \tag{4.111}$$

where, as discussed in Section 4.6, $\widetilde{\Phi}(\gamma)$ is the corresponding nucleation barrier given by

$$\widetilde{\Phi}(\gamma) = \Phi(\gamma) + \frac{3}{2} k_B T \ln \gamma + \frac{1}{2} m(\gamma) \left(\mathbf{v}_c - \mathbf{v}_0 \right)^2 \tag{4.112}$$

which now includes two contributions arising from the effects of the Brownian diffusion of clusters, and the presence of the shear flow.

Both effects lead to an increase of the nucleation barrier and therefore a reduction in the nucleation rate. The underlying explanation of this decrease of nucleation rate lies in the fact that the thermal energy must be used to surmount the activation barrier and to increase the kinetic energy of the cluster. Notice that when velocity of the clusters equilibrates with the imposed velocity, $\mathbf{v}_c = \mathbf{v}_0$, the last term in equation above vanishes and the barrier reduces to the one in the absence of external flow.

Proceeding now along the lines of Section 4.6, the balance equation for the velocity is

$$\frac{\partial f_c \mathbf{v}_c}{\partial t} = -\nabla \cdot \int f \mathbf{u} \mathbf{u} d\mathbf{u} - f_c \left[\overset{\rightrightarrows}{\alpha} + \overset{\rightrightarrows}{\zeta} \nabla \mathbf{v}_0 \right] \cdot (\mathbf{v}_c - \mathbf{v}_0)$$
$$+ \frac{\partial}{\partial \gamma} \left[D_\gamma \int \mathbf{u} \left(f \frac{1}{k_B T} \frac{\partial C}{\partial \gamma} + \frac{\partial f}{\partial \gamma} \right) d\mathbf{u} \right] \tag{4.113}$$

The first term on the right hand side can be related with the pressure tensor of the *gas* of clusters using eq. (4.81). The balance equation for the velocity can then be written as

$$f_c \frac{d\mathbf{v}_c}{dt} = -\nabla \cdot \overrightarrow{P} - f_c \left[\overrightarrow{\alpha} + \overrightarrow{\zeta} \nabla \mathbf{v}_0 \right] \cdot (\mathbf{v}_c - \mathbf{v}_0)$$

$$+ \frac{\partial}{\partial \gamma} \left[D_\gamma \int \mathbf{u} \left(f \frac{1}{k_B T} \frac{\partial C}{\partial \gamma} + \frac{\partial f}{\partial \gamma} \right) d\mathbf{u} \right] \qquad (4.114)$$

Notice that the second term on the right hand side can be identified with the friction force per unit mass exerted on the suspended cluster by the host fluid, where $\overrightarrow{\alpha} + \overrightarrow{\zeta} \nabla \mathbf{v}_0$ plays the role of the friction tensor, which determines again the characteristic time scale for the relaxation of the velocities. In general, such friction tensor can be calculated from hydrodynamics; however, for the sake of simplicity, let us approximate this tensor by the Stokes friction $\overrightarrow{\alpha} + \overrightarrow{\zeta} \nabla \mathbf{v}_0 \simeq \beta \overrightarrow{1}$. For times $t > \beta^{-1}$ the system achieves the diffusion regime. In this situation, we can neglect inertial effects manifested through the presence of the time derivative $\frac{d\mathbf{v}_c}{dt}$. Moreover, the pressure tensor in presence of a shear flow is given by [29]

$$\overrightarrow{P} = \frac{k_B T}{m} f_c \left[\overrightarrow{1} - \left(\beta^{-1} \left(\overrightarrow{1} + \overrightarrow{\zeta} \right) \cdot \nabla \mathbf{v}_0 \right)^s \right] \qquad (4.115)$$

Finally, the rate of relaxation of velocities is usually faster than the relaxation of cluster sizes determined by D_γ. Consequently, we can neglect the contribution arising from the current J_γ in the velocity relaxation . Taking all these considerations into account in eq. (4.114), for the diffusion current of clusters we obtain

$$\mathbf{J}_D = f_c \mathbf{v}_c = -\overrightarrow{D} \cdot \nabla f_c + f_c \mathbf{v}_0 \qquad (4.116)$$

where

$$\overrightarrow{D} = \frac{k_B T}{m} \beta^{-1} \left[\overrightarrow{1} - \left(\beta^{-1} \left(\overrightarrow{1} + \overrightarrow{\zeta} \right) \cdot \nabla \mathbf{v}_0 \right)^s \right] \qquad (4.117)$$

can be identified with the *spatial* diffusion coefficient which in this situation possesses tensorial character. Inserting this expression in the kinetic equation (4.111) one has

$$\frac{\partial f_c}{\partial t} = -\nabla \cdot f_c \mathbf{v}_0 + \nabla \left(\overrightarrow{D} \cdot \nabla f_c \right) + \frac{\partial}{\partial \gamma} \left(f_c \frac{D_\gamma}{k_B T} \frac{\partial \tilde{\Phi}}{\partial \gamma} + D_\gamma \frac{\partial f_c}{\partial \gamma} \right) \qquad (4.118)$$

which is the kinetic equation for nucleation in presence of a shear flow in the *diffusion regime*.

The main effects that the presence of a shear flow exert on the nucleation process can be summarized as follows. On one hand, the flow alters the form of the the Fokker-Planck equation and consequently the evolution of the growing clusters distribution function, which has implications in the effective nucleation and growth rate. On the other hand, we have shown that the consideration of the effects of the motion of clusters leads to modification of the nucleation barrier, as manifested in equation (4.112). Finally, the presence of a shear flow changes the spatial diffusion coefficient of the clusters, as shown in equation (4.117). As stated in Section 4.2.1, the relevance of this modification lies in the fact that the rate of attachment of monomers to a clusters and the exponential prefactor of the nucleation rate in the classical expression, eq. (4.18) strongly depends on the diffusivity of monomers in the surface of a growing cluster. The fact that the diffusion coefficient is altered implies that the nucleation rate is modified. Moreover, the symmetry breaking inherent to the presence of the flow promotes that the diffusion coefficient is no longer a scalar quantity, but has *tensorial* nature. The rate of the process will depend on the direction thus making nucleation and specially growth of crystals no longer isotropic. This feature could share some light to explain some experiments in shear-induced polymer crystallization [30].

As stated previously, the main objective of the section has been to analyze the effects that the flow by itself may induce on the nucleation process. That is the reason why we have assumed isothermal conditions. A more complete and realistic treatment may be carried out incorporating thermal effects, in the way discussed in the preceding section.

4.8 Kinetic Equations for the Crystallinity

The aim of the models we have presented has been mainly the description of the mechanism of nucleation. However their validity is not restricted to nucleation process, as they may also describe the earlier stages of crystal growth. We have seen that nucleation can be considered as an activated process where a free energy barrier has to be surmounted. However, the barrier crossing process is of stochastic nature and in fact recent numerical simulations [18], [19] have revealed that this crossing is highly diffusive which means that the barrier can be surmounted in both directions. Consequently, the fact that a cluster is bigger than the critical size n^* does not guarantee that it must grow indefinitely. Post-critical clusters, which are usually assumed as *stable* crystals, may eventually shrink and disappear. Therefore the diffusive nature of nucleation mechanism implies that the separation between a pure "nucleation" regime (as the formation of critical clusters) and a "growth" regime (for clusters larger than the critical size) is not sharp.

Definitively, the nucleation and the subsequent crystal growth are not intrinsically different processes since, at least in the initial stages, they obey the same underlying physical mechanism. In fact, the same ingredients appearing in phenomenological theories of nucleation namely, the free energy barrier ΔG and even the rate of attachment of monomers $k^+(n)$, remain well defined for sizes larger than the critical size. Therefore, they also describe the kinetics of the growth.

Consequently, the model we have introduced describes nucleation and the initial stages of growth in the crystallization process. However, as most of nucleation-aimed models, our theory is subjected to the following considerations:

- It applies while interactions between clusters are practically negligible. When clusters are large enough, the ideality assumption employed in the expression for the chemical potential fails, since interactions between clusters start to play a significant role. Interactions among clusters emerge as they try to fill the whole space.
- We have not considered the presence of secondary crystallization or perfection which is governed by a different mechanism. This assumption is justified from the fact that secondary crystallization is normally slower, and consequently the overall initial crystallization is usually determined by nucleation and growth.
- We have focused our study in *bulk nucleation*. *Surface* nucleation at the surface of the growing clusters can be categorized as heterogeneous nucleation.
- Finally, for the sake of simplicity, we have assumed that the shape of the clusters is spherical until impingement occurs. Consequently, polymorphic crystallization is not taken into account and anisotropic effects are neglected.

In previous sections, we have modeled the initial stage of crystallization focusing on the evolution of the probability density of clusters of a given size. It remains now to describe the subsequent steps of the process.

It is interesting to realize that our formalism still retains information about the size distribution of crystals of the new phase. Sometimes, however, it is more convenient to work out with a simplified global description in terms of merely the fraction of material crystallized. Accordingly, one introduces a new macroscopic variable, the crystalline fraction or *crystallinity w* defined as the fraction of the volume occupied by crystals. The state $w = 0$ corresponds to a melted polymer whereas $w = 1$ would indicate complete crystallization. The latter situation can barely be achieved. As we have stated previously, the reason is that the resulting "crystal" is in fact a crystalline aggregate with amorphous inclusions of non-crystallized material which get trapped between crystal spherulites.

Let us first introduce the *virtual volume* crystallized V defined as the fraction of volume that the nucleated clusters would occupy at time t if they could grow freely in absence of impingement effects. This quantity differs from the actual fraction of volume crystallized in the fact that overlapping between clusters is allowed. In terms of the size distribution of clusters, it can be expressed as

$$V(\mathbf{x}, t) = \int_{\gamma^*(T)}^{\infty} f_c(\gamma, \mathbf{x}, t) \rho_1 \gamma \, d\gamma \tag{4.119}$$

where ρ_1 is the fraction of volume per monomer of the crystal phase, and γ^* is the size of the critical cluster that strongly depends on temperature.

In the initial stages of nucleation, the crystallinity coincides with the virtual volume crystallized $w \cong V$, because impingement or overlapping between clusters is negligible. At more advances stages of the process, this statement becomes no longer valid since neither nucleation nor growth may take place in an already crystallized region.

Those effects, which can be viewed as interactions between clusters, are difficult to include in nucleation/growth theories to the extent that they are usually not considered. We can then wonder how is possible to describe the space-filling with theories that cannot include interactions or impingement of clusters. The most common answer is to resort to the mathematical theory of Avrami-Kolmogoroff (see Chapters 5 and 6).

4.8.1 Avrami-Kolmogoroff Theory

The theory of Avrami-Kolmogoroff, based upon purely stochastic geometric grounds, aims to describe the space-filling in the crystallization kinetics.

In its formulation, the crystallization process is divided into two basic and well differentiated regimes: nucleation and deterministic growth. Nucleation is viewed as a point process in the space of the sample. That is, nuclei, which are the germ of crystallization, may emerge at random points of the system following a suitable spatial process of the Poisson type. The reason for choosing this particular probability distribution for the appearance of nuclei is, apart from its simplicity, the fact that it is the one assuming less information. The rate at which nuclei appear per unit of volume at position x at time t, denoted by $N(x, t)$, is referred to as the nucleation rate.

After the formation of the nuclei, it is assumed that they grow freely with a radial rate $R(x, t)$. Free growth means that clusters are not aware of the presence of neighbor cluster and overlapping is allowed.

For a thorough presentation of this theory we refer to Chapter 5 in this volume.

In spite of its usefulness, the mathematical theory of Avrami-Kolmogoroff presents some drawbacks:

- It assumes that nucleation is a point process with Poisson-like distribution. In spite that this assumption is the more reasonable one when no further information is available, it may be not valid under realistic conditions. For instance, the presence of inhomogeneities in the system can make Poisson distribution for the nucleation events no longer applicable.

- Nucleation seldom proceeds according to a simple single mechanism as one often assumes in Avrami theory.
- Experiments on crystallization kinetics not always can be reproduced by the Avrami equation. In this context, the value of the Avrami exponent strongly depends on the underlying model for nucleation and growth rates, which must be introduced *a priori*.
- The separation between nucleation and growth regimes is not always well defined. As we have shown, nucleation is a stochastic process of diffusive nature which causes that apparently stable nuclei can eventually shrink and even disappear. It is worth to emphasize that, better than delimiting different regimes, the critical size separates different tendencies (to grow or to shrink). In this line, we must take into account the following consideration.
- The growing of clusters is not deterministic but stochastic. Clusters can shrink or grow, even for sizes bigger than the critical one.

In previous sections, we have developed a general framework and discussed a set of examples to describe the evolution of the distribution of clusters under different realistic conditions. Although our formalism is restricted to the earlier stages of crystallization, its results can be used in the Avrami-Kolmogoroff theory to describe the space-filling. The improvements of our theory arise from the fact that it unifies nucleation and growth, retaining the stochastic nature of these processes, and incorporating possible hydrodynamic effects.

From the kinetic equations we have derived, one can evaluate the nucleation $N(t)$ and growth $R(t)$ rates, which take into account the physical mechanisms and try to reproduce the experimental conditions.

In fact, it is not necessary to calculate separately both rates $N(t)$, and $R(t)$. It is more accurate to directly evaluate the virtual volume crystallized from the cluster size distribution function as shown in equation (4.119).

The expression for the crystallinity introduced in the Kolmogoroff equation (see Equation (5.59) in Chapter 5) can describe the advanced stages of crystallization. In the following subsections, we will briefly illustrate the evolution of the crystallinity in some particular situations analyzed in previous sections. For the sake of simplicity, we will analyze the kinetic equations for the variation of the virtual volume with time which following equation (4.119) is given by

$$\frac{\partial}{\partial t}V(\mathbf{x},t) = \int_{\gamma^*(T)}^{\infty} \frac{\partial f_c(\gamma,\mathbf{x},t)}{\partial t}\rho_1\gamma d\gamma + \frac{\partial\gamma^*}{\partial T}\frac{\partial T}{\partial t}f_c(\gamma^*,\mathbf{x},t)\rho_1\gamma^* \quad (4.120)$$

The first term on the right hand side accounts for nucleation and growth as described by the Fokker-Planck equation; and the last term refers to the appearance of stable clusters due to the change in the critical size originated from the variations of the temperature. This last effect is known as *athermal nucleation*.

Consideration of the different Fokker-Planck dynamics corresponding to the models discussed previously gives rise to different types of kinetic equations for the evolution of the virtual volume.

4.8.2 Case 1: Isothermal Homogeneous Nucleation

In section 4.4.1, we have shown that the Fokker-Planck equation corresponding to this situation is given by

$$\frac{\partial f}{\partial t} = \frac{\partial}{\partial \gamma}\left\{D(\gamma,t)\frac{\partial f}{\partial \gamma} + b(\gamma,t)f\frac{\partial \Phi}{\partial \gamma}\right\} \tag{4.121}$$

By using this result in the general expression for the evolution of the virtual volume (4.120) one obtains

$$\frac{\partial}{\partial t}V(t) = -D^*\left.\frac{\partial f}{\partial \gamma}\right|_{\gamma^*}\rho_1\gamma^* - \int_{\gamma^*}^{\infty}\rho_1\left(D\frac{\partial f}{\partial \gamma} + bf\frac{\partial \Phi}{\partial \gamma}\right)d\gamma$$
$$+\frac{\partial \gamma^*}{\partial T}\frac{\partial T}{\partial t}f(\gamma^*,t)\rho_1\gamma^* \tag{4.122}$$

The right hand side of this expression contains contributions of different types. The first term corresponds to the flux of clusters of critical size due to diffusion at the top of the barrier. This term is, essentially, the nucleation rate times the volume of the critical cluster. Notice that there is not contribution arising from the drift because, by definition $\left.\frac{\partial \Phi}{\partial \gamma}\right|_{\gamma^*} = 0$. The second term corresponds to the usually called "growth regime" of the clusters. It accounts for the increase of the virtual volume due to the overall change in size of clusters bigger than the critical size. It is important to emphasize that the growth term is not deterministic. Contrarily, it retains the stochastic diffusive nature of the crystal growth. Finally, the last term takes into account the *athermal nucleation*.

4.8.3 Case 2: Spatial Inhomogeneous Crystallization

The presence of spatial inhomogeneities significantly modifies the evolution of the virtual volume. By introducing the Fokker-Planck equation (4.86) derived in section 4.6, the evolution of the virtual volume is given by

$$\frac{\partial}{\partial t}V(\mathbf{x},t) =$$
$$-D^*\left.\frac{\partial f_c}{\partial \gamma}\right|_{\gamma^*}\rho_1\gamma^* - \int_{\gamma^*}^{\infty}\rho_1\left(D_\gamma\frac{\partial f_c}{\partial \gamma} + \frac{D_\gamma}{k_BT}f_c\frac{\partial \tilde{\Phi}}{\partial \gamma}\right)d\gamma + \frac{\partial \gamma^*}{\partial T}\frac{\partial T}{\partial t}f_c(\gamma^*,\mathbf{x},t)\rho_1\gamma^*$$
$$+\nabla\left(\nabla\int D(\gamma)f_c\rho_1\gamma d\gamma - \int \xi\beta^{-1}\nabla\ln T\, f_c\rho_1\gamma d\gamma\right) \tag{4.123}$$

where the last term arises due to spatial inhomogeneities and thermal diffusion. One can approximate this term as

$$\nabla \left(\nabla \int D(\gamma) f_c \rho_1 \gamma d\gamma - \int \xi \beta^{-1} \nabla \ln T \, f_c \rho_1 \gamma d\gamma \right) \simeq$$
$$\nabla \left(\langle D \rangle \nabla V(\mathbf{x}, t) - \langle \xi \beta^{-1} \rangle \nabla \ln T \, V(\mathbf{x}, t) \right) \qquad (4.124)$$

where $\langle D \rangle$ and $\langle \xi \beta^{-1} \rangle$ are the size-averaged spatial diffusion and thermal friction coefficient. From this expression, the spatial diffusive behavior of the evolution of the virtual volume becomes clearly manifested.

4.8.4 Case 3: Nucleation in a Shear Flow

When the crystallizing system is under the influence of an external flow, its evolution is governed by the Fokker-Planck equation (4.118) derived in section 4.7. In terms of the virtual volume, its evolution is governed by

$$\frac{\partial}{\partial t} V(\mathbf{x}, t) = -D^* \left. \frac{\partial f_c}{\partial \gamma} \right|_{\gamma^*} \rho_1 \gamma^* - \int_{\gamma^*}^{\infty} \rho_1 \left(D_\gamma \frac{\partial f_c}{\partial \gamma} + \frac{D_\gamma}{k_B T} f_c \frac{\partial \widetilde{\Phi}}{\partial \gamma} \right) d\gamma +$$
$$+ \frac{\partial \gamma^*}{\partial T} \frac{\partial T}{\partial t} f_c(\gamma^*, \mathbf{x}, t) \rho_1 \gamma^* - \nabla \cdot (V(\mathbf{x}, t) \mathbf{v}_0) + \nabla \left(\langle \overrightarrow{D} \rangle \cdot \nabla V(\mathbf{x}, t) \right) \quad (4.125)$$

where again the last term has been approximated employing the size-averaged spatial diffusion coefficient $\langle \overrightarrow{D} \rangle$. The novelties introduced by the presence of the flow concerns the two last terms of that expression, corresponding to convection and spatial diffusion of the virtual volume which has tensorial character.

In the previous examples, we have shown how to formulate differential equations for the virtual volume corresponding to different situations. It becomes then clear that, in practical situations, the evolution of the crystallinity may exhibit a more complex behavior than the one resulting form the mathematical models usually introduced in the Avrami-Kolmogoroff theory. The natural extension of our analysis is to employ those equations to study the behavior of the crystallinity in realistic cases. The evolution of the crystallinity given by equation (??) would constitute a valid result to be checked with experimental data.

4.9 Further Extensions

The situations introduced in which the theory we have presented applies do not completely describe the crystallization process. Our aim in this section is precisely to outline possible extensions of the theory with the main purpose of offering a more general framework in which actual physical situations can be studied.

4.9.1 Crystallization at Advanced Stages

As most of nucleation theories formulated up to now, our theory introduces some assumptions that are only valid in the initial stages. These assumptions are in essence the following:

- The appearance of crystallization nuclei only occurs in a few points of the sample. Therefore, in the nucleation regime, it is reasonable to assume that crystal clusters constitute a dilute system. This means that we can neglect interactions between clusters, and the system behaves as ideal. One also assumes that clusters grow by addition of single monomers. In the dilute regime, the number of monomers is much higher than the number of binary, ternary... clusters and therefore one can neglect collisions between them. As clusters grow, collisions acquire relevance and the notion of dilute regime is no longer valid.
- The analysis of nucleation process is mostly focused on the evolution of the cluster size distribution function. One tends to identify the probability density of clusters of a given size with the number distribution of clusters present at the system. A conservation law similar to equation (4.27) is usually formulated for the evolution of this quantity. Consequently, conservation of the total number of clusters present at the system is assumed. Moreover, as stated in the section 4.2.2, the kinetic equation for the distribution function is formulated within a finite range of sizes. Moreover, the lose of monomers due to the growth of clusters is not taken into account since one assumes an almost infinite number of monomers. However, as they grow reaching macroscopic sizes, the number of monomers (and clusters of small sizes) decreases significantly thus leading to a reduction of the crystallization rate.

These restrictions make nucleation theories, as it is, not valid at advanced stages of crystallization process. This is the reason why these theories are only utilized as models for early stages of crystallization. To describe subsequent space-filling and impingement regimes, one has to use other approaches as the mathematical model of the Avrami-Kolmogoroff.

It would be interesting to extend the validity of nucleation theories to describe latest stages of crystallization process, overcoming their restrictions and retaining the physical nature of crystallization process. The restrictions of nucleation theories come mainly from two features: avoidance of interactions between clusters and conservation of the number of clusters.

In the following subsections we will precisely outline how these difficulties could be avoided within the context of our theory.

Interactions Between Clusters At the earlier stages of the process, nucleation and subsequent formation of clusters may be assumed to take place under ideal conditions for which clusters practically do not interact. The theory we have developed holds for this particular situation. A complete description of the process requires the knowledge of the kinetics at later times when clusters have grown to the extent of making ideal conditions inapplicable.

To accomplish for that situation, two slight modifications of the formalism we have presented can be carried out. On one hand, the interaction potential between clusters, U, must be taken into account in the expression of the chemical potential (for instance in equation (4.46)). One may equivalently introduce an activity coefficient a, defined as

$$a = \exp\left(\frac{U}{k_B T}\right) \qquad (4.126)$$

in such a way that the chemical potential is now given by

$$\mu = k_B T \ln af + \Phi \qquad (4.127)$$

On the other hand, the continuity equation (4.48) must be modified to account for direct interactions. With the introduction of these modification, we could now proceed along the lines indicated in section 4.5 to obtain the kinetic equation governing the system under the presence of interactions.

Non-conservative Equations In addition to the interactions among clusters, the advanced stages of crystallization are also characterized by the decrease of the crystallization rate. The reasons are that nucleation and growth cannot proceed in a region already crystallized, and that the amount of amorphous phase to crystallize progressively decreases. Whereas, the first item can be treated as interactions between clusters of the excluded volume type, the second one must be considered separately.

In order to take into account the reduction of the amorphous phase and the disappearance of small clusters, which end up incorporated to the larger ones, our formalism has to be extended to deal with "open systems". That is, our theory must consider the fact that the total number of cluster is not necessarily conserved.

The simplest way to introduce that feature, is to replace the continuity equation for the evolution of the population of clusters (for instance, equation (4.48) in the case of inhomogeneous nucleation) by a non-conservative one including a source/sink term. This term will be responsible for the appearance/disappearance of monomers/clusters of any size in the system.

The incorporation in our formalism of interactions between clusters and of a non-conservative equation governing the evolution of cluster population may lead to a complete description of crystallization even at latest stages, and may constitute a powerful model for crystallization of polymers alternative to the classical Avrami-Kolmogoroff formalism.

4.9.2 Effect of External Potentials: Multidimensional Theory

The original theory of nucleation, although has been successfully applied to many situations, cannot describe important features observed in actual systems subjected to potential fields, mechanical processing, etc... It would then be desirable to extend nucleation theory accounting for as many external influences as possible. With this aim, A. Ziabicki formulated a generalized nucleation theory [21, 22] incorporating effects such as cluster shape, anisotropy and interaction with external potential fields. This theory, which has been widely reviewed in other chapters of this monograph, conceives nucleation as a diffusion process in the multidimensional space of cluster variables (dimensions, positions, orientations, and internal structure) characterized by a single multidimensional vector.

Within our general framework, a multidimensional formulation is also possible, by considering a multidimensional vector of degrees of freedom similar to the one proposed in the generalized nucleation theory.

Following the procedure detailed in section 4.4.1, and neglecting cross effects one can derive the multidimensional Fokker-Planck equation

$$\frac{\partial f(\gamma, t)}{\partial t} = \sum_{i,j} \underline{\underline{D}}_{ij} \left[\frac{\partial f(\gamma, t)}{\partial \gamma_j} + \frac{f(\gamma, t)}{k_B T} \frac{\partial \Phi(\gamma)}{\partial \gamma_j} \right] \qquad (4.128)$$

where $\Phi(\gamma)$ represents the barrier, defined in a multidimensional space, related to the diffusion process, that must account for any external influence (orientational or spatial potentials, energy of formation of clusters, kinetic energies with respect to the reference state, etc...); and $\underline{\underline{D}}$ is the corresponding multidimensional diffusion tensor that in general may be nondiagonal.

Our formalism provides a simple set of rules to obtain kinetic equations for the evolution of the probability density in terms of any relevant variable considered as an internal degree of freedom. An interesting aspect to emphasize is that our formalism intrinsically includes cross effects leading to kinetic equations more general than the ones obtained by the Generalized Nucleation Theory. In previous Sections, we have discussed some illustrative examples corroborating that assertion.

To summarize, the advantages of our formulation are the following:

– Cross effects can be considered within the nonequilibrium thermodynamics formulation.
– The coupled evolution with the hydrodynamic fields can be taken into account in the kinetics.

4.9.3 Anisotropic Effects

In the situations developed in previous sections, the growing clusters feel the influence of possible external agents through the melted phase in which they are embedded. For example, the kinetic equation governing nucleation in a temperature gradient, given by equation (4.66), differs from its isothermal counterpart, equation (4.30), in terms containing the influence of the melted phase. In its derivation, it is implicitly assumed that the nucleation mechanisms remains unaltered.

In the initial stages of the crystallization, this assumption is fully justified. The length scale in which variations of the fields takes place (the gradient length scale) is significantly bigger than the typical size of the emerging clusters. Consequently, for all practical purposes, these clusters appear and growth under homogeneous conditions. When clusters grow, they can achieve sizes comparable to the gradients length scale. As a consequence, clusters undergo growing processes under inhomogeneous conditions leading to their deformation. In this context, it has been shown that clusters growing in the presence of a temperature gradient elongate in the direction of the gradient.

In order to describe this feature , we must incorporate the effect of the gradients in the kinetics coefficients which control the mechanism of nucleation and growth.

In fact, in section 4.7, we have precisely shown how the presence of a flow alters the rate of attachment of clusters, through modifications in the diffusion coefficient, leading to anisotropic behaviors.

4.10 Discussion and Conclusions

In this chapter, we have introduced a theoretical framework for modeling crystallization processes. The theory presents the following characteristics:

- It aims to describe nucleation and initial stages of growth, which play a very important role in crystallization processes.
- It takes into account the stochastic nature of both nucleation and growth, offering a common treatment for both mechanisms, without introducing a sharp and somehow artificial splitting into two different regimes.
- It considers the influence of different external conditions concerning pressure, temperature, presence of external fields or flows, under which crystallization process develops. These conditions may induce drastic modifications in the crystallization process. In fact, in the examples discussed in this chapter, we have shown that inhomogeneities, gradients and flows induce significant alterations in the evolution of the population of crystals and consequently in the evolution of the crystallinity. Moreover, our formalism is able to provide the coupled evolution of cluster distribution function and state variables, accounting for hydrodynamic and coupled effects which may be very relevant.
- We have developed in detail a set of representative examples, some of them of relevance in situations of practical interest as the case of the shear flow. The range of applicability of our theory is by no means limited to these

few particular cases. It constitutes a general framework in which, following systematically a small set of simple rules, one can derive kinetic equations for crystallization processes aiming to reproduce actual conditions and their influence in the process. Thus, our formalism can be employed to analyze crystallization in presence of gradients, external fields, boundaries, and spatial or orientational dependencies.

– For advanced stages of crystallization we may invoke the classical Avrami-Kolmogoroff theory and its extensions, as presented in other chapters of this volume. However, the ultimate goal would be to be able to describe these stages without using mathematical space-filling models but by implementing a physical kinetic model of crystallization valid even at the final stages of the process. We believe that our formalism, with the proper inclusion of interactions and of non conservation of the number of clusters in the way outlined in the previous section, may become a valuable alternative to describe crystallization process. Work along this lines is in progress.

4.11 Acknowledgments

We would like to thank I. Santamaría-Holek and T. Alarcón for valuable discussions. This work has been supported by DGICYT of the Spanish Government under grant PB98-1258. D. Reguera wishes to thank Generalitat de Catalunya for financial support.

References

1. G. Gomila, A. Pérez-Madrid, and J. M. Rubí. Physica A **233**, 208 (1996).
2. I. Pagonabarraga, A. Pérez-Madrid, and J. M. Rubí. Physica A **237**, 205 (1997).
3. D. Reguera, J. M. Rubí, and A. Pérez-Madrid, Physica A **259**, 10 (1998).
4. I. Pagonabarraga, and J. M. Rubí. Physica A **188**, 553 (1992).
5. K. F. Kelton. Solid State Phys. **45**, 75 (1991).
6. D. T. Wu. Solid State Phys. **50**, 37 (1996).
7. A. C. Zettlemoyer. *Nucleation.* (Marcel Dekker, New York, 1969).
8. A. Laaksonen, V. Talanquer and D. W. Oxtoby. Annu. Rev. Phys. Chem. **46**, 489 (1995).
9. P. G. Debenedetti. *Metastable Liquids : concepts and principles.* Chap. 3. (Princeton University Press, 1996).
10. D. Turnbull and J.C. Fisher, J. Chem. Phys. 17, 71 (1949).
11. L. D. Landau and E. M. Lifshitz, *Course of Theoretical Physics Vol 5* (Statistical Physics Part 1) *and Vol 9* (Statistical Physics Part 2) (Pergamon Press, New York, 1980).
12. K. F. Kelton, A. L. Greer and C. V. Thompson. J. Chem. Phys. **79**, 6261 (1983).
13. Ya. B. Zeldovich, Acta Physicochim. URSS, **18**, 1 (1943).
14. I. Prigogine and P. Mazur. Physica **XIX**, 241 (1953).

15. S. R. de Groot and P. Mazur, *Non−Equilibrium Thermodynamics*, (Dover, New York, 1984).
16. J. M. Rubí, and P. Mazur. Physica A **250**, 253 (1998).
17. J. S. van Duijneveldt and D. Frenkel. J. Chem. Phys. **96**, 4655 (1992).
18. P. R. ten Wolde, M. J. Ruiz-Montero and D. Frenkel. J. Chem. Phys. **104**, 9932 (1996) and references quoted therein.
19. P. R. ten Wolde, M. J. Ruiz-Montero and D. Frenkel. Faraday Discuss. **104**, 93 (1996).
20. D. Reguera, J. M. Rubí, and A. Pérez-Madrid, J. Chem. Phys. **109**, 5987 (1998).
21. A. Ziabicki, J. Chem. Phys. **48**, 4368 (1968).
22. A. Ziabicki, J. Chem. Phys. **85**, 3042 (1986).
23. M.L. Di Lorenzo and C. Silvestre, Prog. Polym. Sci. **24** 917 (1999).
24. A. Pérez-Madrid, J. M. Rubí, and P. Mazur. Physica A **212**, 231 (1994).
25. D. Reguera and J.M. Rub, J. Chem. Phys. (in press).
26. K.J. Zhang *et al.* J. Chem. Phys **111**, 2270 (1999).
27. G. Shi and J.H. Seinfeld, J. Appl. Phys. **68**, 4550 (1990).
28. A. Onuki, J. Phys. Cond. Matt. **9** 6119 (1997).
29. I.Santamaría-Holek, D. Reguera, and J. M. Rubí, Phys. Rev. E **63** 051106 (2001).
30. G. Eder and H. Janeschitz-Kriegl. *Processing of Polymers*, Chapter 5, (VCH, Germany, 1997).
31. G. Eder in *Macromolecular Design of Polymeric Materials*. K. Hatada, T. Kitayama, O. Vogl (Eds.) (Marcel Dekker, New York, in press).
32. U.W. Gedde, *Polymer Physics,* (Chapman&Hall, London, 1995).
33. D. Andreucci, A. Fasano, M. Primicerio, and R. Ricci. Surv. Math. Ind. **6**, 7 (1996).

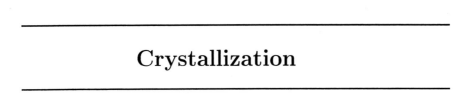

Crystallization

5 Mathematical Models for Polymer Crystallization Processes

Vincenzo Capasso[1], Martin Burger[2], Alessandra Micheletti[1], and Claudia Salani[1]

[1] MIRIAM - MIlan Research Centre for Industrial and Applied Mathematics, University of Milano and Department of Mathematics, University of Milano, Italy
[2] Industrial Mathematics Institute, Johannes Kepler University of Linz, Austria

5.1 Stochastic Models of the Crystallization Process

5.1.1 Introduction

Polymer industry raises a large amount of relevant mathematical problems with respect to the *quality* of manufactured polymer parts. These include in particular questions about the crystallization kinetics of the polymer melt, in presence of a temperature field.

The *final morphology* of the crystallised material is a fundamental factor for the physical properties of a solidified part. Also the long term behaviour of such properties (dimensional stability, physical ageing, ...) is strongly influenced by the microstructure of the crystallized material. For example the field stress σ_γ of the crystallized material is related to the average grain diameter d via the so-called Hall-Petch relation (cf. e.g.[24])

$$\sigma_\gamma = Kd^{-\frac{1}{2}} + \sigma_0$$

where K is a positive constant, and σ_0 is a constant offset, showing that optimal mechanical properties are directly related to optimal morphological properties.

Crystallization is a mechanism of phase change in polymeric materials. If an experiment is started with a liquid (the polymer melt) and the temperature is subsequently decreased below a certain point (the melting point of the material), crystals appear *randomly in space and time* and start to grow. Growth processes may be very complicated, but usually, with polymers we have growth of either spherical crystallites (*spherulites*) or of cylindrical crystallites (so called *shish-kebabs*). In the following the restriction to the case of spherulitic growth is made, which is also a good assumption for the crystallization of relaxed polymer melts. But also in the situation of flowing melts, as occurring in injection moulding, the central core of the solidified part shows spherulitic morphologies.

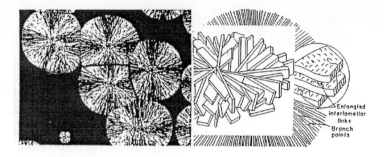

Fig. 5.1. a schematic representation of a spherulite and an impingement phenomenon

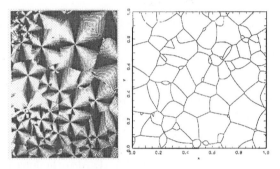

Fig. 5.2. The final Johnson-Mehl tessellation of a crystallization process, both from real and simulated experiments

The spherulites grow until they hit another crystal, which abruptly stops the growth at the interface (Fig.1) This phenomenon, called *impingement*, causes the morphology of the crystallized material; the resulting division of space into cells is called *Johnson-Mehl tessellation.* (Fig.5.2). Because of the randomness in time and location of the birth of crystals, the final morphology is random. A mathematical theory is needed to describe the mean theoretical properties of the process (and possibly the *variability* around the mean behaviour, too), and to predict the final properties of the crystallized material. Tuning of these mathematical models with respect to experimental data may lead polymer industry to optimize the solidification process so to obtain materials with optimal mechanical properties.

It is well known from experiments that the kinetic parameters of nucleation and growth strongly depend upon temperature. Hence, it is relevant to take this dependence into account when dealing with heterogeneous temperature fields.

In order to model the kinetics of the crystallization process in an heterogeneous temperature field we have to introduce the basic mathematical structures representing nucleation and growth of crystals. The nucleation process, being random both in time and space, will be modelled as a *marked point process*, that is a marked counting process, the mark being the (random) spatial location of the germ; this process will be the basis of the building of a *dynamic Boolean model*, representing the growth of crystals in absence of impingement. The dynamic Boolean model, coupled with the concept of causal cone in presence of spatial heterogeneities, provides a generalization of the well known Kolmogorov-Avrami-Evans formula [3, 23, 31] for computing the evolution of the *degree of crystallinity*, i.e., the mean volume fraction of the space occupied by crystals (a rigorous definition of degree of crystallinity will be provided later).

In this section we describe the crystallization process in a given deterministic field of temperature, neglecting at the moment the influence of crystallization due to the release of latent heat during phase change.

5.1.2 The Nucleation Process

The nucleation process is modelled as a stochastic spatially marked point process N on an underlying probability space $(\Omega, \mathcal{A}, P)$, with marks in the physical space E, a compact subset of $\mathbb{R}^d$, $d = 1, 2, 3$ (we shall denote by $\mathcal{B}_{\mathbb{R}_+}$, resp. $\mathcal{E}$, the σ-algebra of Borel sets on $\mathbb{R}_+$, resp. on E). The marked point process (MPP) N is a random measure given by

$$N = \sum_{n=1}^{\infty} \epsilon_{T_n, X_n}$$

where

- T_n is an $\mathbb{R}_+$-valued random variable representing the time of birth of the n-th nucleus,
- X_n is an E-valued random variable representing the spatial location of the nucleus born at time T_n,
- $\epsilon_{t,x}$ is the Dirac measure on $\mathcal{B}_{\mathbb{R}_+} \times \mathcal{E}$ such that for any $t_1 < t_2$ and $B \in \mathcal{E}$,

$$\epsilon_{t,x}([t_1, t_2] \times B) = \begin{cases} 1 & \text{if } t \in [t_1, t_2], x \in B, \\ 0 & \text{otherwise.} \end{cases}$$

In a crystallization process with nucleation events

$$\{(T_n, X_n) \mid 0 \leq T_1 \leq T_2 \leq \ldots\}$$

the crystalline phase at time $t > 0$ is given by a random set

$$\Theta^t = \bigcup_{T_j \leq t} \Theta_j^t,$$

which is the union of all crystals born at times T_j and locations X_j and freely grown up to time t. Note that impingement is obscured by this representation.

The integer valued random variable $N([0,t] \times B)$ counts the (random) number of nucleation events occurred in the time-space region $[0,t] \times B \in \mathcal{B}_{\mathbb{R}_+} \times \mathcal{E}$. We introduce the following

Definition 1. *The (degree of) crystallinity, $\xi(x,t)$, at location $x \in E$ and time $t > 0$ is the probability that, at time t, x is covered (or "captured") by some crystal, i.e.*

$$\xi(x,t) := P(x \in \Theta^t) = E[I_{\Theta^t}(x)].$$

It is possible to show [4] that, under rather general conditions, a compensator ν of N exists, such that

$$N([0,t] \times B) = \nu([0,t] \times B) + \mathcal{M}(t,B), \qquad \forall B \in \mathcal{E}, \ t > 0$$

where $\{\mathcal{M}(t,B)\}_{t \in \mathbb{R}_+}$ is a zero mean martingale, for all $B \in \mathcal{E}$. The (random) measure $\nu(dt \times dx)$ is known as the "stochastic intensity" of the process. It provides the probability that a new nucleation event occurs during the infinitesimal time interval $[t, t+dt[$, in the infinitesimal volume dx around $x \in E$, given the "history" of the process before time t.

For the marked point process (MPP) that models the birth process of crystal nuclei, we assume that the random measure ν is given by

$$\nu(dt \times dx) = \alpha(x,t)(1 - I_{\Theta^{t-}}(x))dt \ dx. \qquad (5.1)$$

where α, known as "nucleation rate", is a real-valued measurable function on $E \times \mathbb{R}_+$ such that $\alpha(\cdot, t) \in \mathcal{L}^1(E)$, for all $t > 0$, and such that, for any $T > 0$,

$$0 < \int_0^T \int_E \alpha(x,t) \ dx \ dt < +\infty,$$

so that the (free) nucleation process is not explosive.

The term $(1 - I_{\Theta^{t-}}(x))$ is responsible of the fact that no new nucleus can be born at time $t > 0$ in a zone already occupied by the crystalline phase. We will denote by

$$\nu_0(dt \times dx) = \alpha(x,t)dt \ dx$$

the "free space" intensity.

The corresponding (deterministic) measure defined by the expected values

$$\Lambda([0,t] \times B) := E[N([0,t] \times B)], \qquad t \geq 0, \ B \in \mathcal{E},$$

is known as the (deterministic) "intensity measure" of the nucleation process N.

By the definitions, the following identities hold

$$\begin{aligned}
\Lambda(dt \times dx) &= E[N(dt \times dx)] \\
&= E[\nu(dt \times dx)] \\
&= E[\alpha(x,t)(1 - I_{\Theta^{t-}}(x))dt\ dx] \\
&= \alpha(x,t)E[(1 - I_{\Theta^{t-}}(x))]dt\ dx \\
&= \alpha(x,t)(1 - \xi(x,t))dt\ dx.
\end{aligned}$$

If we define $\tilde{\nu}(dt) := \nu([t, t+dt] \times E)$, it is always possible to factorize the measure ν in the following way [34]

$$\nu(dt \times dx) = \tilde{\nu}(dt)k(t, dx) \tag{5.2}$$

where the kernel $k(t, \cdot)$ denotes, for any $t \in \mathbb{R}_+$, a probability measure on $(E, \mathcal{E})$ such that

$$k(t, E \setminus \Theta^t) = \int_{E \setminus \Theta^t} k(t, dx) = 1. \tag{5.3}$$

By using Equations (5.2) and (5.1) and under Condition (5.3), we obtain

$$\tilde{\nu}(dt) = \int_E \nu(dt \times dx) = \left(\int_E \alpha(y,t)(1 - I_{\Theta^{t-}}(y))dy \right) dt$$

$$= \left(\int_{\overline{\Theta^{t-}}} \alpha(y,t)dy \right) dt =: \tilde{\alpha}(t, \overline{\Theta^{t-}})dt \tag{5.4}$$

$$k(t, dx) = \frac{\alpha(x,t)(1 - I_{\Theta^{t-}}(x))}{\tilde{\alpha}(t, \overline{\Theta^{t-}})}dx \tag{5.5}$$

where $\overline{\Theta^{t-}}$ denotes the complement of Θ^{t-}.

Remark 3. From Equations (5.4) and (5.5) we can give a physical interpretation to the disintegration of the compensator.

— $\tilde{\nu}(dt)$ is the compensator of the (univariate) counting process

$$\tilde{N}(t): = N([0,t] \times E), t \geq 0,$$

called the *underlying counting process* associated with N; at each time t it counts the number of nuclei born in the whole region E up to time t. Thanks to the assumptions made on the (free) nucleation rate $\alpha(x,t)$, since for large times Θ^t tends to E, the stochastic intensity $\tilde{\alpha}$ of the process $\tilde{N}$ will tend to 0:

$$\lim_{t \to \infty} \tilde{\alpha}(t, \overline{\Theta^{t-}}) = 0.$$

– $k(t, \cdot)$ is the spatial probability measure of a nucleus born during $[t, t + dt[$. Note that with the increase of the crystallized volume, the available space for new nuclei is reduced. In (5.5) one may recognize its density.

The *local intensity measure* of N is given by

$$L(x, t) := \lim_{\lambda^d(B) \to 0} \frac{E[N([0, t] \times B)]}{\lambda^d(B)}$$

$$= \lim_{\lambda^d(B) \to 0} \frac{\int_B \int_0^t \alpha(y, s)(1 - \xi(y, s)) ds \, dy}{\lambda^d(B)}$$

$$= \int_0^t \alpha(x, s)(1 - \xi(x, s)) ds,$$

where λ^d is the Lebesgue measure on $\mathbb{R}^d$ and B is the ball centred at x.

Several special cases of α are of interest. E.g., if time and space are independent coordinates, then $\alpha(x, s) = g(s)f(x)$, and the case $f(x) =$ constant corresponds to a spatially homogeneous nucleation process.

5.1.3 Growth of Crystals

Experimental results and numerical simulations confirm that the crystallization process can be represented by a dynamic Boolean model. It has been introduced in [12, 13, 17] to model isothermal crystallization processes, where nucleation and growth rates are time- but not space-dependent (spatially homogeneous).

The dynamic Boolean model is a dynamical version of the classical definition of Boolean model [19] that takes time evolution into account. We give here the definition of an "heterogeneous version" of the dynamic Boolean model, which may represent a crystallization process with nucleation rate $\alpha(x, t)$ and growth rate $G(x, t)$.

The set Θ^t, representing the crystalline phase at any time $t \in \mathbb{R}_+$, is completely characterized by its hitting functional [36], defined as

$$T_{\Theta^t}(K) := P(\Theta^t \cap K \neq \emptyset),$$

for every compact subset K of E.

In particular for $K = \{x\}$, $x \in E$, and $t \in \mathbb{R}_+$, we have the crystallinity

$$T_{\Theta^t}(\{x\}) = P(x \in \Theta^t)$$

$$= E[I_{\Theta^t}(x)] =: \xi(x, t).$$

The classical Kolmogorov-Avrami-Evans theory for isothermal crystallization heavily used the fact that individual crystals are of a spherical shape if the growth rate is constant; the same is true whenever the growth rate depends upon time only. In the case of non-homogeneous growth, i.e., if the growth rate G depends on both space and time, the shape of a polymeric crystal (in absence of impingement) is no longer a ball centered at the origin of growth. In the case of a growth rate with constant gradient it has been verified that the growing crystal is the union of the *growth lines*, which lead to growth in minimal time (cf.[52]). This principle can be adapted for the case of arbitrary growth rates (cf.[57]).

Assumption 1 (Minimal-time Principle) [11] The curve along which a crystal grows, from its origin to any other point, is such that the needed time is minimal.

The minimal-time principle is obviously satisfied for homogeneous growth, since there the growth lines are just straight lines. The growth of a crystal in $\mathbb{R}^2$ between its origin (x_0, y_0) and another point (x_1, y_1) due to Assumption 1 may be formulated as follows:

$$t_1 = \min_{(x,y,\phi)}$$

subject to

$$\dot{x}(t) = G(x(t), y(t), t) \cos \phi(t), \quad t \in (t_0, t_1)$$

$$\dot{y}(t) = G(x(t), y(t), t) \sin \phi(t), \quad t \in (t_0, t_1)$$

$$x(t_0) = x_0, \quad y(t_0) = y_0$$

$$x(t_1) = x_1, \quad y(t_1) = y_1$$

The necessary first-order conditions for this optimal control problem (cf. e.g.[26]) lead to the following equation for the control variable ϕ:

$$\dot{\phi} = \nabla G(x, y, t).(-\sin \phi, \cos \phi)^T.$$

A similar reasoning is possible also in $\mathbb{R}^3$; we refer to [11] for details.

By eliminating the angle ϕ we may deduce a second-order ordinary differential equation for a growth line given by

$$\frac{d}{dt}\left(\frac{\dot{\mathbf{x}}}{G(\mathbf{x}, t)}\right) = -\nabla G(\mathbf{x}, t) + \left(\nabla G(\mathbf{x}, t). \frac{\dot{\mathbf{x}}}{G(\mathbf{x}, t)}\right) \frac{\dot{\mathbf{x}}}{G(\mathbf{x}, t)}, \qquad (5.6)$$

where $\mathbf{x}$ denotes the vector $(x, y)^T$ in $\mathbb{R}^2$.

The crystal at time t is given as the union of all growth lines, i.e.,

$$\Theta_0^t = \{\mathbf{x}(\tau) \,|\, \mathbf{x} \text{ solves } (5.6), \mathbf{x}(t_0) = \mathbf{x}_0, \tau \in (t_0, t)\}.$$

Each growth line is determined uniquely by the origin of growth $\mathbf{x}(t_0) = \mathbf{x}_0$ and the initial speed of growth, which may be written as

$$\dot{\mathbf{x}}(t_0) = G(\mathbf{x}_0, t_0)\mathbf{n}_0,$$

where $\mathbf{n}_0$ is an arbitrary vector in $\mathbb{R}^2$ with $\|\mathbf{n}_0\| = 1$. Thus, we may introduce a parameterization for the crystal based on the initial direction, namely

$$\Theta_0^t = \{\mathbf{x}(\tau, \gamma) \mid \tau \in (t_0, t), \gamma \in [0, 2\pi)\}, \tag{5.7}$$

where $\mathbf{x}(\tau, \gamma)$ denotes the solution of (5.6) with initial values

$$\mathbf{x}(t_0) = \mathbf{x}_0,$$
$$\dot{\mathbf{x}}(t_0) = G(\mathbf{x}_0, t_0)(\cos\gamma, \sin\gamma)^T.$$

Equation (5.6) yields a description of the crystal growth based on growth lines, which are computed independently. The parameterization introduced in (5.7) also provide another view upon the growing crystal. By fixing the time t we obtain the set

$$\partial\Theta_0^t = \{\mathbf{x}(t, \gamma) \mid \gamma \in [0, 2\pi)\}, \tag{5.8}$$

which is called the *growth front*. In particular (5.6) implies that at any point $\mathbf{x}(t, \gamma)$ of the growth front at time t (initial direction γ) we have

$$\dot{\mathbf{x}}(t, \gamma) = G(\mathbf{x}(t, \gamma), t)\mathbf{n}(t, \gamma), \tag{5.9}$$
$$\dot{\mathbf{n}}(t, \gamma) = -\nabla G(\mathbf{x}(t, \gamma), t) + (\nabla G(\mathbf{x}(t, \gamma), t).\mathbf{n}(t, \gamma))\mathbf{n}(t, \gamma), \tag{5.10}$$

which clearly shows that the growth is determined by the actual normal direction of the growth front as well as by the growth rate and its gradient. The initial values are given by

$$\mathbf{x}(t_0, \gamma) = \mathbf{x}_0, \tag{5.11}$$
$$\mathbf{n}(t_0, \gamma) = (\cos\gamma, \sin\gamma)^T. \tag{5.12}$$

Contrary to the original minimal-time principle, system (5.9), (5.10) needs only information about the shape of the crystal at the actual time, but not about the history of growth. Hence, this description seems to be suitable not only for the case of growth in a given field, but also for growth in interaction with the field.

Figure 5.3 shows a simulation of non-homogeneous crystal growth for which we used a typical parabolic temperature profile (i.e., the solution of the heat equation without latent heat) and data for the growth rate obtained by measurements of isotactic polypropylene (i-PP). The result, presented in Figure 5.3, shows the growth front in the first time steps. The deviation from the spherical shape obviously increases with time, nevertheless the crystals still remain convex and do not produce exotic shapes like dendritic structures.

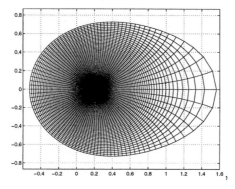

Fig. 5.3. Crystal shapes in a typical temperature field.

5.1.4 Level Sets and Surface Densities

Suppose first that $\alpha(x,t)$ and $G(x,t)$ are given deterministic fields in $E \times \mathbb{R}_+$. In order to make the coupling of the stochastic birth process with the growth of crystals explicit in this case, we refer to an alternative mathematical treatment of the birth-and-growth crystallization process, based on the so called *method of characteristics* [57] or *level set method* [53]. The basic idea is to describe the j-th crystal Θ_j^t via its indicator function $f^j(x,t) := I_{\Theta_j^t}(x)$. The evolution of f^j is determined by the following hyperbolic partial differential equation (cf. [8, 53]), the so-called *level-set equation*

$$f_t^j + G\|\nabla f^j\| = 0, \qquad \text{in } \Omega \times (T_j, \infty) \tag{5.13}$$

supplemented by the additional conditions

$$f^j(x, T_j) = 0 \quad \text{for } x \neq X_j, \tag{5.14}$$
$$f^j(X_j, t) = 1 \quad \text{for } t > T_j, \tag{5.15}$$

having denoted by T_j the (random) time of birth of the j-th crystal. We may also consider the same equation for $t < T_j$ with the additional condition $f^j(X_j, t) = 0$. An inspection of the characteristics of the first-order equation (5.13) (see also Figure 5.4) implies that f^j equals one on the growth lines and zero outside (cf. [53]).

The overall crystalline phase can be described by the indicator function $f(t) = I_{\Theta^t}$, where

$$\Theta^t = \bigcup_j \Theta_j^t. \tag{5.16}$$

The expression of f in terms of the family of indicator functions f^j of all crystals is the following

$$f(x,t) = 1 - \prod_{T_j < t} (1 - f^j(x,t)), \qquad t \in \mathbb{R}^+, x \in E, \qquad (5.17)$$

since f equals zero if and only if all f^j's do. The degree of crystallinity is given by

$$\xi(x,t) = E[f(x,t)], \qquad t \in \mathbb{R}^+, x \in E. \qquad (5.18)$$

Besides the volume distribution f^j of a crystal, we may also define the surface density u^j and the oriented surface density v^j, which are generalized functions, via

$$\langle u^j, \phi \rangle = \int_{\partial \Theta_j^t} \phi(x) \, d\sigma(x), \qquad \forall \, \phi \in C_0^\infty(E, \mathbb{R}) \qquad (5.19)$$

$$\langle v^j, \phi \rangle = - \int_{\partial \Theta_j^t} \phi(x) \mathbf{n}(x) \, d\sigma(x), \qquad \forall \, \phi \in C_0^\infty(E, \mathbb{R}). \qquad (5.20)$$

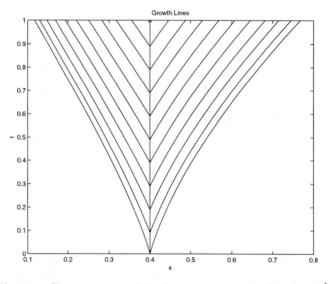

Fig. 5.4. Characteristics of the first-order equation (5.13) in $\mathbb{R}^1$.

In particular, for $B \in \mathcal{B}_{\mathbb{R}^d}$, we have

$$\int_B u^j(x,t) \lambda^d(dx) = \lambda^{d-1}(\partial \Theta_j^t \cap B).$$

Formally

$$u^j(x,t)dx = \lambda^{d-1}(\partial\Theta_j^t \cap dx)$$
$$v^j(x,t)dx = -\mathbf{n}(x)\lambda^{d-1}(\partial\Theta_j^t \cap dx)$$

(note that $v^j(x,t) \cdot \mathbf{n}(x) = -u^j(x,t)$).

It can be shown (cf. [7]) that the generalized functions f^j, u^j and v^j are related via

$$\frac{\partial f^j}{\partial t} = (1-f^j)Gu^j, \tag{5.21}$$

$$\nabla f^j = (1-f^j)v^j, \tag{5.22}$$

Furthermore, u^j and v^j satisfy the compatibility condition

$$\frac{\partial v^j}{\partial t} = \nabla(Gu^j). \tag{5.23}$$

The time derivative of u^j can be calculated as (cf. [7, 11])

$$\frac{\partial u^j}{\partial t} = \text{div } (Gv^j) + S_j^d, \tag{5.24}$$

where S_j^d is the generalized function that describes the (stochastic) event of birth of the j-th crystal in $\mathbb{R}^d$. Formally

$$S_j^d = \frac{\partial}{\partial t}u^j\Big|_{t=T_j},$$

for $t = 1, 2$, so that for any $\phi \in C_0^\infty(E \times \mathbb{R}_+, \mathbb{R})$

$$\langle S_j^1, \phi \rangle = 2\phi_t(X_j, T_j) \tag{5.25}$$

$$\langle S_j^2, \phi \rangle = 2\pi G(X_j, T_j)\phi(X_j, T_j), \tag{5.26}$$

The case $d = 3$ needs more tedious calculations (see [11] for details).

Now we turn our attention to the corresponding global quantities. Besides the volume distribution f, we can define the (stochastic) metric and the oriented free surface density, u and v via

$$u = \sum_{T_j<t} u^j \tag{5.27}$$

$$v = \sum_{T_j<t} v^j. \tag{5.28}$$

The distributions u and v are of special interest for themselves, since they provide additional information about the morphology of the material.

In particular the total surface of nuclei in a domain B in the absence of impingement is given by

$$\int_B u \, dx = \sum_{T_j < t} \lambda^{d-1}(\partial \Theta_j^t \cap B). \tag{5.29}$$

From the results for single crystals we may derive

$$\frac{\partial f}{\partial t} = (1 - f)Gu. \tag{5.30}$$

and

$$\nabla f = (1 - f)v. \tag{5.31}$$

The quantities u and v can be computed directly by solving, for $d = 1, 2$

$$\frac{\partial v}{\partial t} = \nabla(Gu) \quad \text{in } E \times \mathbb{R}_+ \tag{5.32}$$

$$\frac{\partial u}{\partial t} = \text{div}(Gv) + \sum_{T_j = t} S_j^d \quad \text{in } E \times \mathbb{R}_+ \tag{5.33}$$

$$u = 0 \quad \text{in } E \times \{0\} \tag{5.34}$$

$$v = 0 \quad \text{in } E \times \{0\} \tag{5.35}$$

$$u + v \cdot \mathbf{n} = 0 \quad \text{on } \partial E \times \mathbb{R}_+, \tag{5.36}$$

which is an immediate consequence of the fact that u and v are linear superpositions of the functions u^j and v^j, respectively, and the linear equations obtained for the latter.

One may easily recognize that the only source of stochasticity in the above system is due to the birth process described by the term $\sum_{T_j = t} S_j^d$.

For $d = 3$ Equations (5.33) and (5.34) are changed in

$$\frac{\partial^2 u}{\partial t^2} = \text{div}(G_t v) + \text{div}(G v_t) + \sum_{T_j = t} G S_j^d \quad \text{in } E \times \mathbb{R}^+$$

$$u = 0, \quad \frac{\partial u}{\partial t} = 0 \quad \text{in } E \times \{0\}$$

(cf. [11, 7] for further details).

Thanks to the linearity of system (5.32)-(5.36) when $\alpha(x, t)$ and $G(x, t)$ are given deterministic fields in $E \times \mathbb{R}_+$, under suitable regularity assumptions, we may exchange time and space derivatives with expected values to obtain for

$$\tilde{u}(x, t) := E[u(x, t)]$$
$$\tilde{v}(x, t) := E[v(x, t)]$$

the following system of evolution equations:

$$\frac{\partial \tilde{v}}{\partial t} = \nabla(G\tilde{u}) \quad \text{in } E \times \mathbb{R}_+ \tag{5.37}$$

$$\frac{\partial \tilde{u}}{\partial t} = \text{div}(G\tilde{v}) + \mathcal{F}^d(\alpha, G) \quad \text{in } E \times \mathbb{R}_+ \tag{5.38}$$

$$\tilde{u} = 0 \quad \text{in } E \times \{0\} \tag{5.39}$$

$$\tilde{v} = 0 \quad \text{in } E \times \{0\} \tag{5.40}$$

$$\tilde{u} + \tilde{v} \cdot \mathbf{n} = 0 \quad \text{on } \partial E \times \mathbb{R}_+, \tag{5.41}$$

where $\mathcal{F}^d(\alpha, G)$ is the mean value of the source term due to nucleation.

More precisely, if we rewrite Eqn.(5.33) in its weak formulation, we get for a test function $\phi(x, t)$ smooth enough and s.t. $\phi(x, t) = 0$ on ∂B, for any $B \in \mathcal{E}$

$$\int_B \phi(x, t) u(x, t) dx = \int_B \phi(x, 0) u(x, 0) dx$$

$$+ \int_0^t \int_B \nabla(\tilde{G}(T) v(x, s)) \phi(x, s) dx \, ds + \sum_{T_j \leq t} \sum_{X_j \in B} \langle S_j^d, \phi \rangle. \tag{5.42}$$

Substituting u and v with their mean values and defining the function $\mathcal{F}^d(\alpha, G)$ such that

$$E\left[\sum_{T_j < t} \sum_{X_j \in B} \langle S_j^d, \phi \rangle\right] = \int_0^t \int_B \mathcal{F}^d(\alpha, G)(x, s) \phi(x, s) dx \, ds,$$

we obtain system (5.37)-(5.41) in weak sense. In particular, for $d = 1, 2$, we have

$$\mathcal{F}^1(\alpha, G)(x, t) = 2\alpha(x, t)$$

$$\mathcal{F}^2(\alpha, G)(x, t) = 2\pi G(x, t) \int_0^t \alpha(x, s) ds.$$

5.1.5 The Morphology of the Crystallization Process.

We have already announced in the introduction that for industrial processing, the aim is to produce material consisting of homogeneous structures with good mechanical properties. Since the yield stress σ_γ of the crystallized material is related to the average grain diameter d via the so-called Hall-Petch relation (cf. e.g.[24])

$$\sigma_\gamma = Kd^{-\frac{1}{2}} + \sigma_0$$

where K is a positive constant, and σ_0 is a constant offset, optimal mechanical properties are directly related to optimal morphological properties.

The optimality criterion for the crystalline morphology is a fine-grained structure, which is close to spatial homogeneities. In practice, the latter can be realized only on a meso- or macroscale, i.e., local mean densities of $n-$dimensional interfaces (see below) of the final tessellation, in the same way as for the crystallinity ξ (the mean $d-$dimensional interface density) should be homogeneous. We need then to obtain evolution equations for these densities so to establish optimal control problems ([16]).

When the kinetic parameters $\alpha(x,t)$ and $G(x,t)$ are given deterministic fields it is possible to extend the classical theory of Kolmogorov [31] - Avrami [3] - Evans [23] based on the concept of causal cone.

5.2 The Causal Cone

The concept of causal cone was first introduced by Kolmogorov [31] and, more recently, studied by Otha et al. [45, 46], for the case of spatially homogeneous birth-and-growth processes. We will here define and study the causal cone for general heterogeneous but given kinetic parameters.

From the definition of the degree of crystallinity, we have

$$\xi(x,t) = P(x \in \Theta^t)$$

$$= P(x \in \bigcup_{T_n \leq t} \Theta^t_{T_n}(X_n))$$

$$= P(\exists (T_n, X_n) \in [0,t] \times E \,|\, x \in \Theta^t_{T_n}(X_n)). \qquad (5.43)$$

Equation (5.43) justifies the following definition

Definition 2. *The **causal cone** $A(x,t)$ of a point x at time t is the set of points (y,s) in the time space $\mathbb{R}_+ \times E$ such that a crystal born in y at time s covers the point x by time t*

$$A(x,t) := \{(y,s) \in [0,t] \times E \,|\, x \in \Theta^t_s(y)\}$$

where we have denoted by $\Theta^t_s(y)$ the crystal born at $y \in E$ at time $s \in \mathbb{R}_+$, and observed at time $t \geq s$. We denote by $C_s(x,t)$ the section of the causal cone at time $s < t$,

$$C_s(x,t) := \{y \in E \,|\, (y,s) \in A(x,t)\}$$
$$= \{y \in E \,|\, x \in \Theta^t_s(y)\}.$$

Let us suppose that the growth rate of a crystal depends only upon the point (x,t) under consideration, and not upon the age of the crystal, for

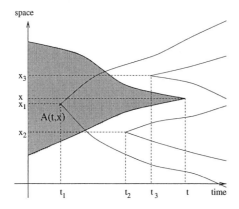

Fig. 5.5. The causal cone of point x at time t.

example. In this case the degree of crystallinity $\xi(x, t)$, i.e., the probability that the point x is covered by time t may be expressed in terms of the probability that no nucleation event occurs inside the causal cone (cf. [11])

$$\xi(x, t) = 1 - P(x \notin \Theta^t) = 1 - P(\text{no nucleation occurs in } A(x, t)). \ (5.44)$$

In order to verify (5.44), we have to show that any nucleation in the causal cone leads to coverage independent of the amount of crystalline phase present at that time. More precisely, we have to show that even if $(t_1, x_1) \in A(x, t)$ is already covered by another crystal, the point x will be covered at time t. This can be done by comparing two crystals - the one already existing before (born at (t_0, x_0) denoted by Θ_0^t) and the (virtual) new-born crystal Θ_1^t. We show that the crystal with later nucleation will always stay 'inside' the other one. Defining the *cone of influence* by

$$\mathcal{I}(x_0, t_0) := \{(x, t) \mid x \in \Theta_0^t\}, \tag{5.45}$$

we can express this statement mathematically in the following lemma:

Lemma 1. *[11, Lemma 3.1]* If $(x_1, t_1) \in \mathcal{I}(x_0, t_0)$, then $\mathcal{I}(x_1, t_1) \subset \mathcal{I}(x_0, t_0)$.

From Lemma 1, there follows that the event "no nucleation takes place in the causal cone $A(x, t)$" has the same probability as the event

$$\{\Theta^s \cap C_s(x, t) = \emptyset, \ s \leq t\} \in \mathcal{F}_t.$$

Conditional upon this event, expression (5.1) for the compensator in the causal cone reduces to

$$\nu_0(ds \times dy) := \nu(ds \times dy \mid \Theta^r \cap C_r(x, t) = \emptyset, \ r \leq s) = \alpha(y, s)ds \ dy, \quad (5.46)$$

for every $(y, s) \in A(x, t)$, so that we have a deterministic compensator (the "free space" nucleation rate).

It is possible to prove [27] that a Poisson process is characterized by a deterministic compensator. Hence, in this case nucleation occurs as an heterogenous Poisson process in space and time. In particular, the number of nuclei in the causal cone $(N(A(x,t)))$ is a Poisson-distributed random variable with intensity

$$\nu_0(A(x,t)) = \int_{A(x,t)} \nu_0(ds \times dy) \tag{5.47}$$

Thus, since

$$\xi(x,t) = 1 - P(N(A(x,t)) = 0),$$

we obtain

$$\xi(x,t) = 1 - e^{-\nu_0(A(x,t))}. \tag{5.48}$$

By considering the factorization of the compensator (5.2) together with equations (5.46) and (5.47), we may write

$$\nu_0(A(x,t)) = \int_0^t \int_{C_s(x,t)} \alpha(y,s)ds\,dy$$

$$= \int_0^t \tilde{\nu}_0(ds) \int_{C_s(x,t)} k_0(s,dy)$$

$$= \tilde{\nu}_0([0,t]) \int_0^t \frac{\tilde{\nu}_0(ds)}{\tilde{\nu}_0([0,t])} k_0(s,C_s(x,t))$$

$$= \tilde{\nu}_0([0,t]) E_S[k_0(S,C_S(x,t))] \tag{5.49}$$

where

$$\tilde{\nu}_0(ds) = \int_E \alpha(y,s)ds\,dy$$

so that $\tilde{\nu}_0([0,t]) = \int_0^t \tilde{\nu}_0(ds)$ is the mean number of nuclei born in the whole space E up to time t; $k_0(s,dy)$ (cf.(5.5)) is the distribution of the nuclei at time s, knowing that no point is born in the causal cone $A(x,t)$ and S is a random variable with distribution

$$P(S \in ds) = \frac{\tilde{\nu}_0(ds)}{\tilde{\nu}_0([0,t])} I_{[0,t]}(s).$$

If we denote by $a(x,t) = E_S[k_0(S,C_S(x,t))]$, the resulting identity

$$\xi(x,t) = 1 - e^{-\tilde{\nu}_0([0,t])a(x,t)} \tag{5.50}$$

may be seen as an extension of the Kolmogorov-Avrami-Evans formula to the heterogeneous case.

As for the spatially homogeneous case we may introduce the quantity $V_{ex}(x, t)$ as the *mean extended volume density*, i.e. the expected value of the spatial density of the sum of the volumes of the grains (crystals) which are born and develop independently of each other with a free space birth rate (see Fig.5.6)

$$V_{ex}(x, t) = \lim_{r \downarrow 0} E \left[\sum_{T_j < t} \frac{\lambda^d(\Theta_j^t \cap B_r(x))}{\lambda^d(B_r(x))} \right].$$

It can be shown that

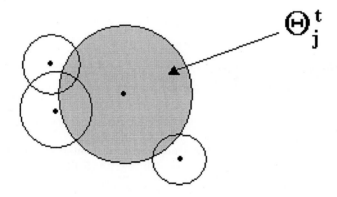

Fig. 5.6. Crystals born and developped independently of each other with a free space birth rate.

$$V_{ex}(x, t) = \nu_0(A(x, t)) = \tilde{\nu}_0([0, t])a(x, t)$$

so that we have also

$$\xi(x, t) = 1 - e^{-V_{ex}(x,t)}.$$

Moreover, it can be shown ([11], [16]) that

$$\frac{\partial}{\partial t} \nu_0(A(x, t)) = G(x, t)\tilde{u}(x, t) \tag{5.51}$$

with

$$\tilde{u}(x,t) = E[u(x,t)] = E\left[\sum_{T_j < t} u_j(x,t)\right]$$

$$= \lim_{r \downarrow 0} E\left[\sum_{T_j < t} \frac{\lambda^{d-1}(\partial\Theta_j^t \cap B_r(x))}{\lambda^d(B_r(x))}\right] =: S_{ex}(x,t)$$

the *mean density of the extended surface* at $(x,t) \in E \times \mathbb{R}_+$.

For the simplest case $d = 1$ the derivation of (5.51) can be made more explicit starting from the integral of $\nu_0(A(x,t))$ in (5.49). In order to compute these integral over $C_s(x,t)$ we introduce the indicator function

$$\chi(x,t,y,s) := I_{C_s(x,t)}(y). \tag{5.52}$$

For the sake of simplicity we restrict our attention to the case of $\Omega = (0,1)$ for the moment, although a similar, but technically more involved argumentation is possible for general domains in higher dimensions, too (cf. [11]). The theory of characteristics of first-order differential equations (cf. [48, 54]) shows that χ satisfies

$$\frac{\partial\chi}{\partial t} - G\frac{\partial\chi}{\partial x} = 0 \qquad \text{in } (0,y) \times (s,\infty) \tag{5.53}$$

$$\frac{\partial\chi}{\partial t} + G\frac{\partial\chi}{\partial x} = 0 \qquad \text{in } (y,1) \times (s,\infty), \tag{5.54}$$

in the sense of distributional derivatives; supplemented by $\chi(y,t,y,s) = 1$ and $\chi(x,s,y,s) = 0$ if $x \neq y$. Now we can rewrite (5.49) as

$$v(x,t) := \nu_0(A(t,s)) = \int_0^t \int_0^1 \chi(x,t,y,s)\ \alpha(y,s)dy\ ds,$$

and derive an evolution equation for v using (5.53) and (5.54). For further details about the derivation we refer to [8], here we just give the result, which is the second-order differential equation

$$\left(\frac{v_t}{G}\right)_t = (Gv_x)_x + 2\alpha \tag{5.55}$$

in $(0,1) \times \mathbb{R}_+$. In an analogous way we may derive the boundary conditions

$$v_t(0,t) = v_x(0,t) \tag{5.56}$$

$$v_t(1,t) = -v_x(1,t). \tag{5.57}$$

Together with the homogenous initial conditions for v and v_t we now have an initial-boundary value problem, from which we may calculate v efficiently. In the multi-dimensional case, similar model equations can be obtained (cf. [11]), which we shall not discuss further here.

5.3 The Hazard Function

Complementary to the degree of crystallinity (or probability of capture) $\xi(x, t)$ is the *survival function* or *porosity* of the point x at time t, which denotes the probability $p_x(t)$ that the point x is not yet crystallized at time t, i.e.,

$$p_x(t) = P(x \notin \Theta^t) = 1 - \xi(x, t) = P(T(x) > t)$$

where we have denoted by $T(x)$ the (random) time of "survival" of point $x \in E$ with respect to its "capture" (occupation) by the crystalline phase.

The probability density $f_x(t)$ of $T(x)$ will be given by

$$f_x(t) = \lim_{\Delta t \to 0} \frac{P(x \in (\Theta^{t+\Delta t} \setminus \Theta^t))}{\Delta t}.$$

As a consequence

$$f_x(t) = -\frac{d}{dt} p_x(t) = \frac{\partial}{\partial t} \xi(x, t)$$

$$= \left[\frac{d}{dt} \nu_0(A(x, t)) \right] p_x(t). \tag{5.58}$$

Definition 3. *The **hazard function** $h(x, t)$ is the rate of capture, i.e., the density of the capture probability, given the history of the process up to time t,*

$$h(x, t) = \lim_{\Delta t \to 0} \frac{P(x \in (\Theta^{t+\Delta t} \setminus \Theta^t)|\mathcal{F}_t)}{\Delta t}$$

Since it can be shown that the crystallization process is a Markov process ([18]), the knowledge of the history up to time t is equivalent to the knowledge of the state of the process at the instant t (see [30], p.120). Furthermore, the hazard function $h(x, t)$ is nonzero only in the case when x has not yet been captured until time t.

Consequently, $h(x, t)\Delta t$ is the probability that x is captured during the time interval $[t, t+\Delta t]$, given that it is not captured before t; thus, considering the limit

$$h(x, t) = \lim_{\Delta t \to 0} \frac{P(x \in (\Theta^{t+\Delta t} \setminus \Theta^t)|x \notin \Theta^t)}{\Delta t}$$

$$= \lim_{\Delta t \to 0} \frac{P(x \in (\Theta^{t+\Delta t} \setminus \Theta^t))}{P(x \notin \Theta^t)\Delta t}$$

$$= \frac{f_x(t)}{p_x(t)}$$

and, using the expression (5.58) for $f_x(t)$, we obtain

$$h(x,t) = \frac{d}{dt}\nu_0(A(x,t)) = -\frac{d}{dt}\ln p_x(t).$$

From the above considerations, we easily derive

$$f_x(t) = p_x(t)h(x,t)$$

i.e.

$$\frac{\partial}{\partial t}\xi(x,t) = (1 - \xi(x,t))h(x,t)$$

$$= (1 - \xi(x,t))\frac{\partial}{\partial t}V_{ex}(x,t) \tag{5.59}$$

$$= (1 - \xi(x,t))G(x,t)S_{ex}(x,t)$$

which is the extension of the Kolmogorov-Avrami-Evans formula to the space-time inhomogeneous case.

We have already introduced the degree of crystallinity $\xi(x,t)$ as

$$\xi(x,t) = P(x \in \Theta^t).$$

It can also be shown [14] that $\xi(x,t)$ is the *mean local density of volume* of the spatial region occupied by the actual crystalline phase at time t, Θ^t:

$$\xi(x,t) = P(x \in \Theta^t)$$

$$= \lim_{r\downarrow 0} E\left[\frac{\lambda^d(\Theta^t \cap B_r(x))}{\lambda^d(B_r(x))}\right] =$$

$$=: V_V(x,t).$$

Correspondingly we may define the mean local surface density of Θ^t, as follows [14]

$$S_V(x,t) := \lim_{r\downarrow 0} E\left[\frac{\lambda^{d-1}(\partial\Theta^t \cap B_r(x))}{\lambda^d(B_r(x))}\right].$$

Under suitable regularity conditions which include the local convexity of Θ^t [15] we may prove that for almost any $x \in E$ and any $t \in \mathbb{R}_+$

$$S_V(x,t) = (1 - \xi(x,t))S_{ex}(x,t)$$

so that

$$h(x,t) = G(x,t)S_{ex}(x,t) \tag{5.60}$$

$$= G(x,t)\frac{S_V(x,t)}{1 - \xi(x,t)}. \tag{5.61}$$

5.4 The Random Tessellation of Space

At any time t during the crystallization process, the available space (and in particular the crystalline phase) is randomly divided into cells due to impingement, forming a so called **(incomplete) Johnson-Mehl tessellation** indexJohnson-Mehl tessellation [29, 42] (the tessellation is **incomplete** at time t if a region with amorphous material is still present). Here we will consider random tessellations in $\mathbb{R}^d$ generated by birth-and-growth processes.

5.4.1 Basic Definitions

Let us denote by $\tau_i(y)$ the time at which a point $y \in E$ is reached by a crystal freely grown (disregarding impingement) from a nucleus $a_i = (X_i, T_i)$, where X_i is the random space location and T_i the random time of birth. We admit $\tau_i(y) = \infty$, which corresponds to the case in which the point y is never covered by the i-th crystal.

Definition 4. *The* **crystal** $C_i(t)$ *of the (incomplete) tessellation, generated by the nucleus* $a_i = (X_i, T_i)$, *at the time of observation* t *is the non-empty set*

$$C_i(t) = \{y \in E | \tau_i(y) \leq t \text{ and } \tau_i(y) \leq \tau_j(y) \; \forall j \neq i\}.$$

We will denote by $C_e(t)$ the **amorphous region** at time t, i.e.

$$C_e(t) := \{y \in E | \tau_i(y) > t, \; \forall i \in \mathbb{N}\}.$$

Let us now introduce a rigorous concept of "interface":

Definition 5. *An* **n-facet** *at time* t *($0 \leq n \leq d$) is the non-empty intersection between* $m+1$ *crystals or between* m *crystals and the amorphous region, i.e.*

$$F_n(t, k_0, \ldots, k_m) = C_{k_0}(t) \cap C_{k_1}(t) \cap \cdots \cap C_{k_m}(t),$$

with $m = d - n$ *and* $k_0, \ldots, k_m \in \mathbb{N} \cup \{e\}$.

Note that in the previous definition

- $d =$ dimension of the space in which the tessellation takes place
- $n =$ dimension of the interface under consideration
- $m + 1 =$ number of crystals (including C_e) necessary for the formation of such an interface

(see also Table 5.1).

Consider now the set of all n-facets at time t

$$\Xi_n(t) = \bigcup_{k_0, \ldots, k_m \in \mathbb{N} \cup \{e\}}^{\neq} F_n(t, k_0, \ldots, k_m),$$

Table 5.1. Specific meaning of "n-facet" in spaces of dimension d

	$d = 1$	$d = 2$	$d = 3$
$n = 0$	vertex=intersection of 2 crystals	vertex=intersection of 3 crystals	vertex=intersection of 4 crystals
$n = 1$	crystal	edge=intersection of 2 crystals	edge=intersection of 3 crystals
$n = 2$		crystal	face=intersection of 2 crystals
$n = 3$			crystal

for $n < d$.

For any Borel set B one can define the n-facet mean content of B at time t as the measure

$$\mathcal{M}_{d,n}(t, B) = \frac{1}{(m+1)!} E \left[\sum_{k_0,\ldots,k_m \in \mathbb{N} \cup \{e\}}^{\neq} \lambda_n(B \cap F_n(t, k_0, \ldots, k_m)) \right], \quad \text{for } n < d$$

$$\mathcal{M}_{d,d}(t, B) = E \left[\sum_{k_0 \in \mathbb{N}} \lambda_d(B \cap F_d(t, k_0)) \right] \tag{5.62}$$

where λ_n is the n-dimensional Hausdorff measure and the symbol $\neq$ denotes that the indices $k_0, \ldots, k_m$ must be distinct (and thus also the corresponding crystals). Note that, with the previous definition, $\mathcal{M}_{d,d}(t, B)$ is the d-dimensional volume of the portion of the set B occupied by crystals at time t.

If the probability distribution of the nuclei a_i which generate the tessellation is non-atomic, i.e., $\mathcal{M}_{d,n}(t, \cdot) \ll \nu_d$, where ν_d is the d-dimensional Lebesgue measure [14], there exists a density $\mu_{d,n}(x, t)$ such that for all Borel sets B in $\mathbb{R}^d$

$$\mathcal{M}_{d,n}(t, B) = \int_B \mu_{d,n}(x, t) dx. \tag{5.63}$$

Definition 6. The function $\mu_{d,n}(x, t)$ defined by (5.63) is called local mean n-facet density of the (incomplete) tessellation at time t .

In particular $\mu_{d,d}(x, t)$ and $\mu_{d,d-1}(x, t)$ are the mean local volume density and the mean local surface density, respectively, of the crystallized region at time t (elsewhere denoted by $V_V(x, t)$ and $S_V(x, t)$, respectively; see e.g.[55] for the nomenclature). The mean local volume density is also the crystallinity. A complete characterization of a crystallization pattern at time t is given in terms of the family of densities $\mu_{d,n}(x, t)$ for $n = 0, 1, \ldots, d$.

Evolution equation for the whole set of spatial densities $\mu_{d,n}(x, t)$ of n-facets have been recently obtained in [16] in term of the hazard function

Theorem 1. *Under sufficient regularity conditions on the kinetic parameters* $\alpha(x,t)$ *and* $G(x,t)$ *(given deterministic fields) we have*

$$\frac{\partial}{\partial t}\mu_{d,n}(x,t) = c_{d,n}\frac{h_{m+1}(x,t)}{(m+1)!}(1-\xi(x,t))[G(x,t)]^{-m} \qquad (5.64)$$

where for $d = 2,3$

$$h_k(x,t) = [h(x,t)]^k, \qquad k = 2,3,$$

and $c_{d,n}$ *are numerical constants depending only upon the dimensions* d *and* n

$$c_{d,n} = \frac{\Gamma(\frac{dm+n+1}{2})\Gamma(\frac{d}{2})^{m+1}}{(m+1)!\Gamma(\frac{dm+n}{2})\Gamma(\frac{d+1}{2})^m\Gamma(\frac{d-m+1}{2})}.$$

Equation (5.64) extends previous results by Moller [43] for the spatially homogeneous case (see also [42]).

System (5.64) is closed by the evolution equation for the hazard function, which is given by (see (5.60))

$$h(x,t) = G(x,t)S_{ex}(x,t).$$

The evolution equation for $S_{ex}(x,t)$ is provided by system (5.37) - (5.41) (note that $S_{ex} = \tilde{u}$).

In the spatially homogeneous case the measure $\mathcal{M}_{d,n}(t,\cdot)$ is translation-invariant, so that

$$\mathcal{M}_{d,n}(t,B) = \mu_{d,n}(t)\nu_d(B),$$

where $\nu_d(B)$ is the d−dimensional volume of B. Then, by the use of the hazard function $h(t)$, the following expression for the densities $\mu_{d,n}$. of an (incomplete) Johnson-Mehl tessellation can be derived [38, 43, 42]:

$$\mu_{d,n}(t) = c_{d,n}\int_0^t\left[s_d\int_0^\tau\left(\int_s^\tau G(u)du\right)^{d-1}\alpha(s)ds\right]^{m+1}G(\tau)p(\tau)d\tau \quad (5.65)$$

where $d = 1,2,3$, $n = 0,\ldots,d$, $m = d-n$ and s_d is the surface of the unit ball in $\mathbb{R}^d$,

5.5 Estimating the Local Density of Interfaces

5.5.1 Estimate of the Density of $d-1$-Facets via the Spherical Contact Distribution Function

For the sake of simple notations let us denote by

$$\Xi(t): = \Xi_{d-1}(t) = \bigcup_i\partial C_i(t)$$

the random set of the boundaries of the cells of a Johnson-Mehl tessellation at time t in $\mathbf{R}^d$ (here we will treat only the cases relevant in the applications, i.e. $d = 2, 3$). Under sufficient regularity conditions on the growth and nucleation rates, this set may be assumed to have Hausdorff dimension $d - 1$.

Definition 7. *The local spherical contact distribution function H_s associated with a random set Σ [14] is*

$$H_s(r, x): \ = P(x \in \Sigma \oplus b(0, r) | x \notin \Sigma),$$

where $\oplus$ is Minkowski addition $(A \oplus B = \{a + b | a \in A, b \in B\}, \forall A, B \subseteq \mathbf{R}^d\})$ and $b(0, r)$ is a d-dimensional ball of radius r centered at the origin.

Since our random set $\Xi(t)$ has Hausdorff dimension $d-1$, for any fixed $x \in \mathbf{R}^d$ we have

$$P(x \notin \Xi(t)) = 1,$$

so that the spherical contact distribution function associated with $\Xi(t)$, that will be denoted by $H_s(r, x, t)$, is

$$H_s(r, x, t): \ = P(x \in \Xi(t) \oplus b(0, r)). \tag{5.66}$$

The estimate of $\mu_{d,d-1}$ is then possible thanks to the following theorem, which has been proven in [39]

Theorem 2. *Let $\Xi(t) = \bigcup_i \partial C_i(t)$ be the set of the boundaries of the cells of an (incomplete) Johnson-Mehl tessellation at time t, having Hausdorff dimension $d - 1$. Let $\Xi_r(t)$ be its parallel set of radius r, i.e.*

$$\Xi_r(t) = \Xi(t) \oplus b(0, r).$$

If the condition

$$\lim_{r \to 0} \frac{\nu_d[\Xi_r(t) \cap b(x, \varepsilon)]}{r} = 2\nu_{d-1}(\partial\Xi(t) \cap b(x, \varepsilon)), \tag{5.67}$$

is satisfied for all $x \in \mathbf{R}^d$ and for all $\varepsilon \in \mathbf{R}_+$ then the function $H_s(r, x, t)$ is differentiable at $r = 0$ for ν_d-almost all x, and its derivative satisfies

$$\frac{d}{dr} H_s(r, x, t)|_{r=0} = 2\mu_{d,d-1}(x, t). \tag{5.68}$$

It can be proven, via the Coarea Formula, that Condition (5.67) is fulfilled by a set $\Xi \subset \mathbf{R}^d$ if the function

$$f: \ \mathbf{R}^d \to \mathbf{R}_+$$
$$x \mapsto d(x, \Xi)$$

(where d is the euclidean distance) is Lipschitz [2].

Formula (5.68) may be used to estimate $\mu_{d,d-1}$ from an estimator of H_s. Suppose to recover a black-and-white image of the boundaries of the simulated crystals, as shown in Figure 5.7. Note from (5.66) that $H_s(r, x, t)$ equals the local volume density of the parallel set $\Xi_r(t)$ of $\Xi(t)$, which is in general nonzero, being $\Xi_r(t)$ a set of Hausdorff dimension d.

The (local) volume density $\xi(x)$ of a random set Σ can be estimated by considering an observation window $W(x)$, centered at x, sufficiently small so that the volume density may be considered constant inside the window, but not too small with respect to the size of the random set (if possible), so that the probability that the window is completely occupied by the set or completely empty is nontrivial, i.e. is not 0 or 1 (usually if the random set Σ is a 2D black and white digitized image, the window $W(x)$ must be chosen much larger than the size of a pixel). Then a grid $x_1, \ldots, x_n$ of n points is overlapped to the window $W(x)$ and the local volume density of Σ is estimated by

$$\widehat{\xi}(x) = \frac{1}{n} \sum_{i=1}^{n} \mathbf{1}_{\Sigma}(x_i).$$

This estimator is unbiased, with variance

$$\sigma^2 = \frac{1}{n^2} \left(n\xi(x)(1 - \xi(x)) + 2 \sum_{i>j} k(r_{ij}) \right),$$

where $r_{ij} = \|x_i - x_j\|$ and k is the covariance function of Σ (see [55] for further details). Thus in our case an unbiased estimator of $H_s(r, x, t)$ in a window $W(x)$, with a grid $x_1, \ldots, x_n$, is

$$\widehat{H}_s(r, x, t) = \frac{1}{n} \sum_{i=1}^{n} \mathbf{1}_{\Xi_r(t)}(x_i).$$

An estimator of $\frac{d}{dr} H_s(r, x, t)|_{r=0}$ may be obtained by numerical approximation

$$\frac{\partial}{\partial r} \widehat{H_s(r, x, t)}|_{r=0} \approx \frac{4\widehat{H}_s(r, x, t) - \widehat{H}_s(2r, x, t)}{2r},$$

for $r > 0$ small, since $H_s(0, x, t) = 0, \forall x \in \mathbf{R}^d, \forall t \in \mathbf{R}_+$ (this is a second order scheme, so that the resulting error is $o(r^2)$). Thus an estimator of the interface density $\mu_{d,d-1}(x, t)$ of the random tessellation is

$$\widehat{\mu}_{d,d-1}(x, t) = \frac{1}{2} \frac{\widehat{H}_s(r, x, t)}{r} = \frac{1}{2nr} \sum_{i=1}^{n} \mathbf{1}_{\Xi_r(t)}(x_i).$$

5.5.2 Numerical Results

In order to test the properties of this estimator, the estimation procedure has been applied to a spatially homogeneous simulated birth-and-growth process, generating an homogeneous Johnson-Mehl tessellation. An explicit formula for the densities $\mu_{d,n}(x,t) = \mu_{d,n}(t)$ (the densities do not depend on the space coordinate because of the spatial homogeneity) is (see Section 5.4)

$$\mu_{d,n}(t) = \bar{c}_{d,n} \int_0^t \left[s_d \int_0^\tau \left(\int_s^\tau G(u)du \right)^{d-1} \Lambda(ds) \right]^{m+1} G(\tau)p(\tau)d\tau \quad (5.69)$$

where $\bar{c}_{d,n}$ are geometric constants, $G(u)$ is the radial growth rate of the cells (the growth is isotropic), Λ is the intensity measure of the birth process and $p(t)$ is the probability that a point (which may be assumed as the origin for the spatial homogeneity) is covered by a cell at time t. A result of a simulation in the square $[0,1] \times [0,1]$ and of an edge detection performed on the image using the Image Processing Toolbox of Matlab 5.2 is shown in Figure 5.7. The radial growth rate and the nucleation rate have been assumed constant, with the following values:

$$G(t) = G_0 t, \quad \text{with } G_0 = 0.02$$
$$\Lambda(ds) = \lambda_0 ds \quad \text{with } \lambda_0 = 30$$

the time step used to perform one iteration of the simulation is $dt = 1$ and the simulation has been stopped after 7 iteration, i.e. at time $t = 7$. The side of one pixel is 0.0026 and we used 381×381 pixels to draw the image in the whole window of observation. The window $[0,1] \times [0,1]$ of observation of the process has been divided into 361 smaller subwindows and in each one the estimator $\hat{\mu}_{d,d-1}(x,t)$ has been computed. The mean and the variance of this estimator has been computed in each subwindow over 46 simulations. The results are shown in Figure 5.8.

The true value of the surface density, computed via Formula (5.69), is

$$\mu_{d,d-1}(7) = \mu_{2,1}(7) = 19.13.$$

The estimated mean surface density varies, spatially, from 14 to 28, but the majority of the estimated values is included between 15 and 20. The estimated variance is mostly of the order of 100 , so that a 95% "local" confidence interval would always include the true value. Let us denote by $\hat{\mu}_{d,d-1}(i), i = 1,\ldots,361$ the mean estimated surface density in each subwindow, then the "total" estimated mean surface density, i.e. the spatial mean of $\hat{\mu}_{d,d-1}$ is

$$\bar{\mu}_{d,d-1} = \frac{1}{361} \sum_{i=1}^{361} \hat{\mu}_{d,d-1}(i) = 19.73,$$

TIME=7 sec

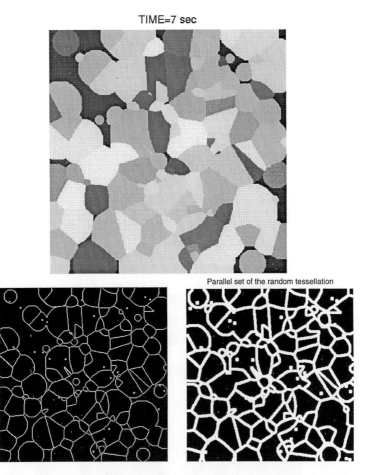

Parallel set of the random tessellation

Fig. 5.7. top figure = the result of a 2D simulation of an homogeneous birth-and-growth process, with constant radial growth rate and constant birth rate; bottom left = the result of an edge detection via an image analiser; bottom right= parallel set computed via an image analiser

which is very close to the true value $\mu_{d,d-1}(7)$, and its spatial standard deviation is

$$\sigma(\hat{\mu}_{d,d-1}) = (\frac{1}{360} \sum_{i=1}^{360} (\hat{\mu}_{d,d-1}(i) - \bar{\mu}_{d,d-1})^2)^{1/2} = 2.47.$$

Obviously these values highly depend on the choice of the growth parameters, of the radius r of the parallel set $\Xi_r(t)$ and on the number and size of the subwindows in which the estimate is performed.

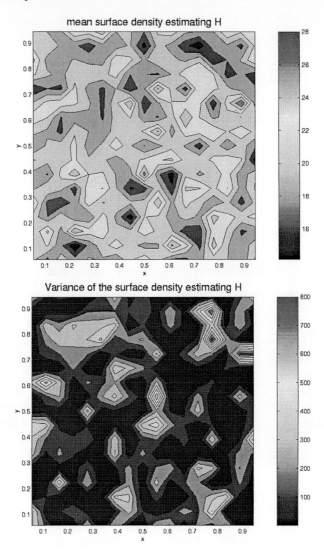

Fig. 5.8. Mean surface density estimated locally over 46 simulations (top figure) and its variance (bottom figure). See also Color Plate 1

5.5.3 Estimate of the Density of n-Facets from a 2D Simulation Under Homogeneous Conditions

In this section we estimate the density of n-facets (at all Hausdorff dimensions $n = 0, 1, 2$) of a 2-dimensional birth and growth process, on the basis of simulations performed under spatially homogeneous conditions, that is with constant nucleation and growth rate

$$\alpha(x, t) = \dot{N}_0$$
$$G(x, t) = G_0.$$

The aim is to estimate, in a window having unit area, the evolution of the *stochastic* local densities of

1. the volume of the crystalline phase,
2. the length of the interfaces,
3. the number of vertices,
4. the length of the outer surface,
5. number of nuclei,

and to see how far they are from their expected values, given respectively by (see Section 5.4)

1.

$$\mu_{2,2}(t) = 3A \int_0^t \tau^2 \exp(-A\tau^3) d\tau$$
$$= 1 - \exp(-At^3), \tag{5.70}$$

2.

$$\mu_{2,1}(t) = 6\dot{N}_0 G_0 A \int_0^t \tau^4 \exp(-A\tau^3) d\tau, \tag{5.71}$$

3.

$$\mu_{2,0}(t) = \frac{3}{2} \pi \dot{N}_0^2 G_0^2 A \int_0^t \tau^6 \exp(-A\tau^3) d\tau, \tag{5.72}$$

4.

$$S_V(t) = \pi \dot{N}_0 G_0 t^2 \exp(-At^3),$$

5.

$$N(t) = \dot{N}_0 \int_0^t \exp(-A\tau^3)) d\tau,$$

where A is the constant

$$A = \frac{\pi \dot{N}_0 G_0^2}{3}$$

Other methods of estimation are possible for digitized images. In particular we propose another method for the estimation of the density of 1-facets, as an alternative to the one presented in the previous section. This second method seems to be more efficient in terms of computational costs of the one presented in the previous section, but it has the disadvantage to be easily applicable only to simulated data, and not to images taken from real transformation.

Suppose to retrieve an history of images of a simulated birth and growth process (for details on the methods used for the simulations see Section 5.7). In every picture the pixels belonging to each crystal assume a different value (we associated the value zero to the pixels which have not yet been captured). In Figures 5.9 and 5.10 we may see a typical result, when the crystallized region is approximately 25%, 50%, 75% and 100% of the available space $E = [0,1] \times [0,1]$.

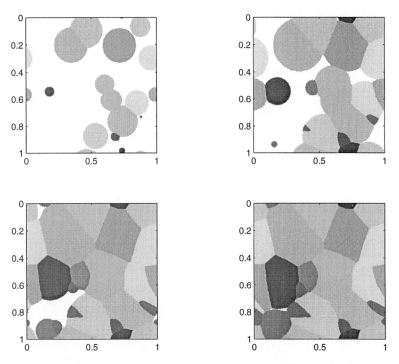

Fig. 5.9. Evolution of the crystalline phase with $\dot{N}_0 = 10$, $G_0 = 0.1$.

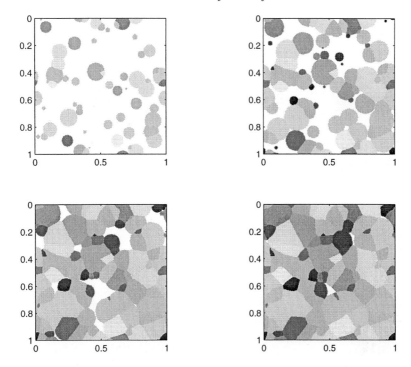

Fig. 5.10. Evolution of the crystalline phase with $\dot{N}_0 = 100$, $G_0 = 0.1$.

The structure of the algorithm for the estimation procedure is the following

- location of the occupied volume: all the pixels with value greater then zero are marked;
- edge detection: the pixels in the crystallized region that have a neighbour with a different value receive a new mark, assuming different values if the neighbour is in the inner or outer part of the crystalline phase;
- estimate of the number of vertices: the pixels on the inner boundary that have neighbours of at least two different colors are signed as vertices.

In this way we obtain the images shown in figures 5.11 and 5.12.

An estimator of the stochastic length of interfaces, obtained from a picture composed by N pixels, in a unitary window of observation W, is

$$b = c\frac{\text{number of pixels which lie on the edges}}{N}, \tag{5.73}$$

where c is a correction factor which we could call *"specific length"* of a pixel, which represents the mean length of the portion of line contained in each pixel (see Figure 5.13).

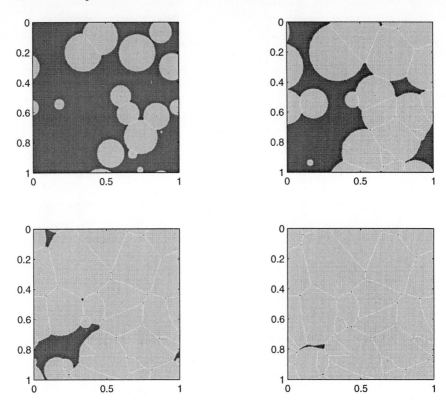

Fig. 5.11. Boundaries and vertices($\dot{N}_0 = 10$, $G_0 = 0.1$).

The value of c has been estimated via a linear regression procedure on a test case: a quarter of circle of radius r has been drawn in a window of the same size and with the same number N of pixels of the picture of the Johnson-Mehl tessellation. Then the length of the border of the circle of radius r has been estimated via the estimator

$$l(r) = \frac{\text{number of pixels which lie on the edge}}{N}.$$

The procedure has been iterated for various values of r, and then a linear regression has been performed of the type

$$l(r) = \beta_1 r + \varepsilon$$

and the coefficient β_1 has been estimated on the basis of the joint sample $(l(r_i), r_i)$. Since the true value of β_1 is $\pi/2$, the ratio

$$c = \frac{\pi/2}{\hat{\beta}_1}.$$

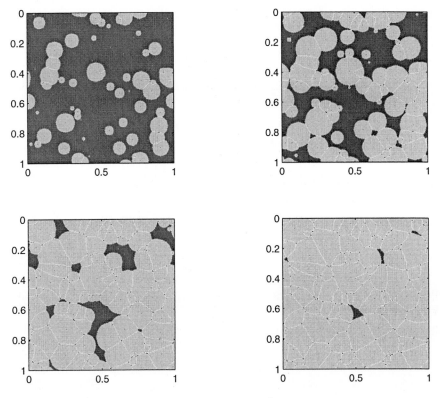

Fig. 5.12. Boundaries and vertices ($\dot{N}_0 = 100$, $G_0 = 0.1$).

gives the correction factor to be used in (5.73). From simulated data we obtained the value

$$c \approx 0.5534.$$

5.5.4 Numerical Results

We run n simulations with parallel processors with the aim to find the mean values of length of interfaces and number of vertices when the crystallization is complete. In Figures 5.14 and 5.15 the output of 64 simulations is plotted.

For any simulation, at the end of the process we estimated the length of interfaces b_j, and the final number of vertices v_j, for $j = 1, \ldots, n$. Then we computed the corresponding mean values, standard deviations and relative deviations:

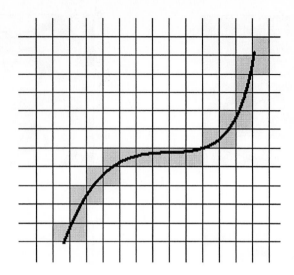

Fig. 5.13. When we draw a line on a screen using pixels, every pixel contains only a portion of the "real" line.

$$\bar{b} = \frac{1}{n} \sum_{j=1}^{n} b_j$$

$$\sigma^b = \sqrt{\frac{1}{n} \sum_{j=1}^{n} (b_j - \bar{b})^2}$$

$$\sigma^b_\% = 100 \frac{\sigma^b}{\bar{b}}$$

and

$$\bar{v} = \frac{1}{n} \sum_{j=1}^{n} v_j$$

$$\sigma^v = \sqrt{\frac{1}{n} \sum_{j=1}^{n} (v_j - \bar{v})^2}$$

$$\sigma^v_\% = 100 \frac{\sigma^v}{\bar{v}}$$

and we made the comparison with their corresponding teorical values $\mu_{2,1}(\infty)$ and $\mu_{2,0}(\infty)$. In Tables 5.2 and 5.3, the results are reported over 64 simulations.

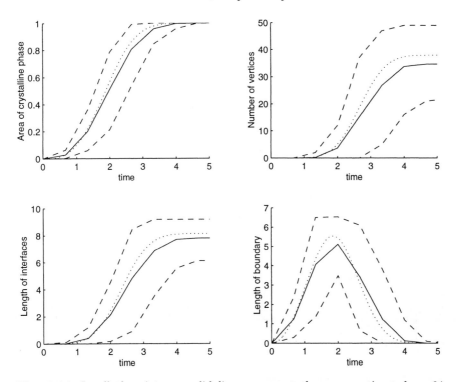

Fig. 5.14. In all the pictures, solid lines represent the mean estimated on 64 simulations, dashed lines are maximum and minimum value of the simulations, dotted lines represent the expected values. ($\dot{N}_0 = 10$, $G_0 = 0.1$).

Table 5.2. Length of the interfaces at the end of the process, with $G_0 = 0.1$, $n = 64$.

N_0	$\mu_{2,1}(\infty)$	$\bar{b}$	σ^b	$\sigma^b_\%$
10	8.13	7.38	0.68	8%
100	17.51	17.66	0.69	3%
1000	37.72	37.37	0.69	1%

Table 5.3. Number of vertices at the end of the process, with $G_0 = 0.1$, $n = 64$.

N_0	$\mu_{2,0}(\infty)$	$\bar{v}$	σ^v	$\sigma^v_\%$
10	37.89	35.30	6.09	17%
100	175.87	175.05	14.37	8%
1000	816.32	816.44	29.96	3%

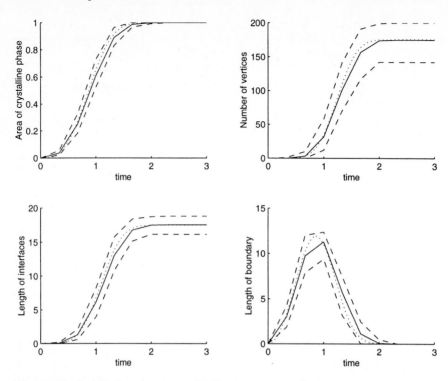

Fig. 5.15. In all the pictures, solid lines represent the mean estimated on 64 simulations, dashed lines are maximum and minimum value of the simulations, dotted lines represent the expected values. ($\dot{N}_0 = 100$, $G_0 = 0.1$).

5.6 Interaction with Latent Heat

The analysis is much more complicated if we want to include the interaction with temperature due to latent heat in the modelling of crystallization processes. This implies randomness in the temperature field due to the intrinsic randomness of the nucleation process.

Indeed nucleation and growth rates depend themselves upon temperature (cf. [28, 47, 50]). This causes problems with the definition of the causal cone and consequently with the related quantities describing the tessellation. As a first approach to circumvent these problems we shall focus on the fact that in a typical industrial situation we face a multiple-scale phenomenon. Under these circumstances averaging is possible by applying "laws of large numbers" as we shall see below.

5.6.1 Random Evolution Equations

In an experimental situation where spatial heterogeneities are caused only by the heat transfer in the material, we may assume that growth and nucleation rates are determined as material functions of the temperature [21].

$$G(x,t) = \tilde{G}(T(x,t)), \tag{5.74}$$

$$\alpha(x,t) = \frac{\partial}{\partial t}\tilde{N}(T(x,t)) + N_0(T(x,t)). \tag{5.75}$$

In typical industrial situations the first term in the representation of α strongly dominates; nucleations due to cooling of the sample (i.e., $\frac{\partial T}{\partial t} < 0$) are numerous, even though in various materials some rare nucleations may still be observed when the cooling is stopped (i.e. $\frac{\partial T}{\partial t} = 0$) at a temperature between the melting and the glass transition point.

It is also often assumed that $\tilde{G}$ and $\tilde{N}$ are exponential functions of the temperature (cf. e.g. [1, 21])

$$\tilde{G}(T) = G_{\text{ref}} \exp(-\ln 10\, y_1(T)), \tag{5.76}$$

$$\tilde{N}(T) = N_{\text{ref}} \exp(-\ln 10\, y_2(T)), \tag{5.77}$$

where $G_{\text{ref}}, N_{\text{ref}}$ are constants depending on the material and $y_1(T), y_2(T)$ are piecewise polynomial curves, usually of degree at most 2, also depending on the specific material. The curves $\tilde{G}(T)$ and $\tilde{N}(T)$ are usually bell shaped (see e.g. Fig. 5.17, for the curves relative to i-PP), but only the left part of the bell can usually be explored experimentally. For example in the case of isotactic Polypropylene (i-PP), the whole sample crystallizes in the temperature range $80 - 160^\circ C$, and only few experimental data are available to estimate the left part of the bell.

Vice versa, the increase of the crystalline phase influences the heat transfer process in the material because of the release of latent heat, i.e., the temperature T is determined by the heat transfer model

$$\frac{\partial}{\partial t}(\rho c T) = \text{div}\,(\kappa \nabla T) + \frac{\partial}{\partial t}(h\rho I_{\Theta^t}), \qquad \text{in } \mathbb{R}^+ \times E, \tag{5.78}$$

$$\frac{\partial T}{\partial n} = \beta(T - T_{out}), \qquad \text{on } \mathbb{R}^+ \times \partial E, \tag{5.79}$$

where $\frac{\partial}{\partial n}$ denotes the (outer) normal derivative on ∂E.

Here $\frac{\partial}{\partial t}(I_{\Theta^t})$ denotes the (distributional) time derivative of the indicator function I_{Θ^t} of the crystalline phase at time t, noting that Equation (5.78) has to be understood in the sense of distributional derivatives. The parm ρ denotes the density, c the heat capacity, κ the heat conductivity, β the heat transfer coefficient and h the latent heat due to the phase change.

The parameters in the heat equation may depend upon the phase, i.e., if ρ_1, c_1, κ_1 and β_1 denote the parameters of the crystallized material and ρ_2, c_2, κ_2 and β_2 the ones of the non-crystallized, we should write

$$\rho = I_{\Theta^t} \rho_1 + (1 - I_{\Theta^t}) \rho_2,$$
$$c = I_{\Theta^t} c_1 + (1 - I_{\Theta^t}) c_2,$$
$$\kappa = I_{\Theta^t} \kappa_1 + (1 - I_{\Theta^t}) \kappa_2,$$
$$\beta = I_{\Theta^t} \beta_1 + (1 - I_{\Theta^t}) \beta_2.$$

Table 5.4. Mean of the measured values for i-PP.

Parameter	Symbol	Value
Density of the crystalline phase	ρ_1	$850 \ [g/m^3]$
Density of the amorphous phase	ρ_2	$940 \ [g/m^3]$
Capacity of the crystalline phase (at $25°C$)	c_1	$1.69 \ [J/gK]$
Capacity of the amorphous phase (at $25°C$)	c_2	$2.14 \ [J/gK]$
Capacity of the crystalline phase (at $180°C$)	c_1	$2.38 \ [J/gK]$
Capacity of the amorphous phase (at $180°C$)	c_2	$2.54 \ [J/gK]$
Conductivity of the crystalline phase	k_1	$0.275 \ [W/mK]$
Conductivity of the amorphous phase	k_2	$0.175 \ [W/mK]$
Latent heat	h	$165 \ [J/g]$

This heat transfer model is a random partial differential equation, since all parameters and the latent heat depend upon the random variable I_{Θ^t}. A direct consequence is the stochasticity of the temperature field, whose evolution influences (via (5.76) and (5.77)) the crystallization process itself. Alternatively, the system (5.78), (5.79) could be reformulated as a moving boundary problem for the heat equation, with a condition for the jump of $\frac{\partial T}{\partial n}$ on the boundary of the crystalline phase (cf. [25] and [10] for further details). For the growth of a single (deterministic) crystal, regularity properties of the solution T and the boundary $\partial \Theta^t$ have been thoroughly analysed in [25].

Considering first the full system obtained by coupling the crystallization process, as described by system (5.30)-(5.36), to the evolution equation for the temperature (5.78)-(5.79), we obtain (for spatial dimension $d = 2$):

$$\frac{\partial}{\partial t}(\rho c T) = \text{div}\,(\kappa \nabla T) + (h\rho f)_t \qquad \text{in } E \times \mathbb{R}^+ \qquad (5.80)$$

$$\frac{\partial f}{\partial t} = \widetilde{G}(T)(1-f)u \qquad \text{in } E \times \mathbb{R}^+ \qquad (5.81)$$

$$\frac{\partial u}{\partial t} = \text{div}(\widetilde{G}(T)v) + \sum_{T_j = t} S_j^d(T) \qquad \text{in } E \times \mathbb{R}^+ \qquad (5.82)$$

$$\frac{\partial v}{\partial t} = \nabla(\widetilde{G}(T)u) \qquad \text{in } E \times \mathbb{R}^+ \qquad (5.83)$$

$$\frac{\partial T}{\partial n} = \alpha(T - T^1) \qquad \text{on } \partial E \times \mathbb{R}^+ \qquad (5.84)$$

$$T = T^0 \qquad \text{in } E \times \{0\} \qquad (5.85)$$

$$f = 0 \qquad \text{in } E \times \{0\} \qquad (5.86)$$

$$u = 0 \qquad \text{in } E \times \{0\} \qquad (5.87)$$

$$v = 0 \qquad \text{in } E \times \{0\} \qquad (5.88)$$

$$u + v \cdot \mathbf{n} = 0 \qquad \text{on } \partial E \times \mathbb{R}_+. \qquad (5.89)$$

An analogous system is obtained in $\mathbb{R}^3$ (see [11]). This system is random since all parameters are random; on the other hand we may notice that the only source of stochasticity is explicitely given by the random measures S_j^d.

5.7 Simulation of the Stochastic Model

We mow turn our attention to an (efficient) simulation of the stochastic model (i.e., (5.78)-(5.79), coupled with the stochastic evolution of the term I_{Θ^t}), which consists of three main parts:

– **Nucleation**, which is the part with intrinsic stochasticity.
– **Growth** of the crystals nucleated at random locations.
– **Heat Conduction**, which is influenced by the crystalline phase.

5.7.1 Nucleation

Since the nucleation rate $\alpha(x,t)$ depends on space and time, the birth of new crystals may be represented by an inhomogeneous space-time Poisson process (see e.g. [18, 10]) having stochastic intensity α (actually $\alpha(1 - I_{\Theta^t})$, since crystals already covered at the time of their birth are not of interest).

A standard algorithm for the simulation of such processes is the *thinning* or *random sampling* method (see [49, p.77]). This method consists in simulating an homogeneous Poisson process with constant intensity $\hat{\alpha}$, satisfying

$$\alpha(x,t) \le \hat{\alpha}, \qquad \forall x \in E, t \in \mathbb{R}_+.$$

In a subsequent thinning step a uniformly distributed random variable $R_i \in [0,1]$ is evaluated for each generated point $\{(t_i, x_i)\}$, which is then retained if $R_i > \alpha(t_i, x_i)/\hat{\alpha}$. For crystallization problems with a large number of nuleation events and a rate α with significant spatial heterogeneities, it turned out that this algorithm enforces unreasonably high computational efforts, since the number of points generated in the first step is usually much larger than the number of points that are finally kept. Therefore, we use a different approach, taking into account the spatial discretization of the domain E. Suppose we have a decomposition

$$E = \bigcup_{i=1}^{m} E_i, \tag{5.90}$$

$$0 = t_0 < t_1 < \ldots < t_n = t_{final} \tag{5.91}$$

with diam E_i and $|t_j - t_{j-1}|$ sufficiently small such that the intensity of the process can be well approximated by a constant in each space-time cylinder $E_i \times (t_j, t_{j-1})$. Then we can simulate the births in the time interval (t_j, t_{j+1}) via the following algorithm

Algorithm 1 (Generation of births in the time interval (t_j, t_{j+1}))
   for $i = 1$ to m do

1. *Compute $\alpha_{ij} := \int_{t_j}^{t_{j+1}} \int_{E_i} \alpha(x, t) dx dt$.*
2. *Generate a random number P_i, Poisson distributed with intensity α_{ij}. This number represents the number of 'virtual' births in the i−th space cell of the discretization.*
3. *generate P_i independent random numbers $U_1, \ldots, U_{P_i}$ having uniform distribution in the cell E_i. These numbers represent the locations of the new 'virtual' nuclei in the cell E_i.*
4. *check if the new 'virtual' nuclei are already covered by the crystalline phase: if yes delete them, otherwise add them to the set of nuclei.*

end

Note that if we are not interested in the evolution of the interfaces between two crystals, but only in the evolution of the crystallinity, we may disregard the effects of impingement and do not need to perform Step 4 in the previous algorithm. Indeed, we have shown that a new 'virtual' crystal that is inside another one will never grow out of it. Hence, the indicator function I_{Θ^t} does not change because of the presence of the new 'virtual' nuclei inside the crystalline phase.

5.7.2 Growth

Together with the nucleation, we have to simulate the growth of crystals and their interaction with heat transfer.

In spatial dimension one, a crystal Θ^k nucleated at (X_k, T_k) is an interval $[c_k, d_k]$, the evolution of the boundary is determined by

$$\frac{\mathrm{d}}{\mathrm{d}t}c_k(t) = -G(c_k(t), t), \qquad \frac{\mathrm{d}}{\mathrm{d}t}d_k(t) = G(d_k(t), t), \qquad (5.92)$$

with initial values $c_k(T_k) = d_k(T_k) = X_k$. Hence, an obvious way to compute the growth of a single crystal is to apply a numerical scheme to the nonlinear ordinary differential equations (5.92). Since the time steps and the increments in G are usually very small in realistic simulations, even an explicit method yields reasonable results. Hence, if impingement is disregarded and we are interested only in the evolution of I_{Θ^i}, growth can be computed according to the following algorithm:

Algorithm 2 (Growth of m one-dimensional crystals in the time interval (t_j, t_{j+1}) without impingement)

for $k = 1$ to m do

1. *If $T_k \in (t_j, t_{j+1})$ set $c_k = d_k = X_k$ and $s = T_k$. Else, set $s = t_j$ if $T_k < t_j$ and $s = t_{j+1}$ if $T_k > t_{j+1}$.*
2. *Compute (or approximate) the growth rate at the boundary points of the $k-$th crystal $[c_k, d_k]$:*

$$G_1 = G(T(c_k, s)), \quad G_2 = G(T(d_k, s))$$

3. *Update the boundary points by*

$$c_k = c_k - G_1(t_{j+1} - s), \quad d_k = d_k + G_2(t_{j+1} - s).$$

end

Of course, also more advanced schemes for the numerical integration of (5.92) can be used in Algorithm 2, and the algorithm can in principle also be carried over to higher dimensions.

   If we are interested also in the effects of impingement, i.e., in monitoring the evolution of the interface densities, Algorithm 2 has several disadvantages. Obviously, in one spatial dimension, one can check if two crystals Θ^j and Θ^k intersect by simply comparing c_k with d_j and d_k with c_j, but if we want to find all interfaces this method is computationally expensive. Therefore it seems necessary to derive a faster numerical method for the simulation of growth, which allows to trace the interfaces efficiently also in higher spatial dimensions.

   The clue for such a method seems to be a change from the Lagrangian perspective induced by (5.92) to an Eulerian approach, i.e., we fix locations in space and watch the crystals arriving. We use this Eulerian viewpoint for a 'pixel colouring' technique, i.e., we assign a number (respectively a colour) to each crystal and then assign this number (colour) to the spatial locations (pixels) covered by the crystal. More precisely, we fix a spatial grid (usually finer than the grid on which the temperature is assigned, because of the larger scale of heat conduction with respect to growth) in the domain E and at every time step we check for each grid point if it is inside a crystal (which is the case if there exist a grid point of radial distance less than $G\Delta t$);. If this is the case, we assign the number of the crystal to this point (respectively colour it). An additional speed-up can be achieved by marking those grid points, which are at the boundary of a crystal and computing the growth only there (all the points in the interior of a crystal do not provide additional information).

If we have K_j marked pixels at time t_j (i.e. the ones which are coloured and at the boundary of a crystal), then we simulate the growth in a time step (t_j, t_{j+1}) in the following way

Algorithm 3 (Growth in a time interval (t_j, t_{j+1}) with impingement)

for $k = 1$ to K_j do

1. *Compute the growth rate at the pixel p_k:* $G_k = G(T(p_k, t_j))$
2. *Compute the radius associated with the pixel p_k:* $\Delta r_k = G_k(t_{j+1} - t_j)$
3. **For** *every pixel p included in a disk centered at p_k having radius Δr_k* **do**
   - *if p is not in the crystalline phase (uncoloured), assign to p the same number of p_k (colouring), mark p and compute the approximate time at which it is captured by p_k:* $\tau_k = \frac{\|p - p_k\|}{G_k}$.
   - *if p is already inside a crystal (i.e. it has been captured at a time τ) and is marked, compute the time at which it is captured by p_k:* $\tau_k = \frac{\|p - p_k\|}{G_k}$. *If $\tau_k < \tau$, assign to p the number (colour) of p_k .*
4. *Unmark pixel p_k.*

end

We note that Step 3 ensures that every pixel assumes the colour of the first crystal by which it is reached, independently from the order of checking of the marked pixels and of the chosen time step. The accuracy of the algorithm increases as the step size in time and the spatial grid size decrease; obviously they cannot be chosen independently, since one should ensure that a crystal can really reach neighbouring grid points in one time step, i.e., if h is the typical grid size and Δt the time step we need at least

$$h < G_0 \Delta t,$$

where G_0 is a typical value of the growth rate.

5.7.3 Temperature Evolution

For the numerical experiments we used data from isotactic Polypropylene (i-PP). Referring to the notations of equation (5.78), the quantities h, ρ, c, k have been assumed to be constant and equal in the amorphous and in the crystalline phase, since their variability for i-PP is relatively small. Thus, equations (5.78)-(5.79) may be rewritten in the form

$$T_t = D\Delta T + L(I_{\Theta^t})_t \qquad\qquad \text{in } E \times \mathbb{R}^+ \qquad\qquad (5.93)$$

$$T = T_{out}, \qquad\qquad \text{on } \partial E \times \mathbb{R}^+, \qquad\qquad (5.94)$$

with appropriately scaled parameters D and L: Since Equation (5.93) should be understood in a weak sense, the discretization can be performed via finite elements. We propose to use linear finite elements on a grid of the type shown in Fig. 5.16. This is similar to a finite difference discretization for the heat

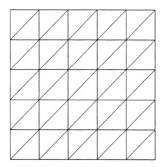

Fig. 5.16. The spatial grid used to solve Equation (5.93) using linear finite elements

equation, while the stochastic term $L\frac{\partial}{\partial t}I_{\Theta^t}$ is "smoothed" by integration on the cells of the grid. Because of the larger scale of heat conduction, we propose to use a coarser grid for the heat equation than for the computation of growth.

With the above approximations we can simulate the complete process as follows:

Algorithm 4 (Stochastic simulation of nonisothermal crystallization)
Start with the initial temperature T^0 and set $\Theta^0 = \emptyset$.
$j = 0$: solve the discretized version of equation (5.93) with $f^0(x) = f^1(x) = 0, \forall x \in E$ to obtain T^1.
for *$j = 1$ to $n - 1$* **do**

1. *Approximate the nucleation rate in $[t_j, t_{j+1})$ by*

$$\alpha \sim \frac{\partial \tilde{N}}{\partial T}(T_j)\frac{T^j - T^{j-1}}{t_j - t_{j-1}}.$$

2. *Simulate the nucleation in $E \times [t_j, t_{j+1})$ according to Algorithm 1 and add the new generated nuclei to $\Theta^{t_{j+1}}$.*

3. *Let the (old and new) crystals grow until time t_{j+1} and update $\Theta^{t_{j+1}}$.*
4. *Compute the indicator function $f^{j+1}(x) = I_{\Theta^{t_{j+1}}}(x)$.*
5. *Solve the discretized heat equation to obtain T^{j+1}.*

end

5.7.4 Numerical Results

The constants D and L of Equation (5.93) have been assumed $D = 0.5\, m^2/sec$ and $L = 50°C$, which correspond to mean values of experimental data for i-PP.

For the nucleation and growth rate we used curves fitted to experimental data from [51] of the form

$$\widetilde{G}(T) = 10^{y(T)}, \qquad \widetilde{N}(T) = N_{\text{ref}}10^{y(T)}$$

where y is a piecewise quadratic spline given by

$$y(T) = \begin{cases} aT + b & T \geq 125°C \\ cT^2 + dT + e & T \leq 125°C. \end{cases} \tag{5.95}$$

with

$$\begin{cases} a = -0.085825 \\ b = 4.062614 \\ c = -0.00133 \\ d = 0.246637 \\ e = -16.716251 \end{cases}$$

$N_{\text{ref}}, G_{\text{ref}}$ are scale factors, taken of the order of 10^7 and 10^{10}, respectively, so that the ratio $N_{\text{ref}}/G_{\text{ref}}$ is usually at least of the order of 10^3 (see Figure 5.17).

All algorithms have been implemented with the Scientific Computing Tool MATLAB, which was also used for the image analysis and visualization of results.

Numerical Results in $\mathbb{R}^1$ The results of a one-dimensional stochastic simulation are shown in Figure 5.18. The simulation has been performed starting from a uniform temperature of $130°C$ in the whole domain $E = (0,1)$ at time $t = 0$ and suddenly cooling the boundary points to $80°C$. The result clearly shows the strong effects on temperature due to the release of latent heat and due to the small diffusion coefficient. The peaks in the temperature profile are caused by new nucleations; they are smoothed in time because of diffusion. If the temperature is kept only slightly below the melting point, the increase of temperature due to the phase change may even stop the growth of nucleated crystals. This is evident if we take into account impingement in the simulation, obtaining thus results as shown in Figures 5.19 and 5.20. In Figure 5.20

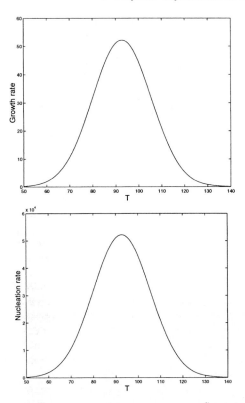

Fig. 5.17. Growth rate $\tilde{G}(T)$ (left) and nucleation rate $\tilde{N}(T)$ (right) as a function of temperature T

the cones of influence of crystals are plotted, i.e., the space-time region occupied by each growing crystal. It can be seen that already before two crystals hit each other, their growth slowed down for a certain time because of the increase of temperature and the consequent decrease of the growth rate.

Numerical Results in $\mathbb{R}^2$ A two-dimensional numerical experiment was performed on a rectangle, with the same data for the parameters as in the one-dimensional case. We cooled the sample starting from a uniform temperature of $120°C$ and then cooling the boundary with constant speed $\frac{\partial T}{\partial t}|_{\partial E} = 0.5°Cs^{-1}$. The results are shown in Figure 5.21, the behaviour is similar to the one-dimensional example; most crystals nucleate close to the boundary, while the temperature in the interior is too high for significant nucleation. The density of boundaries of crystals, $\mu_{d,d-1}$ may be easily estimated from pictures like the ones in Figure 5.21, using e.g. the estimator described in Section 5.5.1.

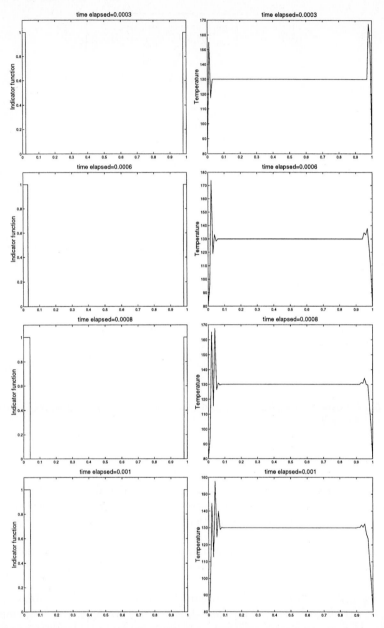

Fig. 5.18. Evolution of the indicator I_{Θ^t} (left) and of the temperature field (right) in a 1D stochastic simulation with growth computed disregarding impingement.

Fig. 5.19. Result of a 1D stochastic simulation with growth computed with the "pixel colouring" technique. The black zones are empty.

5.8 Stochasticity of the Causal Cone

As we have seen in Section 5.2, a possible way of deriving the evolution equations (5.64) for the relevant quantities that we would like to control, is based on the use of the causal cone $A(x,t)$, i.e., the set of all points y and times s, such that nucleation at y at time s would lead to coverage of x at time t. The whole theory of the causal cone (and even its well-definedness) is based on Lemma 1, which can be applied to any situation, where the growth rate $G(x,t)$ is a given field.

In the case of non-isothermal crystallization, this is a very delicate point, since the growing crystals influence temperature and thus nucleation and growth rates, so that, in the case of a substantially non-isothermal situation, they depend upon temperature. Hence, via (5.78), (5.79), the temperature depends upon the evolution of the crystalline phase. As already discussed in Section 2.4, in this case two growing nuclei cannot be compared as in Lemma 1, since both of them influence temperature and consequently, their growth is driven by different growth rates (see e.g. the case described in Figure 5.22). In a situation like that, the causal cone itself cannot be defined independently of the process anymore, since it always depends on the pattern of the crystalline phase via the temperature.

To overcome this problem, one can use certain approximations such as disregarding the effect of latent heat, which is reasonable if L is very small (and therefore rather realistic for other materials than polymers). A more sophisticated approximation that can be found in literature is the so-called *hybrid model*. In the heat equation (5.78), the indicator function of the crystalline phase is approximated by its mean value, the degree of crystallinity

$$I_{\Theta^t}(x) \longrightarrow \xi(x,t).$$

In this way one obtains a modified heat equation that gives a deterministic temperature field. Coupling this equation with the birth and growth process described by the geometric Kolmogorov-Avrami-Evans approach, gives a model to which the causal-cone approach can again be applied. The rationale behind this approximation is that small fluctuations in the source term

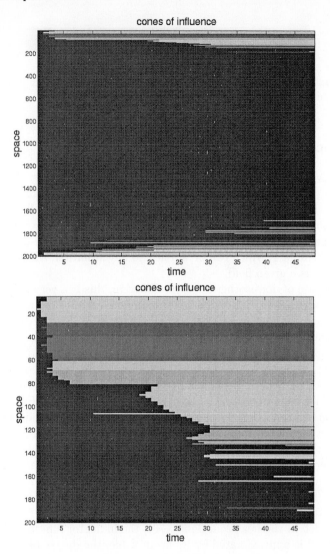

Fig. 5.20. Cones of influence in a 1D stochastic simulation (top) and a zoom of the same image (bottom). Note that the growth of some crystals is stopped before impingement occurs by the increase of temperature due to the release of latent heat

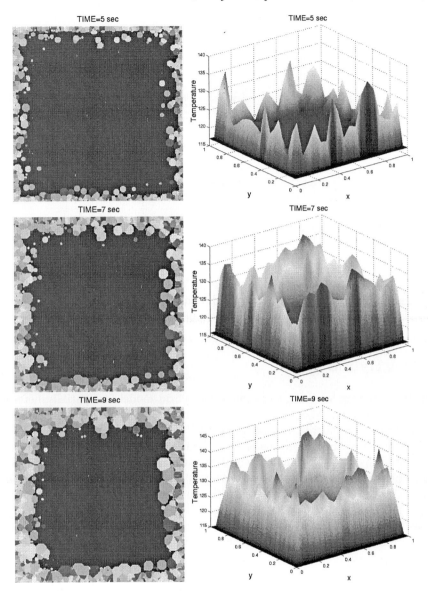

Fig. 5.21. Evolution of the crystals (on the left) and of the temperature field (on the right) in the stochastic simulation with cooling at the boundary.

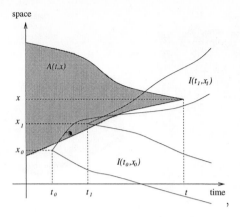

Fig. 5.22. The point (t_1, x_1) belongs to $\mathcal{I}(x_0, t_0)$ but $\mathcal{I}(x_1, t_1) \not\subset \mathcal{I}(x_0, t_0)$, so that (x_1, t_1) is not part of the causal cone in the case that a crystal is born in (x_0, t_0).

are smoothed out by the heat equation (5.78), and therefore we may use the corresponding (deterministic) averaged temperature field in (5.74) and (5.75) to obtain deterministic fields for $G(x, t)$ and $\alpha(x, t)$.

5.9 A Hybrid Model

In the following we want to formulate a hybrid model for spatially heterogeneous crystallization, which leads to evolution equations for the degree of crystallinity and the n-facets densities. This hybrid approach, namely to describe the stochastic process of crystallization via the evolution of a (deterministic) distribution function, was first introduced by Kolmogorov [31] for a homogeneous crystallization process (i.e., with constant growth and nucleation rates), and has been applied with considerable success to more general spatially homogenous processes (cf. e.g. [20]). We are now going to generalize this approach to an heterogenous situation both in space and time; as the main tool for our modelling approach we will use the causal cone, which was introduced above.

For many practical tasks, the stochastic models presented above, which are able to describe the process on microscopic scales, are too sophisticated for typical industrial conditions; it suffices to use averaged quantities on a larger scale. The advantage of using these averaged quantities is that they can be computed with much lower computational costs than a stochastic simulation at the microscale. We will now derive explicitly an hybrid model by means of an averaging procedure to the stochastic model (5.80)-(5.89). In the next chapter, a second macroscopic model is derived by an interacting particle approach.

Under typical industrial conditions, several assumptions can be made, which allow to approximate the stochastic model (5.80)-(5.89) by a deterministic system:

i) The nucleation rate and consequently the number of crystals is very large.
ii) The growth rate and consequently the sizes of the crystals are small.
iii) The typical scale for diffusion of temperature (macroscale) is much larger than the typical crystal size (microscale).

Under these conditions a mesoscale may be introduced, which is sufficiently small with respect to the macroscale of heat conduction and sufficiently large with respect to typical crystal sizes. First this means that temperature may be considered approximately constant on this mesoscale, and secondly that this scale contains a sufficiently large number of crystals to apply a "law of large numbers" . A typical size x_{meso} on this mesoscale satisfies

$$x_{crystal} << x_{meso} << x_{temperature},$$

where $x_{crystal}$ and $x_{temperature}$ are typical sizes for single crystals and for the heat transfer process. If t_0 is a typical time, G_0 and D_0 are typical values for the growth rate and the diffusion coefficient, we may choose

$$x_{crystal} = G_0 t_0, \qquad x_{temperature} = \sqrt{D_0 t_0},$$

which differ by several orders of magnitude under usual processing conditions. This typical feature of the process is illustrated in Figure 5.23.

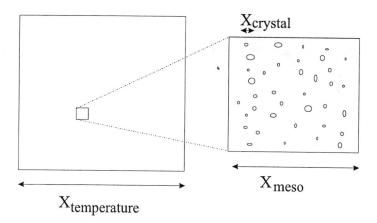

Fig. 5.23. The typical scales in the crystallization process.

At the level of the mesoscale it makes sense to consider a region B small enough that the space variation of temperature inside B may be denied and such that there is a sufficiently high number of "small" crystals in B, so that a law of large numbers allows the approximation

$$\sum_{X_j \in B} \sum_{T_j \leq t} S_j^d \approx E\Big[\sum_{X_j \in B} \sum_{T_j \leq t} S_j^d \Big], \qquad (5.96)$$

i.e. we may approximate the source term by its expected value, which is the mean rate of surface production. This mesoscale B, introduced by our assumptions upon nucleation and growth, is illustrated in Figure 5.23.

As we have already noticed before, for system (5.80)-(5.89) the only source of stochasticity is explicitly contained in Eqn.(5.82) due to the random measures S_j^d which describe the birth process of nucleation. We take now in account the averaging approximation (5.96) for $d = 1, 2, 3$. As in Equation (5.42), we may write in weak form, for a suitable test function ϕ,

$$\sum_{X_j \in B} \sum_{T_j \leq t} \langle S_j^d, \phi \rangle \approx \int_0^t \int_B \mathcal{F}^d[\widetilde{G}, \widetilde{\alpha}, \widetilde{T}](x, t) \phi(x, s) \, dx \, dt. \qquad (5.97)$$

Hence, we can define averaged source terms corresponding to $\sum_j S_j^d$; disregarding thermal nucleation they are given by (cf. [11] for further details)

$$\mathcal{F}^1[\widetilde{G}, \widetilde{\alpha}, \widetilde{T}](x, t) := 2\frac{\partial}{\partial t}\widetilde{\alpha}_t(\widetilde{T}(x, t))$$

$$\mathcal{F}^2[\widetilde{G}, \widetilde{\alpha}, \widetilde{T}](x, t) := 2\pi\widetilde{G}(\widetilde{T}(x, t))\widetilde{\alpha}(\widetilde{T}(x, t))$$

$$\mathcal{F}^3[\widetilde{G}, \widetilde{\alpha}, \widetilde{T}](x, t) := 8\pi\widetilde{G}^2(\widetilde{T}(x, t))\widetilde{\alpha}(\widetilde{T}(x, s)).$$

If we now substitute all stochastic quantities in (5.80)-(5.89) by their corresponding mean values, including approximation (5.97) for the source term $\sum_j S_j^d$, we obtain an initial-boundary value problem for a coupled system of parabolic and hyperbolic partial differential equations:

$$\frac{\partial \xi}{\partial t} = \widetilde{G}(\widetilde{T})(1 - \xi)\widetilde{u}$$

$$\frac{\partial \widetilde{u}}{\partial t} = \text{div } (\widetilde{G}(\widetilde{T})\widetilde{v}) + \mathcal{F}^d[\widetilde{G}, \widetilde{N}, \widetilde{T}]$$

$$\frac{\partial \widetilde{v}}{\partial t} = \nabla(\widetilde{G}(\widetilde{T})\widetilde{u}) \qquad (5.98)$$

$$\frac{\partial}{\partial t}(c\rho\widetilde{T}) = \text{div } (k\nabla\widetilde{T}) + \frac{\partial}{\partial t}(h\rho\xi),$$

in $E \times \mathbb{R}^+$, $d = 1, 2, 3$, supplemented by the boundary conditions

$$\widetilde{u} + \widetilde{v} \cdot \mathbf{n} = 0,$$

$$\frac{\partial \widetilde{T}}{\partial n} = \beta(\widetilde{T} - T_{out}),$$

on $\partial E \times \mathbb{R}^+$ and initial values given by

$$\xi = 0$$
$$\widetilde{u} = 0$$
$$\widetilde{v} = 0$$
$$T = T^0$$

in $E \times \{0\}$, usually with $T^0(x) \geq T_m$ for all $x \in E$. In order to express the essential difference between the (generalized) functions f, u, v and T and their deterministic counterparts in the averaged equations we have now written ξ, $\widetilde{u}$, $\widetilde{v}$ and $\widetilde{T}$.

System (5.98) provides now a deterministic temperature field $\widetilde{T}(x,t)$ in $E \times \mathbb{R}_+$. If we introduce in cascade this temperature field in (5.74) and (5.75), we obtain deterministic growth rates $G(x,t)$ and birth rates $\alpha(x,t)$ for the crystallization process described by system (5.30)-(5.36), that now is again simply stochastic. Also the hazard function $h(x,t)$ may be obtained in terms of these fields $\alpha(x,t)$ and $G(x,t)$ (see Section 5.3), so that the evolution equations (5.64) for the n-facets densities may again be used.

This approach is called "hybrid", since we have substituted the stochastic temperature field $T(x,t)$ given by the full system (5.80)-(5.89) by its "averaged" counterpart $\widetilde{T}(x,t)$ given by (5.98). It is relevant to notice that the hybrid system (5.98) is fully compatible with the rigorous derivation of the evolution equation for ξ, $\widetilde{u}$ and $\widetilde{v}$, in presence of deterministic kinetic parameters. In fact, once we approximate T with its deterministic counterpart $\widetilde{T}$, obtained via system (5.98), we are given deterministic fields for the kinetic parameters

$$\alpha(x,t) = \widetilde{\alpha}(\widetilde{T}(x,t))$$

and

$$G(x,t) = \widetilde{G}(\widetilde{T}(x,t)).$$

With these parameters the evolution equations for $\widetilde{u}$ and $\widetilde{v}$ are exactly the ones already obtained for the expected values of u and v (see equations (5.37)-(5.41)), and the evolution equation for the crystallinity $\xi(x,t)$ is the same as in (5.64) for $\mu_{d,d}(x,t)$, as it should be.

We note that the well-posedness of the initial-boundary value problems (5.98), i.e. the existence and uniqueness of a weak solution, can be shown (cf. [6]) as well as the stable dependence of the solution on the input parameters such as the growth rate $\widetilde{G}$ and the nucleation rate $\widetilde{N}$. For further details on a mathematical theory of the averaged crystallization model we refer to [6].

5.9.1 Numerical Methods for the Hybrid Model

The numerical solution of the averaged model (5.98) has to take into account its special structure, in particular the combination of the parabolic heat equation with the hyperbolic growth model, which can be interpreted as a system of conservation laws with source term. As also for the derivation of analytical results about the model equations (5.98), physical insight into the problem is an important guideline for the development of any numerical method. As we have seen several times in the derivation of the averaged model, the crystallization process can be split into two parts: first of all, temperature causes crystals to nucleate and to grow and secondly, the phase change has a feedback to the diffusion of heat via the latent heat. From this viewpoint it seems to be a good choice to design a numerical method such that in each time step first the growth part is computed and then the heat equation is solved.

An algorithm for the numerical approximation of the averaged model (5.98) has been proposed in [7] and analyzed with respect to convergence (cf. [6]). The basic idea of this algorithm is to perform an explicit time step in the hyperbolic part of the system together with an implicit step in the heat equation, which yields a decoupling and partial linearization in each time step. In semidiscrete form, this strategy for the hyperbolic part reads

$$\xi^{j+1} = \xi^j + \tau^j(1 - \xi^j)\widetilde{G}(T^j)u^j \tag{5.99}$$

$$u^{j+1} = v^j + \tau^j\left(\nabla(\widetilde{G}(T^j)v^j) + \mathcal{F}_d[\widetilde{G}, Nt, T^j]\right) \tag{5.100}$$

$$v^{j+1} = w^j + \tau^j\left(\nabla(\widetilde{G}(T^j)u^j)\right). \tag{5.101}$$

By the index j we denote the function at time t^j and $\tau^j = t^{j+1} - t^j$ is the j-th time step. For the full discretization any appropriate method like Lax's method (cf. [44]) or the Lax-Wendroff method for problems with source terms (cf. [35, 44, 59]) can be used. The stability bound for these methods is given by the Courant-Friedrichs-Levy condition (cf. [32, 44]), i.e.,

$$2(\max G)\tau \leq \Delta x. \tag{5.102}$$

Because of the small size of G, this bound is not very restrictive and still allows a good performance of the algorithm.

For the numerical approximation of the parabolic part one can then use standard discretization methods like the Crank-Nicholson scheme, since the source including ξ is known at time $t = t^{j+1}$. Hence, we can solve the model equations efficiently by the following algorithm:

Algorithm 5 (Numerical Solution of the Averaged Model) *Set the initial values* $\xi^0 = u^0 = v^0 = 0$.

for $j = 0$ **to** $n - 1$ **do**
    0. Use T^j, ξ^j, u^j *and* v^j *as initial values.*
    1. Compute $\mathcal{F}_d[\widetilde{G}, \widetilde{N}, T^j]$.
    2. Compute u^{j+1} *and* v^{j+1} *by an explicit step for*

$$\frac{\partial u}{\partial t} = div\,(\widetilde{G}(T)v) + \mathcal{F}_d[\widetilde{G}, \widetilde{\alpha}, T]$$

$$\frac{\partial v}{\partial t} = \nabla(\widetilde{G}(T)u)$$

in $E \times (t_j, t_{j+1})$.

3. *Compute ξ^{j+1} by an implicit (with respect to u) or explicit step for the equation*

$$\frac{\partial \xi}{\partial t} = \tilde{G}(\tilde{T})(1 - \xi)u \qquad in \ E \times (t_j, t_{j+1}).$$

4. *Solve the heat equation*

$$\frac{\partial}{\partial t}(c\rho T) = div\ (k\nabla T) + \frac{\partial}{\partial t}(h\rho\xi) \qquad in \ E \times (t_j, t_{j+1})$$

using ξ^{j+1} and ξ^j for the right-hand side to obtain T^{j+1}.

end

5.9.2 Numerical Simulations

In our numerical simulations we considered crystallization in a rectangular domain, whose length is twice the width ($\Omega = (0, L) \times (0, 2L)$). We performed simulations with two different choices for the temperature of the cooling material, using a uniform temperature T_{out} (Figs 5.25 and 5.26 - Example1) and a temperature T_{out} which is lower on two boundary segments (Figs 5.28 and 5.29 - Example 2). For all material parameters we used measurements for isotactic polypropylene (cf. [6, 7] for further details)

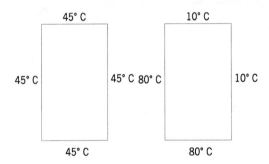

Fig. 5.24. Setup of the numerical examples.

The heat transfer coefficient and the outer temperature were chosen such that the assumptions about nucleation and growth are satisfied, but cooling is still slow compared to industrial conditions. The results clearly show the boundary layer in ξ (Figures 5.26 and 5.29), but there is no such effect in the temperature, which is due to the fact that cooling at the boundary is much stronger than the reheating effect caused by the latent heat. A comparison of the left and right hand side in Figure 5.25 shows that the reheating effect in the interior is more significant, the temperature is not necessarily monotone decreasing in time there.

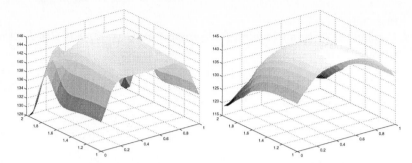

Fig. 5.25. Temperature in Example 1 after 5 and 10 minutes. Because of the symmetry, only the upper part of the rectangle $((0, L) \times (L, 2L))$ is plotted.

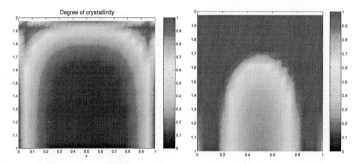

Fig. 5.26. Degree of crystallinity in Example 1 after 5 and 15 minutes. See also Color Plate 2.

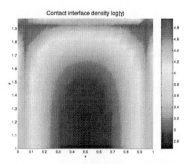

Fig. 5.27. Plot of $\log(\gamma)$, where γ is the mean density of crystals interfaces in $\mathbf{R}^2$, in Example 1 after 5 minutes. See also Color Plate 3.

In Example 2 the temperature after 6 and 9 minutes is shown in Figure 5.28, the evolution of the degree of crystallinity (after 6, 9, 12 and 15 minutes) is plotted in Figure 5.29. Here the crystallization process seems to behave

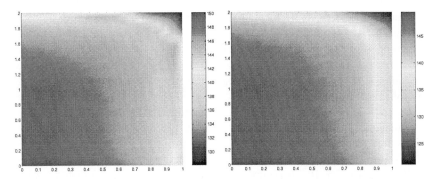

Fig. 5.28. Temperature in Example 2. See also Color Plate 4.

almost like a two-phase problem, the crystalline phase propagates from the corner with lowest temperature to the interior. This corresponds very well to earlier modelling approaches, approximating crystallization by a two-phase process [37].

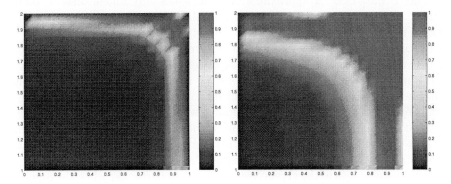

Fig. 5.29. Degree of crystallinity in Example 2. See also Color Plate 5

5.9.3 Comparison between Simulations of the Stochastic and the Hybrid models

A comparison of the results obtained from the stochastic model (5.78) and from the deterministic system (5.98), is shown in Figure 5.30. To compare the two models, 100 stochastic simulations, independent but with the same parameters, have been performed. The crystallinity at time t has then been computed by averaging, both in space and time, of the indicator function I_{Θ^t} computed in each stochastic simulation. The result is compared with the correspondent crystallinity obtained by the deterministic model (left); the same averaging procedure is applied to the temperature field, too. The plot of the deterministic (solid) and averaged stochastic quantities (dashed), shows good agreement in general, in particular for the degree of crystallinity. The results for temperature shows that the deterministic model seems to react more slowly to changes in the source term.

We imposed a ratio between the nucleation rate α and the growth rate $\tilde{G}$ of the order of 10^3 (in many practical examples it is much larger). It can be seen from the pictures that already with this relatively small ratio between the two quantities the numerical results are quite similar. Due to the averaging arguments based on the law of large numbers we expect an improvement for increasing nucleation rate, but so far we were not able to realize such a case in reasonable computational time.

5.9.4 Conclusions and Open Problems

We have shown above how nonisothermal crystallization can be simulated either by a stochastic or a deterministic approach. While the deterministic model is based on an approximation that works only under certain conditions (many and small crystals), the stochastic simulation is expected to yield reasonable results in any case. Although the numerical effort could be reduced significantly by Algorithm 3 with respect to the first approach, it is still very expensive to perform numerical experiments, in particular in $\mathbb{R}^3$. The improvement of the methods used for computing as well as a rigorous mathematical analysis are open problems for future work.

We note that real chemical experiments of the same kind could be performed in a lab, by using thin slices of polymeric material, and cooling them via a thermostat to control the temperature in the whole slice. The disadvantage of this technique is that interfaces are often not clearly visible from microscope images of the crystallized material, in particular when the number of generated crystals is very large. Instead it is rather simple to distinguish and measure the length of the interfaces generated by the simulator, by using some edge detector or image analizer. In practical applications, the approximation by the simulator are much smaller than the information that can be extracted from a picture of a real crystallized sample. Hence, this is a typical example in which simulated data may effectively substitute experimental ones.

Another problem related to the simulation of nonisothermal crystallization is the optimal control of the process. In order to obtain good mechanical properties of the solidified material, a uniform grain size distribution is desired. As we have seen in our numerical experiments, a simple cooling

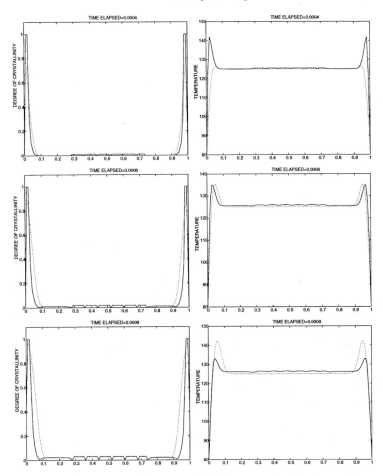

Fig. 5.30. Evolution of the crystallinity (on the left) and of the mean temperature (on the right) in the stochastic and the deterministic models. The solid lines are the solutions of the deterministic model, the dashed lines are the means of the stochastic simulations.

strategy will not yield this result. Therefore, one wants to find a temperature $T^1 = T^1(x,t)$ in the boundary condition

$$\frac{\partial T}{\partial n} = \alpha(T - T^1) \qquad (5.103)$$

such that the densities of the interfaces or the crystal sizes are as much uniform as possible (see Section 5.11.2).

5.10 Scaling Limits

In Section 5.9 we have used the different scales of temperature and crystallinity to derive the deterministic model (5.98). Now we shall discuss the scaling of this models and possible limiting cases of parameters in detail. For the sake of simplicity we restrict our attention to the case of $E \subset R^2$ for the moment, but analogous reasoning is possible in other spatial dimensions, too.

In order to express the dimensionality of the unscaled functions and variables, we will use a bar to mark them; in the following we assume in addition that the parameters in the heat equation are all constant We normalize the minimal length using the new space variable $x = \frac{1}{x_0}\overline{x}$ and introduce a new time variable $t = \frac{1}{t_0}\overline{t}$; the scaled domain will be denoted by D in the following. Temperature and the other free variables are scaled by

$$\theta = \frac{1}{\theta_0}(T - T_{min}) \qquad u = \frac{1}{u_0}\overline{u} \qquad v = \frac{1}{u_0}\overline{v},$$

and the parameter functions are transformed to

$$a(\theta) = \frac{1}{G_0}\widetilde{G}(T) \qquad\qquad ||a||_{C^1} = \mathcal{O}(1) \qquad (5.104)$$

$$b(\theta) = \frac{1}{N_0}\widetilde{N}(T) \qquad\qquad ||b||_{C^0} = \mathcal{O}(1) \qquad (5.105)$$

The arising initial-boundary value problem reads

$$\frac{\partial \theta}{\partial t} = d\Delta\theta + L\frac{\partial \xi}{\partial t} \qquad\qquad \text{in } D \times (0, t_*) \qquad (5.106)$$

$$\frac{\partial \xi}{\partial t} = (1 - \xi)\gamma a(\theta)u \qquad\qquad \text{in } D \times (0, t_*) \qquad (5.107)$$

$$\frac{\partial u}{\partial t} = \epsilon \text{ div } (a(\theta)v) + \delta a(\theta)b(\theta) \qquad \text{in } D \times (0, t_*) \qquad (5.108)$$

$$\frac{\partial v}{\partial t} = \epsilon\nabla(a(\theta)u) \qquad\qquad \text{in } D \times (0, t_*), \qquad (5.109)$$

with boundary conditions

$$\frac{\partial \theta}{\partial n} = \beta(\theta - \theta_{out}) \qquad\qquad \text{on } \partial D \times (0, t_*) \qquad (5.110)$$

$$u = -v.n \qquad\qquad \text{on } \partial D \times (0, t_*), \qquad (5.111)$$

and initial conditions

$$\theta = \theta^0 \qquad\qquad \text{in } D \times \{0\} \qquad (5.112)$$

$$\xi = 0 \qquad\qquad \text{in } D \times \{0\} \qquad (5.113)$$

$$u = 0 \qquad\qquad \text{in } D \times \{0\} \qquad (5.114)$$

$$v = 0 \qquad\qquad \text{in } D \times \{0\} \qquad (5.115)$$

The parameters in the new system are given by

$$\beta = \gamma_0 x_0, \qquad \gamma = G_0 t_0 u_0, \qquad \delta = 2\pi \frac{G_0 N_0 t_0}{u_0},$$
$$\epsilon = \frac{G_0 t_0}{x_0}, \qquad d = \frac{\kappa t_0}{\rho c x_0^2}, \qquad L = \frac{h}{c \theta_0}. \qquad (5.116)$$

Table 5.5. Typical values for isotactic Polypropylene (cf. [56, 58]) .

Parameter	Symbol	Typical Value
Diffusion coefficient	d	10^{-7} m^2 s^{-1}
Latent heat	$\frac{h}{c}$	50 K
Growth rate	G_0	10^{-5} m s^{-1}
Nucleation rate	N_0	10^{12} m^{-2}
Typical length	x_0	10^{-2} m
Typical temperature	θ_0	150 K

It seems reasonable to balance u and the source term in the hyperbolic part of the system, i.e.,

$$1 = \delta = 2\pi \frac{G_0 N_0 t_0}{u_0} \qquad \text{or} \qquad u_0 = 2\pi G_0 N_0 t_0, \qquad (5.117)$$

for the other free parameters we have several choices dependent on which scale we are interested in.

Diffusion Time Scale

In order to balance the heat transfer process, we choose t_0 such that the diffusion coefficient is scaled to 1, i.e.,

$$1 = d = \frac{a_0 t_0}{x_0^2}, \qquad (5.118)$$

which implies a typical time scale of

$$t_0 = \frac{x_0^2}{a_0} \sim 10^3 s. \qquad (5.119)$$

If the latent heat is ignored and the heat transfer coefficient is sufficiently large, t_0 is a good approximation for the time which is needed to decrease the temperature from its initial value to an equilibrium determined by the cooling temperature T_{out}. In the presence of the latent heat as a heating source, the cooling is obviously slower, but the numerical simulations show that this effect is not very strong.

On the diffusion time scale, the hyperbolic part has an almost scaled 'wave number', i.e. $\epsilon \sim 1$, which represents the fact that the growth of nuclei can be important for large time scales. The coefficient γ in equation (5.107) is very large $\gamma \sim 10^{12}$, which means that the degree of crystallinity grows strongly in a typical time interval for diffusion. Solving (5.107) for ξ and using the initial condition $\xi|_{t=0} = 0$ yields

$$\xi(x,t) = 1 - e^{-\gamma \int_0^t au\, d\tau}. \tag{5.120}$$

This means, that with the growth of $\int_0^t au\, d\tau$, the second term tends to zero and ξ tends to 1 very fast, i.e., we have to expect a boundary layer in a small initial time interval.

Growth Time Scale

If one is interested in the effects of the crystal growth on a typical time scale one should rather normalize the parameters γ and δ than the diffusion coefficient d. From $\gamma = \delta = 1$ we obtain

$$t_0 = \frac{1}{G_0\sqrt{2\pi N_0}} \tag{5.121}$$

$$u_0 = \sqrt{2\pi N_0}. \tag{5.122}$$

We note that t_0 is rather small compared to the diffusion time scale. Furthermore, $\epsilon = \frac{1}{x_0\sqrt{2\pi N_0}}$ is very small now ($\epsilon \sim 10^{-5}$). Thus, (5.106)-(5.115) is a singularly perturbed problem, and we have to expect a boundary layer since the boundary conditions are not satisfied by the solution of the reduced problem. In addition, the diffusion coefficient is rather small ($d \sim 10^{-3}$), which expresses the slowness of diffusion.

The asymptotic properties of the hyperbolic part as $\epsilon \to 0$ have been analyzed in [11, 6], it can be shown that the solutions converge to the solution of the reduced problem arising in the limiting case $\epsilon = 0$, i.e,

$$\begin{aligned} u_t &= \delta a(\theta)b(\theta) &&\text{in } D \times (0, t_*) \\ v_t &= 0 &&\text{in } D \times (0, t_*), \end{aligned}$$

Hence, the reduced problem is a straightforward extension of the classical Avrami-Kolmogorov model for spatially homogenous growth, which used the same differential equations (but in their case for global quantities in space). As a consequence of the singular perturbation in the hyperbolic part a boundary layer arises close to ∂R, which is of the size ϵt. In a one-dimensional example, this boundary layer is shown in Figure 5.31.

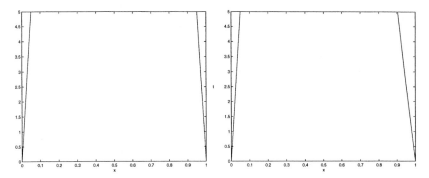

Fig. 5.31. The development of the boundary layer in time for constant growth rate (top) and varying, non-uniform growth rate (bottom).

Geometric Limits

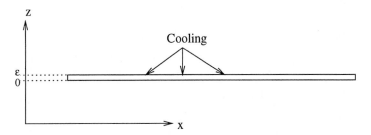

Fig. 5.32. Limit of small height with cooling on top.

As the last limiting case of parameters we want to consider the case of a polymer sample with small height (in the case of $E \subset \mathbb{R}^3$ we will identify the height with the z-variable for convenience) together with cooling on top and bottom (with respect to this small dimension). For the sake of simplicity we restrict our attention to a cylinder $E = E_0 \times (0, Z)$ for the moment.

The crystallization model for $Z \to 0$ can be reduced to the two-dimensional model (5.107)-(5.109) on the domain E_0. The assumption on the heat transfer that yields the equation (5.106) and the boundary condition (5.110) is that the top and bottom of Ω are isolated, i.e.,

$$\frac{\partial T}{\partial n} = 0 \qquad \text{on } E_0 \times \{0, Z\}$$

in the original three-dimensional model. If cooling appears on these parts of the boundary instead (which means that the heat transfer coefficient does not vanish on the top and the bottom), the heat transfer in z-direction will be dominating and we will obtain a completely different heat transfer model on E_0 instead of (5.106). To see this, we assume that T_{out} is constant on $E_0 \times \{0, Z\}$ and that Z is small with respect to the size of E_0. After a preliminary scaling, the dimensionless heat transfer model in E is then of the form

$$\frac{\partial T}{\partial t} = d(T_{xx} + T_{yy} + T_{zz}) + L\frac{\partial \xi}{\partial t} \qquad \text{in } E \times \mathbb{R}^+ \tag{5.123}$$

subject to the boundary condition

$$\frac{\partial T}{\partial n} = \alpha(T - T_{out}) \qquad \text{on } \partial\Omega \times \mathbb{R}^+. \tag{5.124}$$

For small Z, we can eliminate the third spatial dimension by using the mean quantities

$$\tilde{T} = \frac{1}{Z}\int_0^Z T \, dz, \qquad \tilde{\xi} = \frac{1}{Z}\int_0^Z \xi \, dz.$$

If we rescale time via $t = Z\tau$, this yields

$$\frac{\partial \tilde{T}}{\partial t} = dZ(\tilde{T}_{xx} + \tilde{T}_{yy}) + d\left(\frac{\partial T}{\partial z}|_{z=Z} - \frac{\partial T}{\partial z}|_{z=0}\right) + L\frac{\partial \tilde{\xi}}{\partial t}$$

$$= Z(\tilde{T}_{xx} + \tilde{T}_{yy}) + d\alpha\,(T|_{z=Z} + T|_{z=0} - 2T_{out}) + L\frac{\partial \tilde{\xi}}{\partial t}.$$

As $Z \to 0$, the difference between T and the mean quantity $\tilde{T}$ converges to zero and hence, we obtain the reduced problem

$$\frac{\partial \theta}{\partial t} = 2d\alpha\,(\theta - \theta_{out}) + L\frac{\partial \xi}{\partial t} \qquad \text{in } E_0 \times \mathbb{R}^+, \tag{5.125}$$

where θ is the two-dimensional temperature and θ_{out} denotes the outside temperature. An investigation of the reduxed model consisting of (5.125), (5.107) - (5.109) in $E_0 \times \mathbb{R}^+$ shows that all parameters in these evolution equations are now independent of the spatial variable. If furthermore, the initial temperature is uniform, which is a realistic assumption for most crystallization processes, the solutions of the model equations are uniform in space. I.e., we may write $T = T(t)$, $\xi = \xi(t)$ and so on, and these time-dependent quantities will satisfy the following system of ordinary differential equations:

$$\frac{\partial \theta}{\partial t} = 2d\alpha(\theta - \theta_{out}) + L\frac{\partial \xi}{\partial t} \tag{5.126}$$

$$\frac{\partial \xi}{\partial t} = (1 - \xi)\gamma a(\theta)u \tag{5.127}$$

$$\frac{\partial u}{\partial t} = \delta a(\theta)b(\theta), \tag{5.128}$$

the quantity v vanishes and needs not to be considered. One easily verifies that (5.126)-(5.128) is just a scaled version of the extended generalized Avrami-Kolmogorov model, which has been introduced by Eder [20].

5.11 Related Mathematical Problems

In this section we want to discuss some further problems related to the mathematical modelling of polymer crystallization problems. In particular we will investigate the identification of unknown material functions, such as the nucleation rate $\widetilde{N}$, from data that can be obtained in practice and discuss the optimal control of the model, whose motivation is to control the crystallization process such that the final material is optimal with respect to some criterion, e.g. has optimal mechanical properties.

5.11.1 Identification

Parameter identification is a fundamental part of any modelling procedure, since in almost no application the physical parameters that are an essential part of the model, are known or can be determined from direct measurements. The crystallization model for polymers contains a lot of parameters and material functions, whose determination will be discussed in the following

- The parameters in the heat equation c, ρ, κ and β are usually known constant values; which seems to be a reasonable approximation since their variation with respect to temperature and phase seems to be negligible.
- The latent heat L is a known constant value, too.
- The growth rate $\widetilde{G}$ is a material function of temperature, sample values can be determined from experiments, but only with high effort (cf. [47]). Advanced identification procedures could help to speed up the determination of the growth rate.
- The nucleation rate $\widetilde{N}$ is a material function of temperature, too. It can be determined in practice only with extreme effort, namely by counting the final number of crystals in an experiment, where the temperature is quenched to some temperature at the beginning, then kept there for some time and finally decreased suddenly below the melting point. Since the number of crystals is usually very high in such an experiment and automatic image analysis tools have problems to find crystals, it takes a lot of time to determine the value of this material function for only one temperature. Therefore, the identification of nucleation rates from indirect measurements is of major importance.

Identification of Nucleation Rates As explained above it is important to estimate the nucleation rate $\widetilde{N}$ as a function of temperature from indirect measurements that can be obtained in practice. Such measurements are for example the temperature evolution on a part of the boundary ($\Gamma \subset \partial E$ with nonzero $(d-1)$-dimensional Lebesgue measure) in a time interval $I = (0, t_*)$ and an observation of the morphology at the final time t_* in the domain E. In particular we may therefore assume to know

$$T_\Gamma := T_{I \times \Gamma} \in L^2(I \times \Gamma), \qquad \xi_* := \xi_{\{t_*\} \times E} \in L^2(E).$$

It can be shown that these data are sufficient for the identification of the nucleation rate $\widetilde{N}$ in the temperature interval that arises during the experiment, i.e., $\widetilde{N}(\theta)$ is determined uniquely for

$$\theta \in \left(\sup_{t \in I, x \in E} T(x, t), \inf_{t \in I, x \in E} T(x, t) \right),$$

under the assumption that cooling is strong enough (cf. [6]). In practice, however, the measurements are always corrupted by noise, and therefore we have to discuss the stability or possible instability of the identification problem. Indeed, it turns out (cf. [9, 5, 6]) that the identification of nucleation rates from the data (T_Γ, ξ_*) is *ill-posed*, i.e., the solution does not depend on the data in a continuous way, and consequently the precision for computation of the nucleation rates depends heavily on the noise in the data.

We can formulate the identification problem as a least-squares problem, i.e.,

$$\widetilde{N} = arg \min_{\widetilde{N}} ||F(\widetilde{N}) - (T_\Gamma^\delta, \xi_*^\delta)||^2, \tag{5.129}$$

where the index δ announces the noise in the data, which we assume to be bounded by

$$||(T_\Gamma^\delta, \xi_*^\delta) - (T_\Gamma^0, \xi_*^0)|| \leq \delta, \tag{5.130}$$

and F is the *parameter-to-output map* defined by

$$\begin{aligned} F : X &\to L^2(I \times \Gamma) \times L^2(E) \\ \widetilde{N} &\mapsto (T|_{I \times \Gamma}, \xi|_{\{t_*\} \times E}) \end{aligned} \tag{5.131}$$

The pair (T, ξ) solves (5.98) with given nucleation rate $\widetilde{N}$ and X is a function space on an appropriate temperature intervals (cf. [5, 6] for mathematical details), and hence, the evaluation of the parameter-to-output map consists of solving the state equation (5.98) and then evaluating appropriate trace maps on the solution.

In order to obtain a stable approximation of the solution of the least-squares problem (5.129), so-called *regularization methods* have to be used. We tried several iterative regularization methods, where the main regularizing effect comes from an early termination of the iteration procedure in dependence of the noise level δ and the noisy data. In [9] the application of the Landweber iteration

$$\widetilde{N}^{k+1} = \widetilde{N}^k - F'(\widetilde{N}^k)^*(F(\widetilde{N}^k) - (T|_\Gamma^\delta, \xi_*^\delta)). \tag{5.132}$$

was investigated. Since an explicit iteration method like the Landweber iteration is known to be rather slow, Newton-type methods such as the *iteratively regularized Gauss-Newton method*

$$\widetilde{N}^{k+1} = \widetilde{N}^k - (F'(\widetilde{N}^k)^*F'(\widetilde{N}^k) + \alpha_k I)^{-1}F'(\widetilde{N}^k)^*(F(\widetilde{N}^k) - (T|_\Gamma^\delta, \xi_*^\delta)). \tag{5.133}$$

were investigated in [5, 6]. It turned out that results obtained with both methods were of the same quality (see Figure 5.36, but the number of iterations needed until the stopping criteria is satisfied, is significantly smaller for Newton-type methods, which can be seen in Figure 5.33. On the other hand, an iteration step in a Newton-type iteration is by far more expensive than an iteration step for Landweber's method, since the assemblation of the Newton matrix needs several evaluations of $F'(\widetilde{N}^k)$, which is of the same effort as evaluating F (we refer to [5, 6] for further details). It turns out that the total number of floating point operations, which is the effective number characterizing the numerical effort is even lower for the Landweber iteration for many cases - in particular for larger noise levels, which is the more realistic case in practical applications.

The ill-posedness of the identification problem is demonstrated in Figure 5.35, where the development of the error during an iteration procedure is plotted for several Newton-type methods, such as the iteratively regularized Gauss-Newton method (IRGN) and the Levenberg-Marquardt iteration (LM), as well as some Quasi-Newton methods such as a frozen Gauss-Newton method and Broyden's method. It turns out that the error is decreasing only during a small number of iterations, which is also a typical effect for ill-posed problems and causes the need of an appropriate early termination of the iterative method (cf. [22] for an overview of iterative regularization methods, and [5, 6] for their application to the identification of nucleation rates). Finally, the quality of the reconstructed nucleation rates in presence of noise is illustrated in Figure 5.36; the reconstruction is a reasonable approximation already for larger noise level ($\delta = 5\%$) and approaches the exact solution very well for less noise ($\delta = 2\%$).

5.11.2 Optimal Control

Finally, we want to discuss optimal control problems for the crystallization models presented above, which are relevant for industrial applications. The basic variable that can be controlled is the cooling temperature T_{out}, while the possible objectives and constraints vary. Important examples of optimal control problems are:

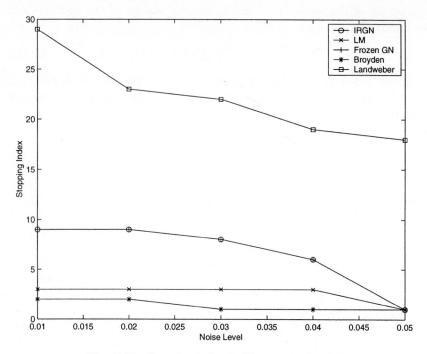

Fig. 5.33. Stopping index $k_*(\delta)$ vs. noise level δ.

– Minimization of the crystallization time until a certain degree of crystallinity is reached in the material, which is of general importance for industrial process optimization.
– Optimization of the morphology; in particular of the distribution of interfaces between the crystals, which is important for the quality of the final
material since it influences the mechanical properties of the material.
– Mathematical characterization and optimization of the crystal size distribution, which is another important factor for the properties of the final
material.

There are still open problems; here we will focus on a particular problem,
namely the optimal control of the morphology, which will be investigated in
the following section.

Optimization of the Morphology In the following we set up the framework for the optimization of the morphology for a crystallization process,
restricting ourselves to the averaged model for simplicity. In principle, the
optimization problem could be formulated for the stochastic model, too, leading to a stochastic optimal control problem, which is investigated in present
and future research.

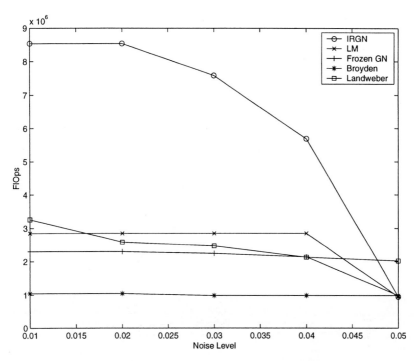

Fig. 5.34. Number of floating point operations $FlOps(k_*(\delta))$ needed until termination of the iteration vs. noise level δ.

As we have noticed above, an optimal morphology with respect to industrial needs is one with many small crystals with almost uniform size. Thus, it seems reasonable to set up an optimal control problem with an objective of the form

$$J(\gamma) = - \int_\Omega \gamma(x, S) \ dx + \frac{\kappa}{2} |\gamma(x, S)|^p \ dx, \qquad (5.134)$$

where γ denotes the density of interfaces [16] and S is the finite time in the crystallization process. The objective J is a pay-off between the first term corresponding to the criteria of many small crystals and the second criteria of uniformity. This objective is to be minimized over all admissible functions γ obtained from a crystallization model, where only the cooling temperature can be controlled over some part of the boundary in the time interval $(0, S)$. A general setup for this problem including also the case of spatial heterogeneity can be found in (cf. [10]), here we restrict our attention to the case of a small height of the sample, using the scaling limit deduced in Section 5.10, i.e., the model (5.126)-(5.128) with vanishing initial values for the degree of crystallinity ξ and the scaled surface density u and appropriately scaled initial values for the temperature θ.

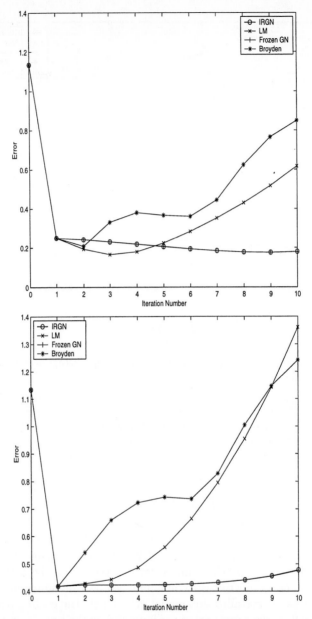

Fig. 5.35. Development of the error $||\widetilde{N}_k^\delta - Nt^0||$ during the iteration (k) for fixed noise level $\delta = 2\%$ (left) and $\delta = 5\%$ (right).

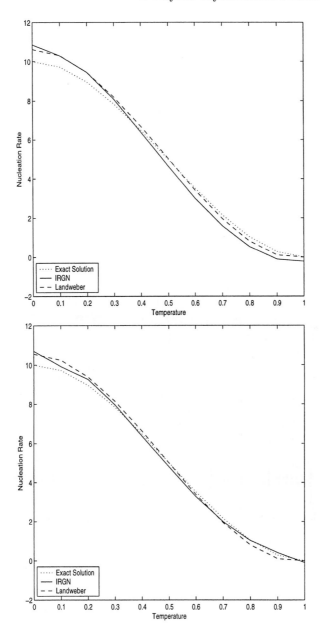

Fig. 5.36. Exact solution (dotted) and closest iterates obtained with the iteratively regularized Gauss-Newton method (solid) and the Landweber iteration (dashed), for noise level $\delta = 5\%$ (left) and $\delta = 2\%$ (right) .

The density of the interfaces in $\mathbb{R}^2$ denoted by $\gamma = \mu_{2,1}$ can be computed by :

$$\gamma(t) = c_{2,1} \int\limits_0^t \left[\int\limits_0^\tau \int\limits_s^\tau G(\sigma) \, d\sigma\alpha(s) \, ds \right]^2 G(\tau)p(\tau) \, d\tau, \qquad (5.135)$$

where $p := 1 - \xi$ is the density of the uncrystallized volume. The objective functional to be minimized can now be reduced due to spatial homogeneity to $J(\gamma) = -\gamma(S)$.

By defining the quantity φ via

$$\varphi(t) := \int\limits_0^t \int\limits_s^t G(\tau) \, d\tau\alpha(s) \, ds, \qquad (5.136)$$

and insert the equations (5.127),(5.128) we may rewrite the objective functional as

$$J(p, \varphi, \theta; u, q) = - \int\limits_0^{S00} p(t)\varphi(t)^2 a(\theta(t)) \, dt, \qquad (5.137)$$

and the equality constraints as

$$\dot{x} := \begin{pmatrix} \dot{p} \\ \dot{\varphi} \\ \dot{\theta} \\ \dot{u} \end{pmatrix} = \begin{pmatrix} -8\pi p\varphi a(\theta) \\ a(\theta)b(\theta) \\ \beta(u - \theta) + 8\pi L p\varphi a(\theta) \\ q \end{pmatrix} := F(p, \varphi, \theta; u, q) \quad (5.138)$$

$$\begin{pmatrix} p(0) \\ \varphi(0) \\ \theta(0) \\ u(0) \end{pmatrix} = \begin{pmatrix} 0 \\ 0 \\ \theta^0 \\ u^0 \end{pmatrix}, \qquad (5.139)$$

where u denotes the outside temperature, q the cooling rate, which is the effective control variable of the system, and $\beta = -2D\alpha$. In addition, there is a constraint for the control variable given by

$$a \leq q = \frac{du}{dt} \leq 0, \qquad (5.140)$$

which is a realistic bound on the cooling speed. We notice that appropriate choices of a and S will guarantee a certain degree of crystallinity at the final time. As an alternative one could also put the additional state constraint

$$p(S) \leq b \qquad (5.141)$$

for some value $b \in (0, 1)$.

The existence of a minimum can be shown by standard techniques for optimal control problem for ordinary differential equations (cf. [10]), further statements on its solution can be made by an inspection of the structure of the problem. The first important note about the structure is that the optimal control problem for the morphology is a separated problem, since the corresponding Hamiltonian

$$H(p, \varphi, \theta, u, q; \lambda) = p\varphi^2 a(\theta) + F(p, \varphi, \theta; u, q)^T \lambda. \qquad (5.142)$$

can be written in the form

$$H(p, \varphi, \theta, u, \lambda) = H_0(p, \varphi, \theta, \lambda) + q\lambda_4. \qquad (5.143)$$

As a direct consequence of the maximum principle in control theory, an optimal control q^* must satisfy

$$q^*(t) = \arg \min_{a \leq r \leq 0} \lambda_4^*(t) r, \qquad (5.144)$$

where λ^* is the corresponding Lagrange-multiplier. Since the optimization problems (5.144) are linear in each time step, we immediately obtain that the minimum is attained at the boundary if $\lambda_4^*(t) \neq 0$, i.e., we have

$$q^*(t) = \begin{array}{l} a \text{ if } \lambda_4^*(t) > 0 \\ 0 \text{ if } \lambda_4^*(t) < 0 \end{array} \qquad (5.145)$$

Hence, the optimal control is of bang-bang type, i.e., in a sequence of time intervals one should either apply maximal or minimal cooling to the sample.

In the absence of latent heat ($L = 0$) it seems obvious from the physics of the problem that maximal cooling ($q^* = a$ in $(0, S)$) leads to the highest density of interfaces, since it causes most nucleations. In presence of latent heat, the situation can differ, since with a too high number of nucleations the reheating effect can dominate in a time period and cause crystals to grow in absence of new nucleations, which leads to a lower density of interfaces. In any case, the result will depend strongly on the model parameters and has therefore to be determined by numerical methods in each specific case.

Acknowledgements

It is a pleasure to acknowledge important stimulating discussions with all members of the ECMI Special Interest Group on "Polymers" (www.mat.unimi.it/~polymer).

Particular thanks are due to Dr. S. Mazzullo of the "G.Natta" Research Centre of MONTELL-Italia. Financial support from the Italian Space Agency (ASI), Italian CNR (National Research Council) under project grant n. 98.03635.ST74, the Austrian Research Fund (FWF) under project grants F 13/08 and P 13478-INF, the EU-TMR Programme *DEIC* under project grant n. FMRXCT97-0117 and the Italian MURST/cofin. Programme " *Stochastic processes with spatial structure*" is gratefully acknowledged.

References

1. G. Alfonso, Polimeri cristallini, in *Fondamenti di Scienza dei Polimeri*, (M. Guaita et al. Eds.), Pacini, Pisa, 1998.
2. L.Ambrosio, N.Fusco, D.Pallara, *Functions of Bounded Variation and Free Discontinuity Problems*, Clarendon Press Oxford, 2000.
3. M.Avrami, Kinetics of phase change. Part I, *J. Chem. Phys.*, **7**, 1103-112 (1939).
4. P.Brémaud, *Point Processes and Queues, Martingale Dynamics*, Springer-Verlag, New York, 1981.
5. M.Burger, *Iterative regularization of an identification problem arising in polymer crystallization*, SFB Report 99-21 (University of Linz, 1999), and SIAM J. of Numerical Analysis (submitted).
6. M.Burger, *Direct and Inverse Problems in Polymer Crystallization Processes*, PhD-Thesis, University of Linz, 2000.
7. M. Burger, V. Capasso, Mathematical modelling and simulation of non-isothermal crystallization of polymers. *Math. Models and Methods in Appl. Sciences*, **11** (2001), 1029–1053.
8. M.Burger, V.Capasso, G.Eder, Modelling crystallization of polymers in temperature fields *Z. Angew. Math. Mech.* **82** (2002), 51-63.
9. M.Burger, V.Capasso, H.W.Engl, *Inverse problems related to crystallization of polymers*, Inverse Problems **15** (1999), 155-173.
10. M.Burger, V.Capasso, A.Micheletti, *Optimal control of polymer morphologies*, *Quaderno n. 11/2002*, Dip. di Matematica, Università di Milano, 2002.
11. M. Burger, V. Capasso, C. Salani, Modelling multidimensional crystallization of polymers in interaction with heat fransfer, *Nonlinear Analysis: Real World Application*, **3** (2002), 139–160.
12. V.Capasso, A.Micheletti, Stochastic geometry of birth-and-growth processes, *Quaderno n. 20/1997*, Dip. di Matematica, Università di Milano, 1997.
13. V.Capasso, A.Micheletti, Birth-and-Growth stochastic processes modelling polymer crystallization *Technical Report* **14**, *Industrial Mathematics Institute, J.Kepler Universität, Linz, 1997.*
14. V. Capasso, A. Micheletti, Local spherical contact distribution function and local mean densities for inhomogeneous random sets, *Stochastics and Stoch. Rep.*, **71**, (2000),51-67.
15. V.Capasso, A.Micheletti, The hazard function of an inhomogeneous birth-and-growth process, *Quaderno n.39/2001*, Dip. di Matematica, Universita' di Milano.
16. V.Capasso, A.Micheletti, M.Burger, Densities of n-facets of incomplete Johnson-Mehl tessellations generated by inhomogeneous birth-and-growth processes. Quaderno n.38/2001, *Dip. di Matematica, Università di Milano.*
17. *V.Capasso, A.Micheletti, M.De Giosa, R.Mininni, Stochastic modelling and statistics of polymer crystallization processes, Surv. Math. Ind., 6, (1996),109-132.*
18. *V.Capasso, C.Salani, Stochastic-birth-and-growth processes modelling crystallization of polymers with spatially heterogeneous parameters, Nonlinear Analysis: Real World Application, 1, (2000) pp.485-498.*
19. *N.A.C. Cressie, Statistics for Spatial Data, Wiley, New York, 1993.*

20. *G.Eder*, Mathematical modelling of crystallization processes as occurring in polymer processing, *Nonlinear Analysis* **30** *(1997), 3807-3815.*

21. *G.Eder, H.Janeschitz-Kriegl*, Structure development during processing: crystallization, in *Materials Science and Technology* , *Vol.18 (H.Meijer, Ed.), Verlag Chemie, Weinheim, 1997.*

22. *H.W.Engl, O.Scherzer*, Convergence rate results for iterative methods for solving nonlinear ill-posed problems, *in: D.Colton, H.W.Engl, J.McLaughlin, A.Louis, W.Rundell, eds., Surveys on Solution Methods for Inverse Problems (Springer, Vienna, New York, 2000).*

23. *V.R.Evans*, The laws of expanding circles and spheres in relation to the lateral growth rate of surface films and the grain-size of metals, *Trans. Faraday Soc.,* **41** *(1945), 365-374.*

24. *L.H.Friedman, D.C.Chrzan*, Scaling theory of the Hall-Petch relation for multilayers, *Phys. Rev. Letters*, **81** *(1998), 2715-2719.*

25. *A.Friedman, J.J.L.Velasquez*, A free boundary problem associated with crystallization of polymers in a temperature field, *Indiana Univ. Math. Journal,* **50** *(2001), 1609–1650.*

26. *R.V. Gamkrelidze, Principles of Optimal Control Theory Plenum Press, N.Y., London, 1976.*

27. *B.G.Ivanoff, E.Merzbach*, A martingale characterization of the set-indexed Poisson Process, *Stochastics and Stochastics Report* **51** *(1994), 69-82.*

28. *H.Janeschitz-Kriegl, E.Ratajski, H.Wippel*, The physics of athermal nuclei in polymer crystallization, *Colloid & Polymer Science* **277** *(1999), 217-226.*

29. *W.A.Johnson, R.F.Mehl*, Reaction Kinetics in processes of nucleation and growth, Trans. A.I.M.M.E.,**135**, *416-458 (1939).*

30. *J.D.Kalbfleisch, R.L.Prentice, The Statistical Analysis of Failure Time Data, John Wiley and Sons, New York, 1980.*

31. *A.N.Kolmogorov*, On the statistical theory of the crystallization of metals, Bull. Acad.Sci.USSR, Math.Ser. **1** *(1937), 355-359.*

32. *D.Kröner*, Numerical Schemes for Conservation Laws, *Wiley & Teubner, Chichester, Stuttgart, 1997.*

33. *G.Knowles*, An Introduction to Applied Optimal Control, *Academic Press, New York, 1981.*

34. *G.Last, A.Brandt, Marked Point Processes on the Real Line. A Dynamic Approach, Springer, New York, 1995.*

35. *R.LeVeque*, Numerical Methods for Conservation Laws, *Birkhäuser, Basel, Boston, Berlin, 1990.*

36. *G.Matheron, Random Sets and Integral Geometry, Wiley, New York, 1975.*

37. *S.Mazzullo, M.Paolini, C.Verdi*, Polymer crystallization and processing: free boundary problems and their numerical approximation, *Math. Engineering in Industry* **2** *(1989), 219-232.*

38. *J.L.Meijering*, Interface area, edge length, and number of vertices in crystal aggregates with random nucleation, Philips Res. Rep., **8**, *270-290 (1953).*

39. *A. Micheletti*,The surface density of a random Johnson-Mehl tessellation, Quaderno n. 17/2001, *Dip. di Matematica, Università di Milano, 2001.*

40. *A.Micheletti, M.Burger*, Stochastic and deterministic simulation of nisothermal crystallization of polymers, J. Math. Chem., **30** *(2001), 169-193.*

41. *A.Micheletti, V.Capasso*, The stochastic geometry of polymer crystallization processes, *Stoch. Anal. Appl.* **15** *(1997), 355-373.*

42. A. Micheletti, V. Capasso, G. Eder The density of the n-facets of an incomplete Johnson-Mehl tessellation. Preprint of the Institute for Industrial Mathematics, Johannes-Kepler University of Linz (Austria), 1997.

43. J. Moller, Random Johnson-Mehl tessellations, Adv. Appl. Prob., **24** , 814-844 (1992).

44. H.Niessner, Stability of Lax-Wendroff methods extended to deal with source terms, ZAMM 77 (1997), Suppl. 2, S637-S638.

45. T.Ohta, Y.Enomoto, R.Kato, Domain growth with time dependent front velocity in one dimension, 1990, preprint.

46. S.Ohta,T.Ohta,K.Kawasaki, Domain growth in systems with multiple degenerate ground states, Physica, **140A**, 478-505 (1987).

47. E.Ratajski, H.Janeschitz-Kriegl, How to determine high growth speeds in polymer crystallization, Colloid Polym. Sci. **274** (1996), 938 - 951.

48. M.Renardy, R.C.Rogers, An Introduction to Partial Differential Equations, Springer, New York, 1993.

49. S.M.Ross, Simulation. 2nd ed. Academic Press, San Diego, 1997.

50. C.Salani, Crystallization of polymers with thermal heterogeneities, ECMI Thesis, Linz (1997).

51. C. Salani, On the mathematics of polymer crystallization processes: stochastic and deterministic models., Ph.D. Thesis, University of Milano, Italy, 2000.

52. G.E.W.Schulze, T.R.Naujeck, A growing 2D spherulite and calculus of variations, Colloid & Polymer Science **269** (1991), 689-703.

53. J.A.Sethian, Level Set Methods. Evolving Interfaces in Geometry, Fluid Mechanics, Computer Vision, and Materials Science, Cambridge Univ. Press, Cambridge, 1996.

54. J.Smoller, Shock Waves and Reaction-Diffusion Equations, Springer, New York, Berlin, Heidelberg, 1983.

55. D.Stoyan, W.S.Kendall, J.Mecke, Stochastic Geometry and its Application, John Wiley & Sons, New York, 1995.

56. P.Supaphol, J.E.Spruiell, Thermal properties and isothermal crystallization of syndiodactic polypropylenes: differential scanning calorimetry and overall crystallization kinetics, Journ. Appl. Polym. Sci. **75** (2000), 44-59.

57. J.E.Taylor, J.W.Cahn, C.A.Handwerker, Geometric models of crystal growth, Acta metall. mater. **40** (1992), 1443-1475.

58. D.W.Van Krevelen, Properties of Polymers, 5th ed., Elsevier, Amsterdam, 1990.

59. Y.Zhang, B.Tabarrok, Modifications to the Lax-Wendroff scheme for hyperbolic systems with source terms, Int. J. Numer. Methods Eng. **44** (1999), 27-40.

6 Polymer Crystallization Processes via Many Particle Systems

Vincenzo Capasso[124], Daniela Morale[13], and Claudia Salani[12]

[1] MIRIAM - MIlan Research Centre for Industrial and Applied Mathematics, University of Milano, via Saldini 50, I-20133 Milano, Italy
[2] Department of Mathematics, University of Milano, Italy
[3] Department of Mathematics, University of Torino, Italy
[4] capasso@mat.unimi.it

6.1 Introduction

In this chapter we introduce a new approach that thanks to the multiple-scale structure, allows us to use mathematical techniques of averaging at the lower scale.

As already mentioned crystallization is considered as a superposition of three processes, each of them being of stochastic nature in principle:

- *Nucleation*, which is the process of appearance of spots in space and time (called nuclei), which form the starting points of growing crystalline objects (called crystallites).
- *Growth*, which is the process of space-filling by individual crystallites.
- *Perfection*, which is the process of improvement of the interior crystalline structure of the crystallites, also called secondary crystallization.

Until now almost always the process of perfection has been neglected in modelling, thus assuming that the crystalline structure of the crystallites remains constant after their formation. This simplifying assumption leads to a one-to-one correspondence between the degree of crystallinity and the degree of space-filling. However, one must keep in mind that, in particular for some polymers, secondary crystallization may not be negligible. Indeed the crystallization of polymers is never complete: the properties of crystalline polymeric materials are in between the ones of a material perfectly ordered and the ones for a completely amorphous material. The fraction of crystallized material defines the *degree of crystallinity*. For typical semi-crystalline polymers the degree of crystallinity is between 20% and 60%.

In the model we propose here we take into account perfection and saturation effects. We propose a stochastic model for the processes of nucleation and crystallization, strongly coupled with the underlying temperature field. Thus we consider again a multiple scale, as in PART III, Chapter 1, according to which the temperature field evolves at a larger scale, according to a classical diffusion equation coupled with a lower scale point processes modelling the crystallization process. A fully deterministic continuum model can be recovered by considering the usual asymptotic case in which nucleation occurs much faster than growth in terms of a suitable scaling parameter. As a consequence of the multiple scales, we may model the crystallization process as a birth-and-growth process localized at points of birth in $\mathbb{R}^d$, $(d = 1, 2, 3)$. In this way an explicit geometrical dimension of growth is lost in the model; it will be recovered by segmentation techniques out of the results of the simulation.

6.2 The Nucleation Process

Let $N \in \mathbb{N}$, a scaling parameter, be fixed. As in Part III, Chapter 1, the nucleation process is modelled as a stochastic spatially marked point process Φ_N on an underlying probability space $(\Omega, \mathcal{F}, P)$ with marks in the physical space E, a compact subset of $\mathbb{R}^d$, $d = 1, 2, 3$.

We denote by $\mathcal{B}(\mathbb{R}^+)$ and by $\mathcal{E}$ the σ-algebra of Borelians on $\mathbb{R}^+$ and E, respectively.

The marked point process (MPP) Φ_N is a random measure given by

$$\Phi_N = \sum_{n=1}^{\infty} \epsilon_{(t_n, X_N^n)}, \tag{6.1}$$

where

- t_n $(t_n < t_{n+1})$is an $\mathbb{R}^+$-valued random variable representing the time of birth of the n-th nucleus (crystal);
- X_N^n is an E-valued random variable representing the spatial location of nucleus born at time s at time t_n,
- $\epsilon_{(t,x)}$,, is the Dirac measure on $\mathcal{B}(\mathbb{R}^+) \times \mathcal{E}$ such that for any $A \in \mathcal{B}(\mathbb{R}^+) \times \mathcal{E}$

$$\Phi_N(A) := \sum_n \epsilon_{(t_n, X_N^n)}(A)$$
$$= \text{card}\{n : (t_n, X_N^n) \in A\}, \tag{6.2}$$

which says that $\Phi_N(A)$ is the random variable which counts the marked points in A. By definition $\Phi_N(\{0\} \times E) = 0$.

The "underlying counting process" associated with Φ_N is denoted by

$$\Lambda_N(t) := \Phi_N([0, t] \times E). \tag{6.3}$$

It counts the total number of nuclei born in the whole space region E up to the time $t \in \mathbb{R}^+$.

The function

$$t \mapsto \Lambda_N(t)$$

takes values in $Z_+ = \{0, 1, 2, \dots\}$ and is right continuous, non decreasing and piecewise constant with jumps of size 1.

We have $t_n = inf\{t \geq 0 : \Lambda_N(t) \geq n\}$; in particular

$$\{\Lambda_N(t) \geq n\} = \{t_n \leq t\}$$
$$\{\Lambda_N(t) = n\} = \{t_n \leq t < t_{n+1}\} \tag{6.4}$$

Now we define the history of the process that reflects the accumulated information available at a certain time $t \in \mathbb{R}_+$, i.e. the following σ-algebra, generated by the process Φ_N:

$$\mathcal{F}_t^{\Phi_N} := \sigma(\Phi_N((a,b] \times B) : a < b \leq t, B \in \mathcal{E}); \tag{6.5}$$

$$\mathcal{F}_{t-}^{\Phi_N} := \sigma(\Phi_N((a,b] \times B) : a < b < t, B \in \mathcal{E})$$
$$= \sigma(\{T \leq s\} \cap \{X \in B\} : s < t, B \in \mathcal{E}) \tag{6.6}$$

$\{\mathcal{F}_t^{\Phi_N}\}_{\{t \geq 0\}}$ represents the history, i.e. the dynamic evolution of the marked point process.

We define

$$A_N(t, B) = \int_B \alpha_N(t, x) dx, \tag{6.7}$$

where $\alpha_N(t, x)$ is the birth rate (per unit time and unit volume) with which a new nucleus is created during the infinitesimal time interval $[t, t+dt[$ in the infinitesimal region $[x, x+dx]$. So

$$A_N(t, B) = \lim_{dt \to 0} \frac{1}{dt} P\big(\Phi_N([t, t+dt) \times B) > 0 | \mathcal{F}_{t-}^{\Phi_N}\big), \tag{6.8}$$

is the birth rate in B. The family $\{A_N(t, B) : t \geq 0, B \in E\}$ is known as the *stochastic intensity* kernel of Φ_N [6].

Formally we can say that $P\big(\Phi_N(dt \times B) > 0 | \mathcal{F}_{t-}^{\Phi_N}\big) = A_N(t, B)dt + o(dt)$.

The random variable $\alpha_N(t, x)$ is a finite $\{\mathcal{F}_t\}$-predictable kernel from $\mathbb{R}^+ \times \Omega$ to E such that, for any $\mathcal{F}_t^{\Phi_N}$-predictable function $f(t, \cdot)$

$$E\left[\int f(t, x)\Phi_N(dt \times dx)\right] = E\left[\int \int f(t, x) A_N(t, dx) dt\right]$$
$$= E\left[\int \int f(t, x)\alpha_N(t, x)\, dt\, dx\right], \tag{6.9}$$

i.e.

$$\nu(dt \times dx) = \alpha_N(t, x)\, dt\, dx \tag{6.10}$$

is a compensator of Φ_N.

We may recall the following martingale characterization of a compensator.

Theorem 3. [6] *Let ν be a predictable measure on $\mathbb{R}^+ \times E$. Then ν is a compensator of Φ_N iff the process $\{\Phi_N((0, t \wedge t_n], B) - \nu(t \wedge t_n, B) : t \geq 0\}$ is $\forall n \in \mathbb{N}, B \in \mathcal{E}$ a uniformly integrable martingale, i.e.*

$$E\left[\Phi_N((s \wedge t_n, t \wedge t_n], B)|\mathcal{F}_s\right] = E\left[\int_{s \wedge t_n}^{t \wedge t_n} A_N(u, B)\, du|\mathcal{F}_s\right]$$

$$= E\left[\int_{s \wedge t_n}^{t \wedge t_n} \int_B \alpha_N(u, x)\, du\, dx|\mathcal{F}_s\right]. \quad (6.11)$$

In our case we have in particular

$$E\left[\Lambda_N(t)\right] = E\left[\Phi_N((0, t] \times E)\right]$$

$$= E\left[\int_0^t \int_E \alpha_N(u, x)\, du\, dx\right]. \quad (6.12)$$

So if we define $\alpha_N(t, x)$ in such a way that the last integral in (6.12) is finite, i.e. in such a way that

$$\alpha_N(u, \cdot) \in \mathcal{L}^1(E), \quad \forall u \in \mathbb{R}^+ \quad (6.13)$$

$$0 < \int_0^t du \int_E \alpha_N(u, x)\, dx < \infty, \quad \forall t > 0, \quad (6.14)$$

we have [6] (Theo. 4.1.10) that the compensator ν of Φ_N is characterized by its predictability and the fact that

$$\{\Phi_N((0, t] \times B) - \nu(t, B) : t \geq 0\} \quad (6.15)$$

is a martingale for any $B \in \mathcal{B}$.

6.3 Growth of Crystals

We now describe the growth process of nuclei as follows: for each $k \leq \Lambda_N(t)$, $k \in \mathbb{Z}_+$ at time t we consider a counting process $g_N^k(t)$ defined by

$$g_N^k(t) := g_N^k([0, t]) = \sum_{m \geq 1} \epsilon_{\tilde{t}_m^k}(]t \wedge t_k, t]), \quad (6.16)$$

where $\{\tilde{t}_m^k\}$ are the times when the growth events occur. So (6.16) counts the number of jumps of the mass $Y_N^k(t)$ of the k-th crystal up to time t at location $X_N^k \in E$, after the time of birth T_k. If we assume that the initial mass is v_0; i.e. $Y_N^k(T_k) = v_0$, and a jump size $1/\lambda_N$

$$Y_N^k(t) = v_o + \frac{1}{\lambda_N} g_N^k(t)$$

$$= \frac{1}{\lambda_N} \sum_{m \geq 1} \epsilon_{\tilde{t}_m^k}(]t \wedge t_k, t]), \quad (6.17)$$

If we denote by $G(t, X_N^k, Y_N^k(t^-))$ the stochastic intensity of the counting process (6.16), we have

$$E\left[\int_{t_k}^t f(s, X_N^k)g_N^k(ds)\right] = E\left[\int_{t_k}^t f(s, X_N^k)G(t, X_N^k, Y_N(s^-))ds\right] \quad (6.18)$$

6.4 The Mass Distribution of Crystals

The macroscopic description of the whole process is expressed in terms of the empirical spatial distribution of crystals $\mathbb{X}_N(t)$ and the empirical spatial distribution of mass $\mathbb{Y}_N(t)$

$$\mathbb{X}_N(t) = \frac{1}{N}\sum_{m=1}^{\Lambda_N(t)} \epsilon_{X_N^m} = \frac{1}{N}\Phi_N([0,t] \times \cdot) \quad (6.19)$$

$$\mathbb{Y}_N(t) = \frac{1}{N}\sum_{m=1}^{\Lambda_N(t)} Y_N^m(t)\epsilon_{X_N^m}$$

$$= Y_N(0) + \frac{1}{N}\sum_{k:X_N^k \in B} \frac{1}{\lambda_N} g_N^k(t) + \frac{1}{N} v_0\, \Phi_N([0,t] \times \cdot), \quad (6.20)$$

where ϵ_x is the localizing measure at x: $\forall B \in \mathcal{B}_{\mathbb{R}^d}$,

$$\epsilon_x(B) = \begin{cases} 1, & \text{if } x \in B, \\ 0, & \text{otherwise.} \end{cases}$$

The "history" of the whole system, that is the σ-algebras generated by the processes (6.1) and (6.16) is

$$\mathcal{F}_t := \mathcal{F}_t^{\Phi_N} \vee \sigma(Y_N^m(s)(B), m \le \Lambda_N(s),\ s \le t, B \in \mathcal{E}); \quad (6.21)$$

$$\mathcal{F}_{t-} := \mathcal{F}_t^{\Phi_N} \vee \sigma(Y_N^m(s)(B), m \le \Lambda_N(s),\ s < t, B \in \mathcal{E}); \quad (6.22)$$

6.5 The Rates

Consider ϕ, ψ and γ probability densities on E, and define their scaled counterparts by

$$\phi_N(x) = N^b\phi(N^{b/d}x), \qquad 0 < b < 1,$$
$$\psi_N(x) = N^{b'}\psi(N^{b'/d}x), \qquad 0 < b' < 1,$$
$$\gamma_N(x) = N^{b''}\gamma(N^{b''/d}x), \qquad 0 < b'' < 1. \quad (6.23)$$

These functions will describe qualitatively the strength and the range of the interactions; because of the scaling, the range of interaction decreases and the strength increases as the scaling parameter N increases. Because of the choice $0 < b, b', b'' < 1$ we say that our system is characterized by moderate interactions: each nucleus is influenced by the others standing within a range which is macroscopically small, but microscopically large. Indeed, any particle can interact with other particles located in a volume of order $\sim N^{-b/d}$, which is sufficiently small with respect to the macroscale (order of the size of the whole space $\sim \mathcal{O}(1)$) but large enough with respect to the microscale (typical distance between particles $\sim N^{-1/d}$). This is the mathematical description of the concept of mesoscale. From (6.23), we have

$$\lim_{N \to \infty} \phi_N = \lim_{N \to \infty} \psi_N = \lim_{N \to \infty} \gamma_N = \delta_0; \qquad (6.24)$$

the range of interaction of each nucleus becomes smaller and smaller as N tends to infinity, but thanks to the scaling of the nucleation and growth rates the convergence is sufficiently slow so that for any N the range is large enough to contain a number of nuclei sufficient to apply a law of large numbers.

The birth rate

We may include the following assumptions for the process of nucleation, the process of appearance of nuclei in space and time, which form the starting points of growing crystals:

i) aggregation: the probability that a spot appears at a certain point depends on the crystallized mass in a small neighborhood; more nuclei, higher is the probability of birth of a new one;
ii) saturation: new nuclei appear till the crystalline mass reaches some saturation level;
iii) spatial heterogeneities exist, due only to the heat transfer in the material and the nucleation rate may depend on the temperature T_N via a suitable function.

Thus the nucleation rate (per unit time and unit volume) is assumed to be of the following form

$$\alpha_N(t, x) = N\kappa((\mathbb{Y}_N(t) * \phi_N)(x))b_b(T_N(t, x)). \qquad (6.25)$$

By $\alpha_N(t, \cdot)$ we denote the rate at time t, conditional upon $\mathcal{F}_{t-}$.

The function $\kappa : [0, \infty) \to [0, 1]$ is such that $\kappa(0) > 0$, and decreases to 0 at infinity. It may be chosen to be increasing in the time interval $[0, \bar{K}]$ and decreasing to 0 in $[\bar{K}, \infty)$, for some $\bar{K} \in \mathbb{R}_+$, in order to model the aggregation and saturation phenomena described at points i) and ii). The function $b_b : \mathbb{R}_+ \to [0, \infty)$ describes the dependence of the nucleation rate on temperature.

Finally, the factor N indicates an increasing speed of nucleation with respect to the typical scale of the process.

The growth rate

As already mentioned, instead of the geometrical growth of a spherulite, i.e. the process of space-filling by each individual nucleus , we consider the increasing of mass. We assume the following assumptions for the process of growth

iv) at birth each nucleus has an initial mass $v_0 > 0$ (its critical size);
v) saturation: crystals grow till there is certain plateau of the mass existing within a small neighborhood;
vi) bigger the crystals are, more often they grow;
vii) the growth rate may depend on the temperature T_N;

Thus, given the history of the process $\mathcal{F}_{t-}$ defined as in (6.22), the mass $v \geq v_0$ of an existing nucleus at point $x \in E$, at time $t > 0$ may grow by a jump $\dfrac{1}{\lambda_N}$ with a conditional rate (per unit time)

$$G_N(t, x, v) = \lambda_N \beta((\mathbb{Y}_N(t) * \psi_N)(x), v) b_g(T_N(t, x)), \qquad (6.26)$$

where $\lambda_N \gg N$.

We assume that the function $\beta : (x, v) \in \mathbb{R}_+ \times [0, \infty) \to [0, 1]$ is decreasing with respect to x and increasing with respect to v to model v) and vi). The function $b_g : \mathbb{R}_+ \to [0, \infty)$ describes the dependence of the growth rate on temperature.

The assumption $\lambda_N \gg N$ makes the increments of the processes very small, so eventually the birth process is, for large N, faster than the growth of mass.

The temperature

Both rates (6.25) and (6.26) depend on the underlying temperature field T_N. So we close the system by considering the time evolution of the temperature.

We assume that for every N, the temperature T_N is a random function satisfying

$$\frac{\partial}{\partial t} T_N(t, x) = \sigma \Delta T_N(t, x) + a_g \, b_g(T_N(t, x))$$

$$\times \frac{1}{N} \sum_{k=1}^{\Lambda_N(t)} Y_N^k(t) \, \tilde{\beta}((\mathbb{Y}_N(t) * \psi_N)(X_N^k)) \gamma_N(x - X_N^k). \quad (6.27)$$

Equation (6.27) derives from (5.78) by considering all parameters being constant (a better modelling should involve the dependence of the diffusion coefficient $\sigma = \kappa/\rho c$ on the process itself). Here the growth process, as a variation of the crystalline phase, is responsible of the production of latent heat. In (6.27) the source term considers the production of latent heat at $x \in E$ due to the phase change in a suitable neighborhood (via the kernel γ_N) of the same point x.

Mathematical assumptions

All the processes describing nucleation and crystallization phenomena are cadlag(right continuous with left limits).

For any fixed time t, conditional upon the past history, the different elementary events described above, like birth and growth of a crystal, are independent of each other in the infinitesimal time interval $[t, t + dt[$. As a consequence, almost surely, two or more elementary events cannot happen simultaneously, and at the same point.

Furthermore, let us consider the following assumptions on the functions in (6.25) and (6.26)

$$b_b, b_g \in C_b^\infty(\mathbb{R}), \tag{6.28}$$

$$\kappa \in C_b^\infty(\mathbb{R}_+), \tag{6.29}$$

$$\beta \in C^\infty(\mathbb{R}_+ \times \mathbb{R}_+). \tag{6.30}$$

We consider the following factorization for $\beta(u, v)$

$$\beta(u, v) = \tilde{\beta}(u)\tilde{g}(v). \tag{6.31}$$

In particular we choose

$$\beta(u, v) = \tilde{\beta}(u)v, \tag{6.32}$$

where $\tilde{\beta}(u) \in C_b^\infty(\mathbb{R}_+)$.

Furthermore we consider the following assumption for $\tilde{\beta}(u)$

$$\tilde{\beta}(u) = \mathcal{O}\left(\frac{1}{u}\right), \quad u \in \mathbb{R}_+, \tag{6.33}$$

so that $\lim_{u\to\infty} \tilde{\beta}(u) = 0$.

Different modelling choices may be done that do not affect the mathematical treatment that follows.

6.6 The Time Evolution of the Processes Λ_N and $\mathbb{Y}_N$

Let us rescale the counting process (6.3) with respect the scaling parameter N, i.e.

$$Z_N(t) := \frac{\Lambda_N(t)}{N}. \qquad (6.34)$$

We obtain a jump process with jumps of size $1/N$ occurring at a rate $\Lambda_N(t)$ defined as in (6.7), i.e.

$$E\left[\int_0^t Z_N(s)ds\right] = \frac{1}{N}E\left[\int_0^t \int_E \alpha_N(u,x)\,du\,dx\right]. \qquad (6.35)$$

Obviously $Z_N(t)$ is coupled with the process $Y_N(t)$, as $\Lambda_N(t)$ was. We need to analyze the joint time evolution of the process

$$(Z_N(t), Y_N(t)) \qquad (6.36)$$

coupled with the continuous field $T_N(t)$ at N fixed. Note that (6.35) expresses the coupling of a rescaled counting process with an empirical measure which gives an Eulerian description of the system, i.e. instead of considering the single nuclei with their mass, we focus our attention on the mass distribution of crystals.

In the next section we see how this evolution changes as the size of the system N increases to infinity.

6.6.1 The Time Evolution of $\mathbb{Z}_N$

From (6.15) we have that

$$\{\Lambda_N(t) - \nu(t, E) : t \geq 0\} \qquad (6.37)$$

is a martingale. We may rewrite

$$Z(t) = Z(0) + \int_0^t \int_E \frac{1}{N}\alpha_N(s,x)dxds + M_N(t), \qquad (6.38)$$

where $M_N(t)$ is a zero mean martingale [2], [3] and its quadratic variation is given by the compensator $\int_E \alpha_N(s,x)dsdx$, i.e.

$$E\left[M_N^2(t)\right] = \int_0^t \int_E \frac{1}{N^2}\alpha_N(s,x)dxds. \qquad (6.39)$$

6.6.2 The Time Evolution of $\mathbb{Y}_N$

For any $k = 1, 2, \ldots, \Lambda_N(t)$, from (6.20) for $f \in C_b^\infty(\mathbb{R}^d) \cap L^2(\mathbb{R}^d)$, we have

$$\int_E f(x)\mathbb{Y}_N(t)(dx) = \int_E f(x)\mathbb{Y}_N(0)(dx)$$

$$+ \int_0^t \left[\frac{1}{N}\sum_{k=1}^{\Lambda_N(s)} f(X_N^k)\frac{1}{\lambda_N}g_N^k(ds) + \int_E f(x)v_0\frac{1}{N}\Phi_N([0,t] \times \cdot)\right]. \qquad (6.40)$$

We obtain from (6.9) and (6.18)

$$\int_E f(x) \mathbb{Y}_N(t)(dx) = \int_E f(x) \mathbb{Y}_N(0)(dx)$$

$$+ \int_0^t ds \left[\frac{1}{N} \sum_{k=1}^{\Lambda_N(s)} f(X_N^k)\tilde{\beta}((\mathbb{Y}_N(s) * \psi_N)(X_N^k))Y_N^k(s)b_g(T_N(t, X_N^k)) \right.$$

$$\left. + \int_E v_0 f(x)\kappa((\mathbb{Y}_N(s) * \phi_N)(x))b_b(T_N(s, x))dx \right] + M_{2,N}(f, t), \quad (6.41)$$

where $M_{2,N}(f, t)$ is the martingale term whose quadratic variation is

$$E\left[M_{2,N}^2(f, t)\right] = \int_0^t \left[\frac{1}{(\lambda_N N)^2} \sum_k |f^2(X_N^k)| G_N(t, X_N^k, Y_N^k(t)) \right.$$

$$\left. + \frac{v_0^2}{N^2} \int_E f^2(x)\alpha_N(s, x)dx \right] ds. \quad (6.42)$$

6.6.3 Simulation of the Many-particle Model

We performed stochastic simulations of the many-particle system both in the isothermal ($a_g = 0$) and in the not isothermal ($a_g \neq 0$) cases, in the window $E = [0, 1] \times [0, 1]$. Dependence on temperature is chosen in the following way [5]:

$$b_b(T) = N_{\text{ref}} \exp(-\beta_N(T - T_{\text{ref}})), \quad (6.43)$$

$$b_g(T) = G_{\text{ref}} \exp(-\beta_G(T - T_{\text{ref}})), \quad (6.44)$$

and, since the functions of y decrease to zero when y goes to 1, we choose the functions κ and $\tilde{\beta}$

$$\kappa(y) = \begin{cases} y^2 - 2y + 1 & 0 \leq y \leq 1 \\ 0 & y > 1 \end{cases}; \quad (6.45)$$

$$\tilde{\beta}(y) = \begin{cases} 1 - y & 0 \leq y \leq 1 \\ 0 & y > 1 \end{cases} \quad (6.46)$$

with coefficients as in Table 6.1.

In the isothermal case even for a low number of particles ($N = 650$) we can see a uniform nucleation with uniform increase of mass, as well (cf. Fig. 6.1). The colour segmentations shows both geometrical and density growth of crystals.

Table 6.1. Parameters values used in simulations.

Symbol	Value	Symbol	Value
N_{ref}	$20\ [s^{-1}]$	σ	$0.1\ [m^2/s]$
G_{ref}	$5\ [m/s]$	a_g	$2500\ [^\circ C]$
T_{ref}	$0\ [^\circ C]$	v_0	0.01
β_N	$-0.1\ [^\circ C^{-1}]$	$T_{\max}$	$100\ [^\circ C]$
β_G	$-0.1\ [^\circ C^{-1}]$	$T_{\min}$	$50\ [^\circ C]$

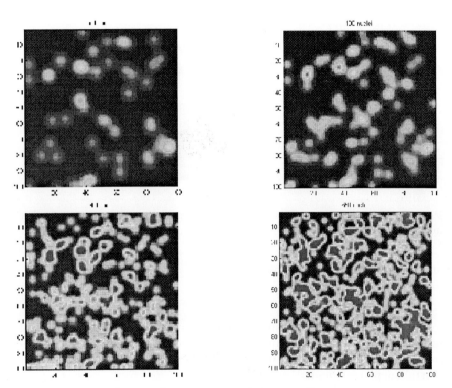

Fig. 6.1. Particle System. Time evolution of the crystallization in the isothermal case. See also Color Plate 6

If we consider the non isothermal case, the situation changes. Indeed, we simulated the crystallization process by a many-particle system of size $M = 10^4$ in the domain $E = [0,1] \times [0,1] \subset \mathbb{R}^2$, by finite difference methods. The initial condition for temperature is

$$T(0, x_1, x_2) = T_{\max} - x_1(T_{\max} - T_{min}),$$

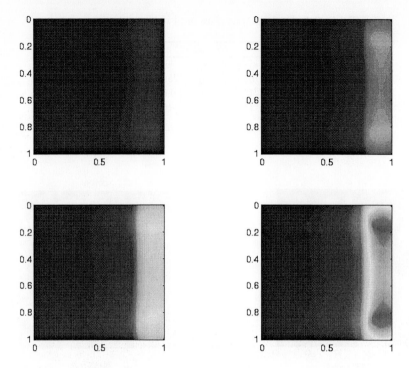

Fig. 6.2. Particle System. Time evolution of the crystallization in the not isothermal case. See also Color Plate 7

and Dirichlet boundary condition

$$T(t, x_1, x_2) = T_{max} - x_1(T_{max} - T_{min}), \qquad t > 0, (x_1, x_2) \in \partial E. \qquad (6.47)$$

From (6.2), (6.3) clearly shows the qualitative behaviour of the crystallization process. Crystallization starts in the coolest part of the sample and from there it expands in the remaining available space. The advance of the crystalline phase is under investigation.

6.7 Heuristic Derivation of a Continuum Dynamics

In the previous sections both (discrete) Lagrangian and Eulerian descriptions of the nucleation and growth processes coupled with the temperature field have been presented. Now we show how an asymptotic (with respect to N) continuum description can be obtained. Indeed, under typical industrial conditions we may assume that nucleation occurs much faster than growth. This is important since in order to justify a continuum model the number of nuclei (particles) should be large enough so that at the mesoscale level we

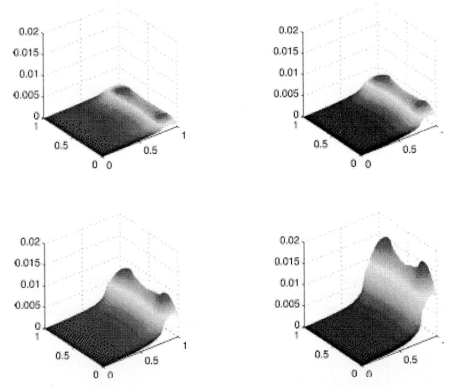

Fig. 6.3. Particle System. Time evolution of the crystallization in the not isothermal case: density profile.

can perfom a "law of large number". (For m athematical details we refer to [8]).

As a prerequisite for the convergence of the processes $Z_N(t)$ and $Y_N(t)$ to deterministic limits, we have to assure that also the initial measures $Z_N(0)$ and $Y_N(0)$ converge to non random limits. We assume

$$\lim_{N \to \infty} Z_N(0) = z_0 \tag{6.48}$$

$$\lim_{N \to \infty} \langle Y_N(0), f \rangle = \langle y_0, f \rangle, f \in C_b(\mathbb{R}^d), \tag{6.49}$$

where $z_0 \geq 0, \lambda \in Z^+$ and y_0 is an $\mathbb{R}^d$-valued regular function.

These conditions include "crude" convergence; for a rigorous derivation further assumptions specification on the type of convergence are required. In this formal setting, let us assume that $Z_N(t)$ and the measure $\mathbb{Y}_N(t)$ admit a limit as $N \to \infty$. Furthermore the two limit measures are assumed to be absolutely continuous with respect to the Lebesgue measure, i.e.

$$\lim_{N\to\infty} Z_N(t) = z(t) \qquad\qquad t > 0, \qquad\qquad (6.50)$$

$$\lim_{N\to\infty} \langle \mathbb{Y}_N(t), f \rangle = \langle \mathbb{Y}(t), f \rangle = \langle y(\cdot, t), f \rangle, \qquad t > 0, \quad f \in C_b(\mathbb{R}^d). \quad (6.51)$$

For N tending to infinity, the martingales M_N, $M_{2,N}$ in (6.38) and (6.41) vanish. Indeed from (6.28)-(6.30), (6.39) and Doob's inequality [10] we have that

$$E\left[\sup_{t\le T} |M_N(t)|\right]^2 \le CE\left[|M_N(T)|^2\right]$$

$$= CE\left[\int_0^T \left[\frac{1}{N} \int_E \kappa((\mathbb{Y}_N(s) * \phi_N)(x)) b_b(T_N(x,s)) dx \ ds\right]\right.$$

$$\le \frac{CT}{N}, \qquad\qquad\qquad\qquad (6.52)$$

so the martingale term vanishes in probability as $N \to \infty$. Hence the process $Z_N(t)$ becomes deterministic in probability as $N \longrightarrow \infty$.

Again we have that in the limit the martingale $M_{2,N}(f, t)$ vanishes in probability. In fact, by (6.28)-(6.32), (6.33) we have

$$E\left[\sup_{t\le T} |M_{2,N}(f,t)|\right]^2 \le CE\left[|M_{2,N}(f,T)|^2\right]$$

$$= CE\left[\int_0^T \left[\frac{1}{(N\lambda_N)^2} \sum_{k=1}^{\Lambda_N(s)} |f(X_N^k)|^2 \lambda_N \tilde{\beta}((\mathbb{Y}_N(s) * \psi_N)(X_N^k))\right.\right.$$

$$Y_N^k(s) b_g(T_N(X_N^k, s)) ds$$

$$\left.+ \frac{v_0^2}{N} \int_{\mathbb{R}^3} |f(x)|^2 \kappa((\mathbb{Y}_N(s) * \phi_N)(x)) b_b(T_N(x,s)) dx \ ds\right]$$

$$\le CE\left[\int_0^T \left[\frac{1}{\lambda_N N^2} \sum_{k=1}^{\Lambda_N(s)} |f(X_N^k)|^2 \frac{Y_N^k(s)}{(\mathbb{Y}_N(s) * \psi_N)(X_N^k)}\right] ds\right.$$

$$\left.+ T\|f\|_2^2 \|k\|_\infty \|b_b\|_\infty \frac{v_0^2}{N}\right]$$

$$\le CE\left[\int_0^T \left[\frac{Z_N(s)}{\lambda_N N^b \psi(0)} \|f\|_\infty^2\right]\right] ds + TC\frac{v_0^2}{N}$$

$$\leq \frac{C}{\lambda_N N^{1+b}\psi(0)} E\left[\int_0^T \int_E \alpha_N(s,x)dxds\right] + TC\frac{v_0^2}{N}$$

$$\leq CT\left(\frac{1}{\lambda_N N^b} + \frac{v_0^2}{N}\right) \tag{6.53}$$

This is the substantial reason for the asymptotic deterministic dynamics. The last inequality in (6.53) derives from (6.29) and (6.35).

We may calculate the limit of the two equations as follows.

Let be T the solution of (6.27) when the coefficients get deterministic. Formally, because of (6.24), (6.50) and (6.51) we have

$$\lim_{N\to\infty} \langle \mathbb{Y}_N(s), \tilde{\beta}((\mathbb{Y}_N(s) * \psi_N)b_g(T_N(s,\cdot))f \rangle$$

$$= \lim_{N\to\infty} a_g\, b_g(T_N(t,x)) \times \frac{1}{N} \sum_{k=1}^{\Lambda_N(t)} Y_N^k(t)\, \tilde{\beta}((\mathbb{Y}_N(t) * \psi_N)(X_N^k))\gamma_N(x - X_N^k)$$

$$= \langle \tilde{\beta}(y(s,\cdot))\, y(s,\cdot)b_g(T(s,\cdot)), f \rangle. \tag{6.54}$$

Since the martingale term vanishes as $N \to \infty$, from (6.27), (6.53), and (6.54), we obtain the weak form of the following PDE system for the mass and the temperature of the process

$$\begin{cases} \dfrac{\partial}{\partial t}y(x,t) = \tilde{\beta}(y(x,t))b_g(T(x,t))y(x,t) + v_0\kappa(y(x,t))b_b(T(x,t)) \\[3mm] \dfrac{\partial}{\partial t}T(x,t) = \sigma\Delta T(x,t) + a_g\tilde{\beta}(y(x,t))b_g(T(x,t))y(x,t). \end{cases} \tag{6.55}$$

6.7.1 Simulation of the Macroscopic Model

System (6.55) is solved in a domain $E = [0,1] \times [0,1] \subset \mathbb{R}^2$ by finite difference methods, with initial conditions

$$\begin{aligned} \rho(0,x_1,x_2) &= 0, \\ y(0,x_1,x_2) &= 0, && x \in E \qquad (6.56) \\ T(0,x_1,x_2) &= T_{\max} - x_1(T_{\max} - T_{min}), \end{aligned}$$

and boundary condition for the temperature (6.47).

We choose the same functions as for the particle system (cf. (6.43)-(6.44)).

Crystallization starts in the colder side of the sample and from there diffuses in all the available space (see Fig. 6.4), reaching saturation.

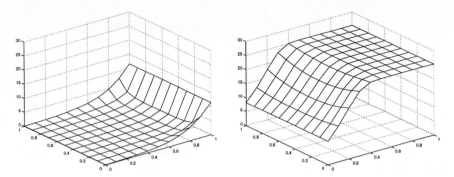

Fig. 6.4. Continuum model. The mass density $\rho(t, x) * 10^3$ at time $t = 100$, $t = 10000$.

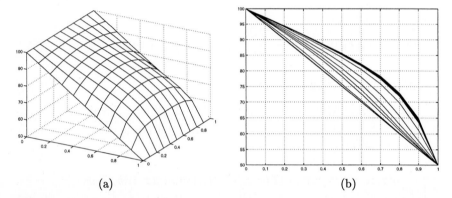

(a)   (b)

Fig. 6.5. Temperature profile.
(a)Particle model. The temperature in presence of 10000 nuclei
(b)Continuum model. Time evolution of the temperature profile.

Even the results for the temperature are qualitatively comparable. Indeed the temperature in the particle case is shown in Fig. 6.5a. We see the effect of latent heat of crystallization, that causes a re-heating in the middle of the sample (in absence of the latent heat, the temperature would be a sloping plane). For the continuum model the crystallization rate becomes smaller getting close to saturation: we may see, in Fig. 6.5b , that its effect on temperature, via enthalpy, is quite strong at the beginning of the process and becomes almost negligible when the process is close to the end.

Acknowledgements

It is a pleasure to acknowledge important stimulating discussions with all members of the ECMI Special Interest Group on "Polymers"
    (www.mat.unimi.it/~polymer).

Particular thanks are due to Dr. Karl Ölschläger of the IWR-Heidelberg for stimulating discussions. Financial support from the Italian Space Agency (ASI), Italian CNR (National Research Council) under project grant n. 98.03635.ST74, the Austrian Research Fund (FWF) under project grants F 13/08 and P 13478-INF, the EU-TMR Programme *DEIC* and the Italian MURST/cofin. Programme " *Stochastic processes with spatial structure*" is gratefully acknowledged.

References

1. G.Alfonso, Polimeri cristallini, In *Fondamentals in Polymer Science*, (.Guaita et.al. Eds), Pacini Ed., Pisa, 1998 (in Italian).
2. P.K. Andersen, O. Borgan, R. Gill, N. Keiding, *Statistical Models Based on Counting Processes*, Springer-Verlag, New-Jork, 1993.
3. P. Bremaud, *Point Processes and Queues. Martingale Dynamics*, Springer-Verlag, New-York, 1981.
4. M. Burger, V. Capasso, C. Salani, *Modeling multi-dimensional crystallization of polymers in interaction with heat transfer.*, *Nonlinear Analysis: Real World Application*, **3** (2002), 139–160.
5. G.Eder, H.Janeschitz-Kriegl, *Structure Development During Processing: Crystallization*, in Materials Science and Technology, Vol.18 (edited by H.Meijer), Verlag Chemie, Weinheim, 1997.
6. G. Last, A. Brandt, *Marked Point Processes on the Real Line. The Dynamic Approach*, Springer-Verlag, New York Berlin Heidelberg, 1995.
7. A.Micheletti, M.Burger, Stochastic and deterministic simulation of nisothermal crystallization of polymers, *J. Math. Chem.*, **30** (2001), 169-193.
8. D. Morale, V. Capasso A *"law of large numbers" for a birth and growth process coupled with a diffusive field*, 2001. In preparation.
9. K.Oelschläger, *Many-Particle Systems and the Continuum Description of their Dynamics* Habilitationsschrift, Faculty of Mathematics, University of Heidelberg, Germany, 1989.
10. A.N. Shirayev *Probability*, Springer-Verlag, New York Inc., 1989.

Manufacturing

Manufacturing

7 Modelling of Industrial Processes for Polymer Melts: Extrusion and Injection Moulding*

Fons van de Ven

Eindhoven University of Technology
P.O.Box 513; 5600 MB Eindhoven, The Netherlands

7.1 Introduction

Polymer research is divided over such diverse fields as there are:
- modelling of cristallization processes;
- derivation of constitutive theories from molecular models;
- modelling of industrial processes for the manufacture of polymeric (plastic) products.

In this chapter, we will deal with some aspects of the latter point. There are various ways in which polymers are processed, depending on the nature of the polymer and the application for which it is intended. Millions and millions of tons of thermoplastic polymers are produced per year and manufactured to plastic products as fibers, sheets, tubes, compact discs and many many others. The mathematical modelling and numerical simulation of the underlying industrial processes is a main activity within the research on polymers. Examples of these industrial processes are: extrusion, injection moulding, blow moulding and fiber spinning. In all these processes, granular polymeric material is first melted in an extruder and then pressed out of it via a die or into a mould to achieve its final shape. A typical extruder consists of a hopper, a screw section, and a die section. Polymer granules are inserted into the hopper and enter the screw section. The screw rotates causing transportation of the granules. They gradually melt during this transport. The energy required for this comes from both external heating and from internal friction. A high pressure forces the melt to flow into the die or the mould, thus acquiring its intended final shape.

The resulting product must fulfil specific requirements. For instance, films are often used in packaging where the film must be transparent and smooth and possess a certain strength. At the same time, economic stimuli demand fast operation with high screw speeds. Unfortunately, it is often found that an increase in extrusion speed causes a decrease in final product properties.

---

* Projects are supported by industrial companies: DOW Benelux, Terneuzen; DSM-Research, Geleen and AXXICON Moulds, Eindhoven; all in the Netherlands.

It is evident that, in view of the huge output of plastic products, even very small improvements with respect to this aspect will yield great economical profits.

Many types of distortions are found to occur in the processes mentioned above. Examples of these are:

- sharkskin or spurt in extrusion;
- surface distortions in cable coating;
- front instability in injection moulding.

Often these distortions are referred to as instabilities, although this term does not always seems to be justified (sometimes they are merely periodic disturbances). Another common classification is *melt fracture*, covering all types of distortions in extrusion processes. This phenomenon has been investigated by many researchers, both in industry and in academia, driven by scientific and economic interests. However, to describe and predict the characteristics of the phenomenon as a function of material, geometrical and operational parameters is still a challenge. This is partly due to the large number of variables playing a part in the phenomenon. Much of the complexity arises from the variation in polymer structure and the nonlinear response of polymers to applied deformation histories and stress fields.

Before entering into more detail two specific processes, we present a short list of the main industrial processes for polymeric products:

- *Extrusion.*
  In extrusion, a polymer melt is pressed from the extruder into a die. This die can be a circular die or capillary (to produce e.g. fibers), an annular die (for tubes) or a slit die (for films or sheets). After leaving the die, the polymer solidifies in air. Instabilities in extrusion are captured under the common name melt fracture ([1], [2], [3], [4], [5]).
- *Injection Moulding.*
  In injection moulding, a hot polymer melt is injected into a cold mould, and the polymer solidifies within the mould before it is ejected. The shape of the mould can be rather simple (e.g. a thin disc-like shape for the production of CD's) or very complex. In the latter case, complete filling of the mould can become the main problem. Thermal effects often play an important role in injection moulding ([6], [7]).
- *Blow Moulding.*
  In blow moulding, a polymer melt is forced through an annular die in order to produce a hollow cylindrical tube called a parison. Blow moulding is an industrial process in which different hollow products can be made, e.g. bottles, cans, air ducts and so called Car Velocity Joints Boots. The length of the parison and its thickness distribution are of central importance for the quality of the final product. Two competitive effects are observed, commonly known as *sag* and *swell*. Sag, is caused by gravitational forces. Swell, is due to elastic properties of the viscoelastic melt. Both sag and swell

lead to a non-uniform thickness distribution of the parison. To be able to control the shape of the parison, one has to know how the geometric and material parameters of the extrusion process influence sag and swell ([8], [9]).

- *Cable Coating.*

  In a cable coating process, a solid (e.g. copper) wire is covered with a thin shield of polymer (e.g. for electrical insulation). In this process, surface distortions of a very special kind were observed: these distortions of the extrudate showed up as two sets of spirals running in two directions, making angles of approximately 45° with the axis of the wire ([10], [11]).

- *Fiber Spinning.*

  In a fiber spinning process, polymer is extruded at high speed via a converging die through a very narrow capillary into free air. Using a take-up device, the thin thread filament or fiber is drawn down by a constant tensile force . The take-up speed is greater than the extrudate velocity, so the filament is elongated uni-axially. Special effects under study in a fiber spinning process are non-uniform elongation and die swell ([12], [13]).

For all of the processes, listed above, mathematical models are constructed, and based on these models both analytical and numerical simulations of the processes can be derived. These simulations can serve us to suggest ways for improving the processes (better final products, shorter production times). For validation of the obtained, theoretical, results by means of experimental data, industrial contacts are indispensable; for the research topics to be discussed in this chapter, we have contacts with companies as DOW, DSM and AXXICON.

In the remainder of this chapter, we will consider flow instabilities in extrusion in Section 7.2 and injection moulding especially with regard to temperature distributions in injection moulding of compact discs, in Section 7.3.

7.2 Flow Instabilities in the Extrusion of Polymer Melts

7.2.1 Introduction

Extrusion of polymeric melts is employed to produce, e.g., plastic wires, pipes, sheets and plates. The principle of extrusion is that the polymeric melt is forced to flow through a die, e.g., by the action of a driving pressure gradient or a moving piston. In extrusion of plastic fibers flow instabilities can occur when increasing the rate of production. The distortions due to these instabilities make the final product worthless. So, it is of great importance for industries to know how these instabilities can be prevented, or at least how the critical rate of production can be increased. This can be done by adapting the geometry of the apparatus and/or the constitution of the polymer.

In order to find out how this is best achieved, a simulation of the processes under consideration is needed. In two ways, such a simulation is searched for:

1. On the basis of a full 3-dimensional theory using a nonlinear viscoelastic constitutive model for the polymer (constitutive instabilities; Aarts [1]);
2. On the basis of a 1-dimensional, discrete model incorporating either stick-slip or spurt-flow (discrete melt fracture models; Den Doelder [2]).

The 3-dimensional model shows the existence of persistent periodic oscillations related to flow instabilities.

The discrete model describes relaxation or spurt oscillations in good correspondence with experimental results (performed at DOW).

7.2.2 Problem Description

This section deals with the modelling of the extrusion process. The extruder consists of a wide cylindrical barrel connected to a narrow cylindrical capillary; see Fig. 7.1. Thus, the radius R of the capillary is small compared to the radius R_b of the barrel. The centerlines of the barrel and the capillary coincide. Cylindrical coordinates (r, θ, z) are introduced with the z-axis along the centerline of the extruder, and $z = 0$ corresponding to the position where the barrel is connected to the capillary.

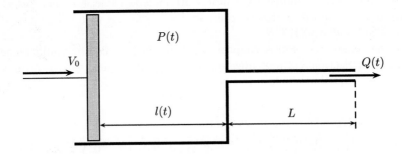

Fig. 7.1. The extruder which consists of a wide barrel and a narrow cylindrical capillary. The melt in the barrel is compressed by a plunger moving with constant speed V_0 in the positive z-direction

The polymeric melt in the barrel is compressed by a plunger, moving at constant speed V_0 in the positive z-direction. Due to the plunger movement a pressure P is built up inside the barrel, and the melt is forced to leave the barrel at $z = 0$ and to flow into the capillary, with volumetric flow rate Q. At the end of the capillary, i.e. at $z = L$ where L denotes the capillary length, the melt leaves the capillary and the extrudate is formed. Since the flow in the barrel and in the capillary are of essentially different types, the two flows are also modelled in different ways. In the main part of the barrel the flow is aligned along the z-axis, and the flow is an almost uniform compression flow. The pressure becomes very high here due to the narrow inlet of the capillary, but the velocity is rather low because the barrel is very wide (as compared to the capillary). Hence, in the barrel the flow is compression dominated, and shear is negligible here. Thus, the compressibility of the melt inside the barrel must be taken into account, and the melt density ρ is variable. Since the flow in the barrel is uniform, P and ρ are only time-dependent, i.e. $P = P(t)$ and $\rho = \rho(t)$. On the other hand, the flow in the capillary is strongly shear dominated (due to the no-slip condition at the wall and the relatively high velocity compared to that in the barrel) and then the influence of compressibility is small, and in fact negligible. Therefore, the melt flowing through the capillary is assumed incompressible. Hence, the melt flows through the whole capillary with volumetric flow rate $Q = Q(t)$.

We first consider the flow inside the barrel. Taking into account the global mass balance in the barrel, together with a linearly elastic compressibility law for the pressure, one can derive the following relation between the pressure in the barrel and the volumetric flow rate flowing into the capillary (cf. [14], [15])

$$\frac{dP(t)}{dt} = -\frac{K\,[Q(t) - Q_i]}{Al(t)}, \quad t > 0, \tag{7.1}$$

where $Q_i = AV_0$, the inlet flow rate at the plunger, and K is the compression or bulk modulus of the polymeric melt.

Next we consider the flow in the capillary, where the polymeric melt may be considered incompressible. This flow is governed by the equation of motion, and the incompressibility condition:

$$\nabla \cdot \mathbf{v} = 0, \tag{7.2}$$

where $\mathbf{v}$ is the fluid velocity.

For the strongly viscous shear flow we consider here the inertia terms in the equations of motion may be neglected. Moreover, body forces are absent here, so the equation of motion reduces to

$$\nabla \cdot \mathcal{T} = 0, \tag{7.3}$$

where $\mathcal{T}$ is the total stress tensor. With the flow aligned along the z-axis, the flow parameters in the capillary are independent of the axial coordinate z and the azimuthal coordinate θ. Under the condition that the flow starts from rest at time $t = 0$, the velocity takes the form

$$\mathbf{v} = v(r, t)H(t)\mathbf{e}_z, \tag{7.4}$$

where H is the (Heaviside) step function and $\mathbf{e}_z$ is a unit vector in the positive z-direction. The conservation of mass is now automatically satisfied.

The no-slip boundary condition at the wall of the capillary and the regularity of the velocity at the axis require

$$v(R, t) = 0, \tag{7.5}$$

and

$$\frac{\partial v}{\partial r}(0, t) = 0, \tag{7.6}$$

respectively.

The characteristic response of the material to a deformation is described by the constitutive equation for the stress. For viscoelastic fluids with fading memory, the stress depends on the deformation history. If a polymer solution contains a small-molecule solvent, this solvent will generally respond in a viscous manner to any applied force or deformation, separately from the elastic response due to the dissolved polymer. Therefore, it is assumed that the extra stress tensor $\mathcal{S} := \mathcal{T} + p\mathcal{I}$ in the fluid consists of a Newtonian viscous component and an isotropic elastic one, namely

$$\mathcal{S} = 2\eta_s \mathcal{D} + \mathcal{S}_p, \quad \Rightarrow \mathcal{T} = -p\mathcal{I} + 2\eta_s \mathcal{D} + \mathcal{S}_p. \tag{7.7}$$

Here, p is the pressure, $\mathcal{I}$ the unit tensor, and η_s the solvent viscosity. The rate-of-deformation tensor $\mathcal{D}$ is defined by

$$\mathcal{D} = \frac{1}{2}(\mathcal{L} + \mathcal{L}^T), \qquad \mathcal{L} = \operatorname{grad} \mathbf{v} = \frac{\partial \mathbf{v}}{\partial \mathbf{x}} (\equiv (\nabla \mathbf{v})^T). \tag{7.8}$$

With (7.4), we see that the only non-zero components of $\mathcal{L}$ and $\mathcal{D}$ are L_{rz} and $D_{rz} = D_{zr}$, where

$$D_{rz} = D_{zr} = \frac{1}{2}L_{rz} = \frac{1}{2}\frac{\partial v}{\partial r}. \tag{7.9}$$

The elastic part $\mathcal{S}_p$, which characterizes the polymer contribution, is assumed to be described by the constitutive model (KBKZ or JSO). For the constitutive models we consider, $\mathcal{S}$ is related to $\mathcal{D}$, and then, just as $\mathcal{D}$, $\mathcal{S}_p$ only depends on r and t. In that case the axial component of the equation of motion (7.3) reads

$$\frac{\partial T_{rz}}{\partial r} + \frac{T_{rz}}{r} - \frac{\partial p}{\partial z} = 0. \tag{7.10}$$

Since T_{rz} is independent of z, the solution for the pressure p takes the form

$$p(r, z, t) = -f(t)z + p_0(r, t). \tag{7.11}$$

Here, f is the pressure gradient driving the flow, and p_0 is a further irrelevant pressure term. Coupling between the flow in the capillary and the flow in the barrel is achieved by equating the pressure terms in the barrel and in the capillary at $r = 0$, $z = 0$. We assume that the pressure outside the capillary is at level zero, so that $p(0, L, t) = 0$. Then it follows that

$$P(t) = Lf(t), \quad \Rightarrow \quad f(t) = \frac{P(t)}{L}. \tag{7.12}$$

Integration of (7.10) with use of (7.11) leads to the following expression for the shear stress (here, $S_{rz} = (S_p)_{rz}$)

$$T_{rz} = \eta_s \frac{\partial v}{\partial r}(r, t) + S_{rz}(r, t) = -\frac{1}{2} r f(t), \quad 0 \le r \le R, \quad t > 0. \tag{7.13}$$

Finally, the volumetric flow rate Q in the capillary is defined by

$$Q(t) = 2\pi \int_0^R v(r, t) r \, dr. \tag{7.14}$$

7.2.3 Models for Constitutive Instabilities

The aim of the theory presented in this section is to get a better insight into the relation between the characteristic behaviour of polymeric melts and the flow instabilities.

The results presented here are mainly from the PhD-thesis of Aarts, [1]. In this thesis, especially the flow instability 'spurt' is investigated.

Spurt in *pressure-driven flows* is experimentally observed through a substantial increase of the volumetric flow rate at a slight increase of the pressure gradient beyond a critical value, while spurt in *piston-driven flows* is accompanied by persistent oscillations in the pressure.

Spurt is explained here in terms of constitutive instabilities (mechanical failure of the polymeric fluid itself), while the no-slip boundary condition at the wall of the die is maintained. The explanation is based on balance laws combined with either of two constitutive models: (1) KBKZ or (2) JSO. To account for the response of a small-molecule solvent, an extra Newtonian viscous term is added to the constitutive model employed.

Pressure-driven Extrusion

For a pressure-driven flow of a KBKZ-fluid through a cylindrical capillary, the occurrence of spurt is demonstrated. It is shown that the steady state solution is not unique if the steady state pressure gradient exceeds a critical value. The asymptotic stability of the possible steady states is established. Numerical computations determine which specific steady state the fluid attains. Phenomena like shape memory and hysteresis are explained (see also [16], [17] or [18]).

The axisymmetric laminar flow of an incompressible fluid through a cylindrical tube, radius R, is considered. The flow starts from rest at $t = 0$, driven by a prescribed constant pressure gradient $\overline{f}$. Under the neglect of inertia forces the balance of linear momentum for the shear stress T_{rz} yields (see (7.13))

$$T_{rz} = -\frac{1}{2}r\overline{f}. \tag{7.15}$$

The characteristic response of the material is described by a constitutive equation for S_p according to the KBKZ-model. The total shear stress T_{rz} is given by

$$T_{rz}(r,t) = \eta_s \frac{\partial v}{\partial r}(r,t) - \mu\lambda \int_{-\infty}^{t} \frac{\gamma(r,t,\tau)}{c+3+\gamma^2(r,t,\tau)} e^{-\lambda(t-\tau)}d\tau, \tag{7.16}$$

where η_s is the solvent viscosity, μ and c are elastic constants and λ is the relaxation rate. The shear strain γ is defined by

$$\gamma(r,t,\tau) = \Gamma(r,t) - \Gamma(r,\tau), \quad \text{where} \quad \Gamma(r,t) = -\int_0^t \frac{\partial v}{\partial r}(r,s)ds. \tag{7.17}$$

After the substitution of (7.16) into (7.15) and making the resulting equation dimensionless, we arrive at the following integrodifferential equation for $\Gamma(r,t)$

$$\varepsilon\frac{\partial\Gamma}{\partial t}(r,t) + h(\Gamma(r,t))e^{-t} + \int_0^t h(\gamma(r,t,\tau))e^{-(t-\tau)}d\tau = \frac{1}{2}r\overline{f}, \tag{7.18}$$

where the function h is defined by $h(x) = x/(1+x^2)$. Once $\Gamma(r,t)$ for each radial coordinate $r\epsilon[0,1]$ is solved from (7.18), the (dimensionless) volumetric flow rate $Q(t)$ can be calculated from (7.14)

$$Q(t) = \int_0^1 r^2 \frac{\partial\Gamma}{\partial t}(r,t)dr. \tag{7.19}$$

The latter result is derived after one integration by parts with the aid of the no-slip condition (7.5) and the definition (7.17).

Steady State Flow The steady state velocity gradient, defined by

$$\omega(r) = \lim_{t\to\infty} -\frac{\partial v}{\partial r}(r,t), \quad \omega(0) = 0, \tag{7.20}$$

satisfies the steady state equation (for each $r\epsilon[0,1]$)

$$\mathcal{F}(\omega(r)) := \varepsilon\omega(r) + \int_0^\infty h(\omega(r)s)e^{-s}ds = \frac{1}{2}r\overline{f} =: F(r). \tag{7.21}$$

For $0 < \varepsilon < 0.0289$ the function $\mathcal{F}(\omega)$ is nonmonotone (see Fig. 7.2). In that case equation (7.21) has three solutions if $F_m < F < F_M$. Stability analysis reveals that the steady state in which $\omega_M < \omega < \omega_m$ is unstable, whereas the other two, $\omega < \omega_M$ and $\omega > \omega_m$, are asymptotically stable. If $\overline{f} > \overline{f}_{\mathrm{crit}} := 2F_M$ (supercritical flow) then ω suffers a jump; if $0 \leq r < r_M := 2F_M/\overline{f}$ then $0 \leq \omega(r) < \omega_M$, whereas if $r_M < r \leq 1$ then $\omega(r) > \tilde{\omega}_M (> \omega_M)$ (see Fig. 7.2). Hence, the velocity gradient has a jump at $r = r_M$, resulting in a kink in the steady state velocity profile $\overline{v}(r)$, as shown in Fig. 7.2. This phenomenon is called *spurt*.

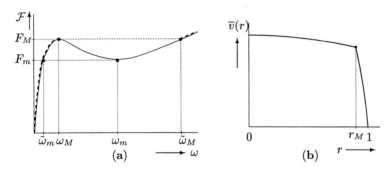

Fig. 7.2. The function $\mathcal{F}(\omega)$ if $0 < \varepsilon < 0.0289$. The *dashed line* represents the possible ω-solution in supercritical flow **(a)**, resulting in a steady state velocity profile, where the spurt layer is located in $r_M \leq r \leq 1$ **(b)**

In the spurt layer (i.e. $r_M \leq r \leq 1$) the magnitude of the velocity gradient is very large, in fact

$$\omega(r) = \frac{\overline{f}r}{2\varepsilon}\left[1 - \frac{4\varepsilon}{(\overline{f}r)^2}\log(\frac{\overline{f}r}{2\varepsilon}) + O(\varepsilon)\right], \quad \varepsilon \to 0. \tag{7.22}$$

These enormous shearing rates near the wall give rise to a dramatic increase in the stationary volumetric flow rate $\overline{Q} := \lim_{t \to \infty} Q(t)$ if the pressure gradient $\overline{f}$ exceeds the critical value $\overline{f}_{\mathrm{crit}}$ (see also Fig. 7.3).

Loading and Unloading; Hysteresis and Shape Memory Consider an experiment in which the flow is initially in a steady state, corresponding to a forcing $\overline{f}_0$, and the forcing is suddenly changed to $\overline{f} = \overline{f}_0 + \Delta\overline{f}$. If $\Delta\overline{f} > 0$ we call this process *loading* and if $\Delta\overline{f} < 0$ *unloading*. Which steady state eventually will be reached after the forcing is changed, depends on the initial state $\overline{f}_0$ and follows from numerical calculations performed on the integrodifferential equation (7.18). If the load is gradually increased from $\overline{f} = 0$ up to $\overline{f} = \overline{f}_{\mathrm{crit}} = 2F_M$ (subcritical flow) the entire flow is classical: The velocity gradient satisfies $\omega(r) < \omega_M$ and is continuous in r for all $r\epsilon[0,1]$. As soon as $\overline{f} > \overline{f}_{\mathrm{crit}}$ (supercritical flow) a kink in the velocity profile at $r = r_M = \overline{f}_{\mathrm{crit}}/\overline{f}$ turns up and spurt occurs. This spurt causes an enormous increase of the volumetric rate $\overline{Q}$, as depicted in Fig. 7.3 (trajectory BC). Let the loading trajectory finish at $\overline{f} = \overline{f}_{\max}(> \overline{f}_{\mathrm{crit}})$; the spurt layer $r^* \leq r \leq 1$, where $r^* = \overline{f}_{\mathrm{crit}}/\overline{f}_{\max}$, is of maximum thickness then. From this point the unloading is started. At first the spurt layer remains fixed between $r = r^*$ and $r = 1$. This phenomenon is called *shape memory* (trajectory CD in Fig. 7.3). During this unloading the magnitude of the shear stress $|T_{rz}| = F$ at r^* decreases according to $F^* := F(r^*) = r^*\overline{f}/2$. If F^* falls below F_m, that is if $\overline{f} < (F_m/F_M)\overline{f}_{\max}$, the boundary $r = r_m := 2F_m/\overline{f}$ of the spurt layer moves back to the wall for further decreasing $\overline{f}$ (trajectory DE in Fig. 7.3). The spurt layer disappears for $\overline{f} = 2F_m$ and in the final unloading path EA the flow is classical again. The phenomenon that no part of the loading curve in Fig. 7.3 is retraced until the flow has become entirely classical again, is typical for the occurrence of *hysteresis* in this process.

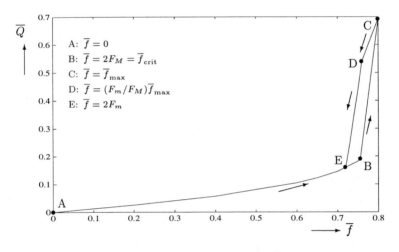

Fig. 7.3. The flow rate $\overline{Q}$ as function of the forcing $\overline{f}$ in a loading-unloading process. AB: subcritical loading (classical flow), BC: supercritical loading (with thickening spurt layer); CD unloading under shape memory; DE unloading with diminishing spurt layer; EA: subcritical unloading (classical flow)

Newtonian, Latency and Spurt Phase In computing the time dependent velocity gradient, we observe distinct time phases before the steady state is reached. These phases are clearly distinguishable in a (S, N)-plane, where S is defined by (see (7.18))

$$S(r,t) = h(\Gamma(r,t))e^{-t} + \int_0^t h(\gamma(r,t,\tau))e^{-(t-\tau)}d\tau, \qquad (7.23)$$

and N, which corresponds to the first normal stress difference $T_{zz} - T_{rr}$, is defined by

$$N(r,t) = g(\Gamma(r,t))e^{-t} + \int_0^t g(\gamma(r,t,\tau))e^{-(t-\tau)}d\tau, \qquad (7.24)$$

where $g(\gamma) = \gamma h(\gamma)$. From (7.18) with (7.21) it follows that

$$S(r,t) = F(r) + \varepsilon \frac{\partial v}{\partial r}(r,t). \qquad (7.25)$$

For $t \to \infty$ these two functions tend to their stationary values $\overline{S}(r) = J(\omega(r))$ and $\overline{N}(r) = 1 - L(\omega(r))$, where $\omega(r)$ is a solution of (7.21) and

$$J(\omega) = \int_0^\infty h(\omega s)e^{-s}ds \quad \text{and} \quad L(\omega) = 1 - \int_0^\infty g(\omega s)e^{-s}ds. \qquad (7.26)$$

According to (7.21), $\overline{S} = F - \varepsilon\omega$, implying that

$$\overline{N} = 1 - L(\frac{F - \overline{S}}{\varepsilon}). \qquad (7.27)$$

For a supercritical flow such that $F > F_M$, there exist three distinct time phases; an initial phase, a pseudo steady state and the *spurt phase* in which the flow becomes stationary. At $t = 0$, $S = N = 0$ and $\partial v / \partial r = -F/\varepsilon$. Hence, the solution starts in the origin of the (S, N)-plane and changes on an $O(\varepsilon)$-time scale until $S = F + O(\varepsilon)$. The period of time during which this occurs is referred to as the *Newtonian phase* $(0 \le t < t_N)$. During this phase the following relation holds

$$N = \frac{1}{2}(1 - \sqrt{1 - 4S^2}) + O(\varepsilon), \quad \varepsilon \to 0. \qquad (7.28)$$

At $t = t_N$ the velocity gradient $\partial v / \partial r$ has become $O(1)$, and this remains so for some time $t > t_N$. Then (7.25) implies that $S = F + O(\varepsilon)$ and hence S is almost constant. This pseudo steady state that precedes spurt is called the *latency phase*. During this phase, N steadily increases until it is sufficiently close to the line $N = 1 - L((F-S)/\varepsilon)$. After that point, S suddenly decreases and spurt ensues, until the steady state $(\overline{S}, \overline{N})$ is reached. Asymptotics for small ε reveal that the stationary values in the spurt phase satisfy (cf. [17])

$$\overline{S} = \frac{\varepsilon}{F}\log\frac{F}{\varepsilon} - \frac{\varepsilon C}{F} + o(\varepsilon), \qquad \overline{N} = 1 - \frac{\varepsilon\pi}{2F} + O(\varepsilon^2\log\varepsilon), \qquad (7.29)$$

where $C = 0.57721...$ is Euler's constant.

Conclusions The KBKZ -model supplied with an extra viscous term has been used to describe the flow of a viscoelastic polymeric fluid through a capillary of an extruder. Internal material properties of the fluid itself account here for the spurt phenomenon and not a global external effect as wall slip. In this section only the flow in the capillary is described. A formulation, in which also the process in the barrel (from which the polymer melt is extrudated into the capillary) is considered, is presented in the next section.

Piston-driven Extrusion

For a piston-driven flow of a JSO-fluid through a cylindrical capillary, spurt accompanied by persistent oscillations in the pressure gradient is found for a bounded range of prescribed flow rates. The influence of compression on the onset of spurt is investigated for an extrusion process that is modelled by the flow of a JSO-fluid through a contraction from a wide barrel into a narrow cylindrical capillary. The fluid in the barrel is compressed by a moving plunger. Numerical computations disclose that persistent oscillations in the pressure as well as in the volumetric flow rate occur for a bounded range of prescribed plunger speeds (see also [1] or [17]).

The elastic part $\mathcal{S}_p$ is described by the constitutive JSO-model. In the JSO-model, $\mathcal{S}_p$ is determined by the following nonlinear differential equation (see Tanner [19, p. 207]):

$$\frac{d\mathcal{S}_p}{dt} - \mathcal{L}\mathcal{S}_p - \mathcal{S}_p\mathcal{L}^T + (1-a)(\mathcal{D}\mathcal{S}_p + \mathcal{S}_p\mathcal{D}) + \lambda\mathcal{S}_p = 2\mu\mathcal{D}, \qquad (7.30)$$

where d/dt denotes the material derivative. The relaxation rate λ, the slip parameter $a \in (-1, 1)$, and the shear modulus μ are material parameters.

With the velocity given by (7.4), $L_{zr} = \partial v/\partial r$ is the only non-zero component of $\mathcal{L}$ and the components S_{ij} of $\mathcal{S}_p$ are functions of r and t only. As a result, the material derivative $d\mathcal{S}_p/dt$ is equal to the partial time derivative $\partial \mathcal{S}_p/\partial t$, and the components $S_{r\theta}, S_{\theta\theta}$ and $S_{\theta z}$ are identically zero here. Then, the JSO-model transforms into the following equations:

$$\frac{\partial S_{rr}}{\partial t} + (1-a)S_{rz}\frac{\partial v}{\partial r} + \lambda S_{rr} = 0,$$

$$\frac{\partial S_{zz}}{\partial t} - (1+a)S_{rz}\frac{\partial v}{\partial r} + \lambda S_{zz} = 0, \qquad (7.31)$$

$$\frac{\partial S_{rz}}{\partial t} - \frac{1}{2}(1+a)S_{rr}\frac{\partial v}{\partial r} + \frac{1}{2}(1-a)S_{zz}\frac{\partial v}{\partial r} + \lambda S_{rz} = \mu\frac{\partial v}{\partial r}.$$

From the first two equations of (7.31) it follows that $(1+a)S_{rr}+(1-a)S_{zz} = 0$. Introduction of the new variables

$$S := -S_{rz}, \qquad\qquad Z := (1+a)S_{rr} = -(1-a)S_{zz}, \qquad (7.32)$$

in (7.31) yields for S and Z the differential equations

$$\frac{\partial Z}{\partial t} - (1-a^2)S\frac{\partial v}{\partial r} + \lambda Z = 0,$$

$$\frac{\partial S}{\partial t} + Z\frac{\partial v}{\partial r} + \lambda S = -\mu\frac{\partial v}{\partial r}. \qquad (7.33)$$

In terms of S and Z, the stress components T_{ij} of $\mathcal{T}$ according to (7.7) become

$$T_{rr} = -p + \frac{1}{1+a} Z(r,t), \qquad\qquad T_{\theta\theta} = -p,$$

$$T_{zz} = -p - \frac{1}{1-a} Z(r,t), \qquad\qquad T_{rz} = \eta_s \frac{\partial v}{\partial r}(r,t) - S(r,t), \qquad (7.34)$$

$$T_{r\theta} = T_{\theta z} = 0,$$

where $p = p(r,z,t)$. The first and second normal stress difference $N_1 :=$ $T_{zz} - T_{rr}$, and $N_2 := -T_{\theta\theta} + T_{rr}$ are determined by

$$N_1 = -\frac{2}{1-a^2} Z(r,t), \qquad\qquad \frac{N_2}{N_1} = -\frac{1-a}{2}. \qquad (7.35)$$

Hence, Z is related to the first normal stress difference, and the ratio of the two normal stress differences is constant.

The equations describing the extrusion process are made dimensionless by appropriate scaling. Then (7.1), governing the flow in the barrel, transforms into its dimensionless form:

$$\frac{dP(t)}{dt} = -\frac{1}{\chi} \frac{Q(t) - Q_i}{l(t)}, \quad t > 0, \qquad (7.36)$$

where the dimensionless parameter χ is given by

$$\chi = \frac{8 A l_0 \mu L}{K \pi R^4}. \qquad (7.37)$$

Furthermore, (7.12) transforms into its dimensionless form

$$f(t) = 8P(t). \qquad (7.38)$$

Equations (7.33) and (7.13), governing the flow in the capillary, transform into

$$\frac{\partial S}{\partial t} = -S + w(1+Z), \qquad \frac{\partial Z}{\partial t} = -Z - wS, \qquad 0 \le r \le 1, \quad t > 0, \qquad (7.39)$$

and

$$\varepsilon w(r,t) + S(r,t) = \frac{1}{2} r f(t), \qquad 0 \le r \le 1, \quad t \ge 0. \qquad (7.40)$$

Here, the velocity gradient, or shear rate, w is defined by

$$w(r,t) = -\frac{\partial v}{\partial r}(r,t), \qquad (7.41)$$

and the dimensionless parameter ε is given by

$$\varepsilon = \frac{\eta_s \lambda}{\mu}. \qquad (7.42)$$

Finally, the volumetric flow rate passes into its dimensionless form

$$Q(t) = 2 \int_0^1 v(r,t)r\,dr. \tag{7.43}$$

The boundary conditions (7.5) and (7.6) read in dimensionless form

$$v(1,t) = 0, \qquad w(0,t) = 0, \qquad t > 0. \tag{7.44}$$

After one integration by parts with the aid of the no-slip boundary condition at the wall, the volumetric flow rate Q can be expressed in terms of the velocity gradient w by

$$Q(t) = \int_0^1 r^2 w(r,t)\,dr. \tag{7.45}$$

Elimination of w by means of (7.40), transforms (7.45) into the following (implicit) relation between the pressure P and the volumetric flow rate Q:

$$P(t) = \varepsilon Q(t) + \int_0^1 r^2 S(r,t)\,dr. \tag{7.46}$$

Thus, the extrusion process driven by a plunger moving at constant speed, is described by the following system of equations:

$$\frac{\partial S}{\partial t} = -S + w(1+Z), \qquad \frac{\partial Z}{\partial t} = -Z - wS,$$

$$\varepsilon w(r,t) + S(r,t) = 4rP(t), \qquad P(t) = \varepsilon Q(t) + \int_0^1 r^2 S(r,t)\,dr, \quad (7.47)$$

$$\frac{dP(t)}{dt} = -\frac{1}{\chi}[Q(t) - Q_i], \qquad 0 \leq r \leq 1, \quad t > 0.$$

For $t < 0$ the fluid is at rest, and at $t = 0$ the flow is suddenly started up by letting the plunger move at constant speed V_0. The polymer melt in the barrel is thus compressed, because at $t = 0^+$ there is no melt flowing into the die yet ($Q(0^+) = 0$). The pressure in the barrel is then steeply built up, started from $P = 0$, following the fifth equation of (7.47). Also the stresses S and Z in the die start at value zero. Although the acceleration of the plunger at $t = 0$ is infinite, the acceleration of the melt in the die remains finite, and it can be shown by dimensional analysis that the inertia terms for the flow in the die remain negligibly small (see [1, Section 2.1]). The plunger movement induces the constant inlet flow rate Q_i.

The initial conditions for P, S and Z, which are supposed to be continuous at $t = 0$, are given by

$$P(0) = 0, \qquad S(r,0) = 0, \qquad Z(r,0) = 0, \qquad 0 \leq r \leq 1. \tag{7.48}$$

Substitution of (7.48) into (7.47) yields the initial values

$$Q(0) = 0, \qquad w(r,0) = 0, \qquad 0 \le r \le 1. \tag{7.49}$$

Equations (7.47), governing the flow in the extruder, can be viewed as a continuous family of quadratic (i.e. containing terms like wZ and wS) ordinary differential equations coupled by the non-local constraint that determines the flow rate, and the non-local ordinary differential equation that describes the compression in the barrel. The material parameters of the polymeric melt, the plunger speed V_0, and the dimensions of the extruder are included in the three dimensionless parameters ε, χ and Q_i. Notice that ε contains only the material parameters of the polymeric melt, whereas Q_i and χ depend on both the material parameters and the geometry of the extruder. The parameter χ is proportional to the melt compressibility $1/K$.

Steady State Flow In this section we investigate the steady state reached by the flow as $t \to \infty$. The steady state flow, driven by the constant inlet flow rate Q_i, is described in terms of the steady state variables

$$\overline{P} = \lim_{t \to \infty} P(t), \qquad \overline{f} = \lim_{t \to \infty} f(t), \qquad \overline{Q} = \lim_{t \to \infty} Q(t),$$
$$\overline{w}(r) = \lim_{t \to \infty} w(r,t), \qquad \overline{S}(r) = \lim_{t \to \infty} S(r,t), \qquad \overline{Z}(r) = \lim_{t \to \infty} Z(r,t), \tag{7.50}$$

under the assumption that these limits exist. For $t \to \infty$, the equations (7.38) and (7.47) reduce to

$$0 = -\overline{S} + w(1 + \overline{Z}), \qquad 0 = -\overline{Z} - w\overline{S}, \qquad \overline{Q} - Q_i = 0,$$
$$\varepsilon w(r) + \overline{S}(r) = 4r\overline{P}, \qquad \overline{P} = \varepsilon \overline{Q} + \int_0^1 r^2 \overline{S}(r)\,dr, \qquad \overline{f} = 8\overline{P}. \tag{7.51}$$

Hence, $\overline{Q} = Q_i$ and the solutions of $(7.51)^{1,2}$ expressed in terms of w read

$$\overline{S}(r) = \frac{w(r)}{1 + w^2(r)}, \qquad \overline{Z}(r) = -\frac{w^2(r)}{1 + w^2(r)}. \tag{7.52}$$

On substitution of $(7.52)^1$ into $(7.51)^4$, we find that the steady state velocity gradient can be determined for each $r \in [0,1]$ by solving $w = w(r)$ from the equation

$$\mathcal{F}(w(r)) = F(r), \tag{7.53}$$

where the steady state shear stress F is defined by

$$F(r) = 4r\overline{P} = \frac{1}{2}r\overline{f}. \tag{7.54}$$

and the function $\mathcal{F}$ is defined by

$$\mathcal{F}(\omega) = \varepsilon\omega + \frac{\omega}{1+\omega^2}. \tag{7.55}$$

For a given inlet flow rate Q_i, the velocity gradient ω must satisfy the constraint

$$Q_i = \int_0^1 r^2\omega(r)dr, \tag{7.56}$$

obtained by letting $t \to \infty$ in (7.45). The steady state velocity profile $\overline{v}(r)$ is obtained by integration of $\overline{v}'(r) = -\omega(r)$ using the boundary condition $\overline{v}(1) = 0$ at the wall.

For $\varepsilon < 1/8$ the function $\mathcal{F}$ is nonmonotone in ω. In Fig. 7.4 the function $\mathcal{F}(\omega)$ is plotted for a specific value of ε with $0 < \varepsilon < 1/8$. Since the Newtonian viscosity η_s is small in comparison to the shear viscosity μ/λ, we will henceforth assume that $0 < \varepsilon = \eta_s\lambda/\mu < 1/8$.

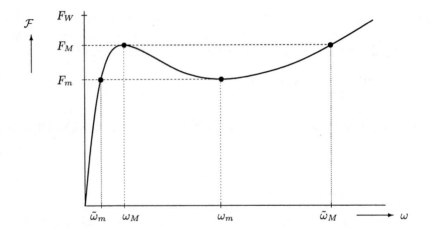

Fig. 7.4. The function $\mathcal{F}(\omega) = \varepsilon\omega + \omega/(1+\omega^2)$, when $0 < \varepsilon < 1/8$. In steady state flow the velocity gradient ω satisfies $\mathcal{F}(\omega) = F$, where $F = 4r\overline{P}$ is the steady state shear stress

The steady state shear stress F is linear in r and has its maximum at the wall $r = 1$. If this maximum, denoted by $F_W = 4\overline{P}$, remains below the minimum F_m, then (7.53) has a unique solution $\omega(r) < \tilde{\omega}_m$ for each radial coordinate r. Clearly, $\omega(r)$ is continuous in r, leading to a smooth steady state velocity profile $\overline{v}(r)$, and the flow is referred to as classical. If the maximum $F_W = 4\overline{P}$ exceeds the minimum F_m, equation (7.53) has

– one solution if $0 \le F < F_m$;

– three solutions if $F_m < F < F_M$;
– one solution if $F_M < F \leq F_W$.

If $F(1) > F_M$, i.e. if $\overline{P} > F_M/4 =: \overline{P}_{crit}$ (supercritical), the steady state velocity gradient $\omega(r)$ has at least one jump at some radial coordinate r. In case of exactly one jump we denote the radial coordinate at which the jump occurs by r^* ($r^* < 1$), and we refer to the flow as spurt flow. Hence, in spurt flow $\omega(r) < \omega_-^*$ for $0 \leq r < r^*$, whereas $\omega(r) > \omega_+^*$ for $r^* < r \leq 1$, with $\omega_-^* \leq \omega_M$ and $\omega_+^* \geq \omega_m$. The jump in ω results in a kink in the steady state velocity profile $\overline{v}(r)$ at $r = r^*$, and a spurt layer with large velocity gradients forms near the wall.

In conclusion, for a prescribed constant inlet flow rate Q_i we have for a possible steady state:

– If $0 \leq Q_i \leq \overline{Q}_0$, the steady state is unique; classical flow occurs .

– If $\overline{Q}_0 < Q_i \leq \overline{Q}_{crit}$, the steady state is not unique; either classical flow or spurt flow occurs.

– If $Q_i > \overline{Q}_{crit}$, the steady state is not unique; spurt flow occurs.

Notice that the results derived in this section are only valid in case the steady state does indeed exist.Results of numerical computations as presented in the next section will show whether or not the flow tends to a steady state as $t \to \infty$.

Transient Flow Behaviour In this section we compute for $t > 0$ the transient flow, starting from rest at time $t = 0$ and driven by the constant inlet flow rate Q_i due to the plunger movement. The flow is governed by the system of equations (7.47), with initial conditions (7.48) and (7.49). From the numerical results we infer whether the flow reaches a steady state, and we determine the steady state variables. The main interest goes to the relationship between Q_i and the steady state pressure $\overline{P}$. In the case of a classical steady state this relationship is one-to-one ,whereas in the case of a spurt steady state, $\overline{P}$ is not uniquely determined by just Q_i. Whether the flow tends to a steady state, is found to depend on the values of Q_i and the dimensionless parameters ε and χ. For example, for $\varepsilon = 0.005$, in case $\chi = 1$ or $\chi = 2$ we find a range of Q_i-values for which the flow shows so-called persistent oscillations that do not die out, so that no steady state is attained. For $\varepsilon = 0.02$, $\chi = 1$ and $Q_i = 0.1$ (subcritical flow) we observe that $P(t)$ and $Q(t)$ are monotone and smooth functions of t. After sufficient time the flow reaches a classical steady state with a continuous steady state velocity gradient.

To investigate supercritical flow ($Q_i > \overline{Q}_{crit} = 0.1714$) for $\varepsilon = 0.02$ and $\chi = 1$, we take $Q_i = 0.6$. In Fig. 7.5, the pressure $P(t)$ and the volumetric flow rate $Q(t)$ are plotted as functions of time t. We observe that oscillations in $P(t)$ and $Q(t)$ appear. These oscillations die out and after sufficient time a steady state is reached. The computations for $Q_i = 0.6$ reveal that the steady state shows a discontinuous velocity gradient $\omega(r)$ with exactly one jump. To

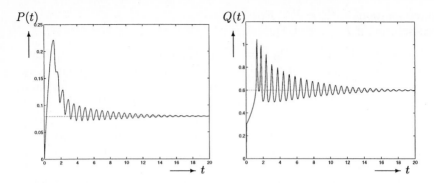

Fig. 7.5. The pressure $P(t)$ and the volumetric flow rate $Q(t)$ as functions of time t, for $\varepsilon = 0.02$, $\chi = 1$ and $Q_i = 0.6$. The *dotted lines* correspond to the steady state values

investigate whether for $\varepsilon = 0.02$, $\chi = 1$ and a given flow rate Q_i the flow starting from rest reaches a steady state, we compute the transient flow for several flow rates, varying from $Q_i = 0$ to $Q_i = 4.0$. The result is that for all values of Q_i considered, a steady state is reached. In Fig. 7.6a. the steady state pressure $\overline{P}$ attained is plotted versus Q_i, for $\varepsilon = 0.02$ and $\chi = 1$; the plot is drawn as a solid curve. The $\overline{P}$ versus Q_i curve is called the *flow curve*. The flow curve shows a kink at $Q_i = \overline{Q}_{crit}$ and is S-shaped.

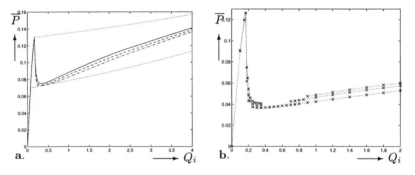

Fig. 7.6. The flow curves of the steady state pressure $\overline{P}$ versus the inlet flow rate Q_i, for: **a.** $\varepsilon = 0.02$ and $\chi = 1$ (*solid curve*), $\chi = 2$ (*dashed curve*) and $\chi = 4$ (*dashed-dotted curve*). The dotted curves correspond to $\overline{P} = P_{clas}(Q_i)$ (classical flow), $\overline{P} = P_{bottom}(Q_i)$ (bottom-jumping) and $\overline{P} = P_{top}(Q_i)$ (top-jumping). The flow curves show a kink at $Q_i = \overline{Q}_{crit} = 0.1714$, and: **b.** $\varepsilon = 0.005$ and $\chi = 1, 2,$ 6. The points $(Q_i, \overline{P})$ marked by the crosses ($\times$) correspond to computed steady states. The gaps in the flow curves for $\chi = 1$ and $\chi = 2$ correspond to flow rates Q_i for which persistent oscillations occur

To investigate the influence of the parameter χ, we compute the transient flow for $\chi = 2$ and $\chi = 4$, keeping $\varepsilon = 0.02$, and we compare the numerical results to those obtained in the case $\chi = 1$, $\varepsilon = 0.02$. The numerical computations disclose that for all values of Q_i considered, a steady state is reached. In Fig. 7.6a. the flow curves are plotted for $\varepsilon = 0.02$ and $\chi = 2$ (dashed curve), and for $\varepsilon = 0.02$ and $\chi = 4$ (dashed-dotted curve). We observe that the flow curves for $\chi = 2$ and $\chi = 4$ are S-shaped, just like the flow curve for $\chi = 1$. Furthermore, the flow curves show a kink at $Q_i = \overline{Q}_{crit}$ independent of χ. Notice that for $Q_i < \overline{Q}_{crit}$ the three flow curves for $\chi = 1, 2$ and 4 coincide, whereas at a fixed supercritical flow rate $Q_i > \overline{Q}_{crit}$, the steady state pressure $\overline{P}$ becomes smaller if χ is changed from $\chi = 1$ to the larger values $\chi = 2$ or $\chi = 4$.

Next, we investigate the influence of the parameters ε and χ on the transient flow behaviour and the steady state values attained. To that end, we compute the transient flow for several values of Q_i, when $\varepsilon = 0.005$ and $\chi = 1, 2$ and 6. In Fig. 7.7, Fig. 7.8 and Fig. 7.9 the pressure $P(t)$ and the volumetric flow rate $Q(t)$ are plotted as functions of time t, for $\varepsilon = 0.005$, $\chi = 1$, and $Q_i = 0.2, 0.6$ and 2.0, respectively. We observe that oscillations in $Q(t)$ appear. For $Q_i = 0.6$ and $Q_i = 2.0$, also oscillations in $P(t)$ appear. For $Q_i = 0.2$ and $Q_i = 2.0$, the amplitudes of these oscillations decay, and the oscillations damp out. Hence, after sufficient time a steady state is reached, if $Q_i = 0.2$ or $Q_i = 2.0$. The steady state variables $\omega(r)$, $\overline{S}(r)$ and $\overline{Z}(r)$ are found to be discontinuous at $r = r^*$.

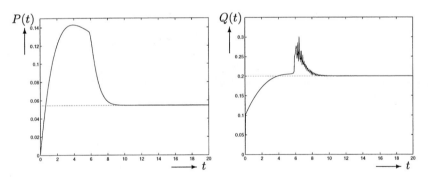

Fig. 7.7. The pressure $P(t)$ and the volumetric flow rate $Q(t)$ as functions of time t, for $\varepsilon = 0.005$, $\chi = 1$ and $Q_i = 0.2$. The *dotted lines* correspond to the steady state values

For $Q_i = 0.6$, however, we observe in Fig. 7.8 that the amplitude of the oscillations in $P(t)$ and $Q(t)$ fails to decay and remains constant after a certain instant. Hence, for $Q_i = 0.6$, the functions $P(t)$ and $Q(t)$ do not settle to a stationary value within the time interval of computation, indicating that no steady state is attained. Instead, $P(t)$ and $Q(t)$ show so-called persistent

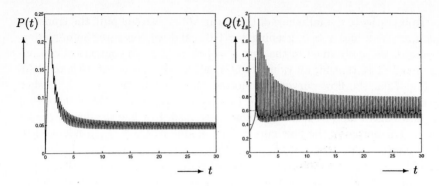

Fig. 7.8. The pressure $P(t)$ and the volumetric flow rate $Q(t)$ as functions of time t, for $\varepsilon = 0.005$, $\chi = 1$ and $Q_i = 0.6$

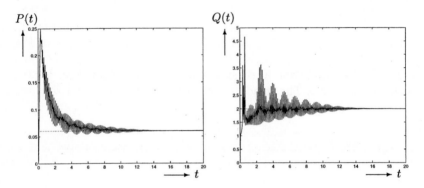

Fig. 7.9. The pressure $P(t)$ and the volumetric flow rate $Q(t)$ as functions of time t, for $\varepsilon = 0.005$, $\chi = 1$ and $Q_i = 2.0$. The *dotted lines* correspond to the steady state values $\overline{P} = \lim_{t \to \infty} P(t)$ and $\overline{Q} = \lim_{t \to \infty} Q(t)$

oscillations. To establish for which inlet flow rates Q_i a steady state is attained when $\varepsilon = 0.005$ and $\chi = 1, 2, 6$, we compute the transient flow for several flow rates, varying from $Q_i = 0$ to $Q_i = 3.0$. The outcome of the computations for $\chi = 1$ is that a steady state is reached for $Q_i \leq 0.35$ and for $Q_i \geq 0.90$. This steady state is classical when $Q_i \leq \overline{Q}_{crit} = 0.1678$, and corresponds to spurt flow when $\overline{Q}_{crit} < Q_i \leq 0.35$ or $Q_i \geq 0.90$. For values of Q_i close to 0.35 or 0.90, the time interval within which the flow settles to a steady state, becomes very large. For $0.40 \leq Q_i \leq 0.80$, however, the functions $P(t)$, $Q(t)$, $w(r,t)$, $S(r,t)$ and $Z(r,t)$ show persistent oscillations and fail to settle to stationary values within the time interval of computation. The computations for $\varepsilon = 0.005$ and $\chi = 2$ disclose that the transient flow behaviour is similar to that observed when $\varepsilon = 0.005$ and $\chi = 1$. For $0.40 \leq Q_i \leq 0.65$ persistent oscillations occur, and the flow fails to settle to a steady state. For $\varepsilon = 0.005$ and $\chi = 6$, the numerical computations reveal that a steady state is reached after sufficient time, for each value of the inlet flow rate Q_i considered.

In Fig. 7.6b. the flow curve of the steady state pressure $\overline{P}$ versus Q_i is plotted for $\varepsilon = 0.005$ and $\chi = 1, 2, 6$. The computed points $(Q_i, \overline{P})$ are marked by a cross ($\times$). The gaps in the flow curves for $\chi = 1$ and $\chi = 2$ correspond to flow rates Q_i for which persistent oscillations occur and no steady state is attained. We observe that the gap becomes smaller if χ is changed from $\chi = 1$ to the larger value $\chi = 2$, and has disappeared when $\chi = 6$.

Conclusions

We have considered 3-dimensional constitutive models for polymeric fluids. We have found that the stationary shear curve can be nonmonotone, yielding a non-unique solution for the stationary shear rate $\omega(r)$. This nonmonotony is crucial for the existence of

- a sudden increase in the flow rate in pressure-driven extrusion at a certain critical pressure gradient;
- persistent oscillations in pressure and flow rate in piston-driven extrusion.

Both phenomena are typical for the flow instability called spurt. Here, we have maintained the no-slip boundary condition at the wall of the capillary. Accordingly, spurt flow is associated with internal properties of the polymeric melt, and is therefore referred to as a *constitutive instability*.

From the numerical results presented here we infer that a bounded range $\mathcal{R} = (\overline{Q}_m, \overline{Q}_M)$ of inlet flow rates Q_i exists, for which persistent oscillations occur and no steady state is attained. At $Q_i = \overline{Q}_m$ and $Q_i = \overline{Q}_M$, the transition from a steady state to a state of persistent oscillations and vice versa takes place. The size of $\mathcal{R}$ depends on the value of ε as well as on the value of χ: for $\varepsilon = 0.02$ and $\chi = 1, 2, 4$, and for $\varepsilon = 0.005$ and $\chi = 6$, the range $\mathcal{R}$ is empty, whereas for $\varepsilon = 0.005$ and $\chi = 1, 2$, the range $\mathcal{R} = (\overline{Q}_m, \overline{Q}_M)$ is not empty. For $\varepsilon = 0.005$ and $\chi = 1$, the range $\mathcal{R}$ has transition points $\overline{Q}_m$ between 0.35 and 0.40, and $\overline{Q}_M$ between 0.80 and 0.90. For $\varepsilon = 0.005$ and $\chi = 2$, the range $\mathcal{R}$ has transition points $\overline{Q}_m$ between 0.35 and 0.40, and $\overline{Q}_M$ between 0.65 and 0.70. Hence, we conclude that the range $\mathcal{R}$ becomes smaller with increasing χ.

Finally, we mention some further results from [1],

- Based on the numerical results presented in Fig. 7.11, the frequency of the persistent oscillations is calculated by means of a Fourier spectral analysis. These frequencies are compared with the frequencies that come up as a solution of a linear stability analysis of the steady state solution. This

comparison shows a perfect correspondence between the numerical and the analytical results.
- The influence of the deformation history has been elucidated by numerical simulations of loading and unloading processes. The following peculiarities are observed:
  - the boundary of the spurt layer remains fixed also after unloading (shape memory);
  - persistent oscillations may occur after unloading to supercritical as well as to subcritical flow rates;
  - the deformation history of the melt affects the transient flow behaviour and the steady state attained.
- The models with one relaxation rate and supplied by a Newtonian term as considered here, are compared with models having two, sufficiently widely spaced, relaxation rates. Similar results are obtained for these latter models.

7.2.4 Discrete Melt Fracture Models

The results presented in this section are mainly from the PhD-thesis of Den Doelder, [2].
The main types of melt fracture, showing up as extrudate distortions, can be related to a specific region in the processing geometry:

- surface distortions: initiated near the die exit region,
- spurt distortions: initiated near the die land region,
- volume distortions: initiated near the die entry region.

A simple mathematical model for *spurt* is presented. Spurt is a type of instability occurring in the extrusion of polymer melts. It shows up by periodical
jumps in the volumetric flow rate and by relaxation oscillations in the pressure. In capillary flow, spurt is characterized by the occurrence of a thin
layer of very high shear rates near the capillary wall (the spurt zone). By
splitting up the flow region in a (narrow) spurt zone, where the viscosity
is very small, and a kernel, with $O(1)$-viscosity, a discrete model for spurt
flow is constructed. At the wall a no-slip condition is maintained (cf [20]).
The model is compared with a model allowing slip, and a complete equiva-
lence is shown (cf [15] or [21]). On the basis of the model, spurt or relaxation
oscillations are found, which are, qualitatively, in correspondence with experimental observations. When an extrusion process is in the spurt regime, the
extrudate shows alternating smooth and distorted regions. In this stage, the
volumetric flow rate Q periodically jumps between a lower value ($\Rightarrow$ *smooth
surface: classical flow*) and a much higher value ($\Rightarrow$ *distorted surface: spurt
flow*). Also the pressure gradient, driving the capillary flow, shows oscillations. These oscillations are many times observed in experiments, and look

like relaxation oscillations. Therefore, we shall refer to them as *spurt* or *relaxation oscillations*. In *spurt*, as it is to be understood here, the flow profile looks strongly different in two regions:

- in the inner region (the kernel) the velocity gradient is small and the flow profile is flat, looking very much like cork flow; in this inner region the flow is similar to classical Poiseuille flow;
- in a very small region close to the wall the velocity profile is very steep, yielding very large values for the velocity gradient; this region is called the spurt zone.

The extrusion device we consider consists of a huge barrel filled with polymeric melt, closed at one side by a movable plunger, and on the other side connected to a narrow cylindrical capillary (see Fig. 7.1). The moving plunger compresses the polymeric melt in the barrel and forces it to flow into the capillary. In the barrel a uniform state is assumed, and the polymeric melt in it is taken to be compressible. The unknowns here are the pressure $P(t)$ and the volumetric flow rate $Q(t)$ flowing from the barrel into the capillary. The relations to be used here are:

- a global balance law for the total mass of the melt in the barrel;
- a constitutive (linearly elastic) compressibility law, relating the pressure to the density.

This results in the following relation, in dimensionless form, (see (7.1))

$$\frac{dP(t)}{dt} = -(Q(t) - Q_i). \tag{7.57}$$

For the shear dominated flow in the capillary we assume:

- laminar incompressible flow;
- pressure linear in the axial coordinate z;
- inertia term negligible.

The axial component of the equation of motion then yields the following relation for the, dimensionless, shear stress τ (see (7.13) and (7.38); $T_{rz} \rightarrow -\tau$)

$$\tau(r,t) = 4rP(t), \qquad 0 \leq r < 1. \tag{7.58}$$

These relations must be supplemented by

- a constitutive equation for the shear stress τ ;
- a relation for the volumetric flow rate (see (7.45))

$$Q(t) = \int_0^1 r^2 w(r,t)dr. \tag{7.59}$$

Two results of the theory for constitutive instabilities motivated us to propose a discrete model consisting of two distinct Newtonian fluids in concentric die regions. These results are

- The non-monotonicity of the stationary shear curve (total shear stress at the wall versus shear rate), having as a consequence that the apparent viscosity in the spurt zone is much smaller than the one in the kernel.
- The shape memory, implying that the thickness of the spurt layer remains constant during spurt.

Discrete Model

Our model is based on the possible existence of a spurt zone in capillary flow. The existence of such a spurt zone is regulated by a switch function. During spurt, the spurt zone reaches from $r = r^* < 1$, to $r = 1$, $(1 - r^* \ll 1)$, whereas in classical flow no spurt zone exists ($r^* = 1$). The discrete model consists of a set of three equations for the pressure $P(t)$, the volumetric flow rate $Q(t)$ and the thickness of the spurt zone $(1 - r^*(t))$. This set can be solved analytically. The analytical calculations make use of asymptotics that are based on different typical time scales during distinct phases of the process (spurt flow versus classical flow). The obtained analytical results clearly predict relaxation oscillations.

The difference between classical flow and spurt flow is essentially expressed in τ. Both in the classical region as well as in the spurt zone a linear (Newtonian) relation for the shear stress is assumed, however, with different values for the viscosity.

1. In classical flow we take (the normalized viscosity is one, here) $\tau = w$, yielding

$$\tau(r, t) = w(r, t) = 4rP(t). \tag{7.60}$$

2. In the spurt zone, due to the very high shear rate, the viscosity is much smaller; so there

$$\tau(r, t) = \varepsilon w(r, t) = 4rP(t), \qquad 0 < \varepsilon \ll 1. \tag{7.61}$$

3. Transitions from classical to spurt flow, and vice versa, are very fast.
4. Shape memory causes the thickness of the spurt zone to remain constant during spurt.

Our discrete model is formulated in the following three sets of equations:
A. The spurt zone reaches from $r = r^* \leq 1$ to $r = 1$, where $(1 - r^*) \ll 1$. If $r^* = 1$: no spurt zone exists (classical flow). Hence,

$$w(r, t) = 4rP(t), \quad 0 \leq r < r^* ; \quad w(r, t) = \frac{4rP(t)}{\varepsilon}, \quad r^* < r \leq 1. \tag{7.62}$$

B. The volumetric flow rate, (7.59), can be evaluated into

$$Q(t) = P(t) \left((r^*)^4 + \frac{(1 - (r^*)^4)}{\varepsilon} \right) \approx P(t) \left(1 + \frac{4}{\varepsilon}(1 - r^*) \right), \qquad (7.63)$$

since $(1 - r^*) \ll 1$.

We define the normalized thickness of the spurt zone by

$$R(t) = (1 - r^*)/\varepsilon, \quad \Rightarrow \quad Q(t) = P(t) + 4R(t)P(t). \qquad (7.64)$$

C. Finally, we propose an evolution equation for $R(t)$ as

$$\frac{dR}{dt} = -\lambda[R(t) - \alpha H(P - B(Q))], \qquad (7.65)$$

with λ and α material parameters, H the Heaviside function, and $B(Q)$ a switch curve, as defined in Fig. 7.10. Since, by assumption, the transition from classical to spurt flow, and vice versa, is very fast, we have $\lambda \gg 1$.

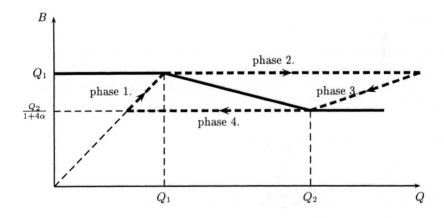

Fig. 7.10. The switch curve $B(Q)$ (*bold line*). *Dashed line* is relaxation loop

Recapitulating, we have the following discrete model:

– Unknowns: $P(t), \quad Q(t), \quad R(t)$.

– Equations

$$\frac{dP(t)}{dt} = Q_i - Q(t), \qquad (7.66)$$

$$Q(t) = (1 + 4R(t))P(t), \qquad (7.67)$$

$$\frac{dR(t)}{dt} + \lambda R(t) = \alpha\lambda H(P - B(Q)). \qquad (7.68)$$

– Initial conditions: $P(0) = R(0) = 0$.

Comparison of Spurt and Slip Model

The results of the theory of constitutive instability are compared to those for a stick-slip model, based on a model for slip at the wall, in agreement with a slip law (cf. [4] or [15]). In this slip model, $v(1,t)$ can be either larger than zero or equal to zero, depending on whether slip occurs or not. When slip starts, the flow rate suddenly increases. The slip velocity at the wall is taken proportional to the maximum shear stress at the wall, which at its turn is proportional to the pressure $P(t)$. The appearance of slip is governed by an evolution equation of the same type as the one for $R(t)$. Thus, in this model, the velocity at the wall is governed by the relation

$$v(1,t) = G(t)P(t), \tag{7.69}$$

where $G(t)$ satisfies a similar evolution equation as $R(t)$ in (11). Stick or slip occurs accordingly to $G(t) = 0$, or $G(t) > 0$, respectively.
Here, everywhere in the flow $\tau = w = 4rP(t)$, yielding for the flow rate

$$Q(t) = 2\int_0^1 rv(r,t)dr = v(1,t) + \int_0^1 r^2 w(r,t)dr = (1 + G(t))P(t). \tag{7.70}$$

Hence, when $G(t) \to 4R(t)$ exactly the same model as (7.66)–(7.68) follows. For both models relaxation oscillations of similar type in both the pressure $P(t)$ and the volumetric flow rate $Q(t)$ are found. Hence, from a mathematical point of view, the two models are equivalent.

Analysis of the Discrete Model

The nonlinear system (7.66)–(7.68) can be solved analytically if we make use of asymptotics that are based on $\lambda \gg 1$ (fast transitions). For this, we distinguish four phases, of which phase 1. represents classical flow, phase 3. spurt flow (or *slip*), while the phases 2. and 4. are transition phases from classical to spurt flow (*stick to slip*) and vice versa, respectively. Hence, in phase 1., $R(t) = 0$, whereas in phase 3., $R(t) = \alpha$. During the very short $(O(\lambda^{-1}))$ phases 2. and 4., $R(t)$ jumps from $0 \to \alpha$, and from $\alpha \to 0$, respectively. Since $R(t)$ is constant during the phases 1. and 3. (shape memory), the system (7.66)–(7.68) is linear then and can easily be solved. At the other hand, the transition phases are so short $(O(\lambda^{-1}))$ that the changes in $P(t)$ according to (7.66) are also of $(O(\lambda^{-1}))$ and, hence, negligible in an approximation for $\lambda \gg 1$. Thus, $P(t)$ may be taken constant during these phases, and again the system (7.66)–(7.68) is linear then and easy to solve.
Therefore, we distinguish the following four phases:

Phase 1. $\underline{0 < t < t_1}$; classical flow.

In this initial phase $P < B(Q)$ and, hence, $H(P - B(Q)) = 0$. Then (7.68), with $R(0) = 0$, yields $R(t) = 0$. This reduces (7.66)–(7.68) to

$$\frac{dP}{dt} = Q_i - Q(t), \qquad P(0) = 0,$$
$$Q(t) = P(t), \tag{7.71}$$

the solution of which reads

$$P(t) = Q(t) = Q_i(1 - e^{-t}). \tag{7.72}$$

The end of phase 1. is at $t = t_1$, where t_1 is the time where $P(t)$ reaches for the first time the switch curve $B(Q)$. Hence (see Fig. 7.10), $P(t_1) = Q_1$, yielding

$$t_1 = \ln(\frac{Q_i}{Q_i - Q_1}), \tag{7.73}$$

provided $Q_i > Q_1$.
We note that $t_1 = O(1)$, which justifies our time scaling.

Phase 2. $t_1 < t < t_2$: transition from classical to spurt flow.

Since $P(t)$ crosses $B(Q)$ from below at $t = t_1$, we assume that during this phase $P > B(Q)$, so $H(P - B(Q)) = 1$.
For this phase we introduce a new time scale: $\tau = \lambda(t - t_1)$, such that (7.66)–(7.68) becomes ($R = R(\tau)$, etc.)

$$\frac{dR}{d\tau} + R(\tau) = \alpha, \qquad R(0) = 0,$$
$$\frac{dP}{d\tau} = \frac{1}{\lambda}(Q_i - Q(\tau))(= O(\lambda^{-1})), \quad P(0) = Q_1,$$
$$Q(\tau) = (1 + 4R(\tau))P(\tau). \tag{7.74}$$

The solution of this system reads

$$P(\tau) = P(0)(1 + O(\lambda^{-1})) \approx Q_1,$$
$$R(\tau) = \alpha(1 - e^{-\tau}) \ (\to \alpha),$$
$$Q(\tau) = [1 + 4\alpha(1 - e^{-\tau})]Q_1 \ (\to (1 + 4\alpha)Q_1). \tag{7.75}$$

We define the end of phase 2. as $\tau = \tau_2$, such that $e^{-\tau_2} = \lambda^{-1}$. This yields

$$\tau_2 = \ln \lambda, \quad \text{or} \quad t_2 = t_1 + \frac{\ln \lambda}{\lambda} = t_1(1 + O(\lambda^{-1})). \tag{7.76}$$

So, indeed $\tau_2 = O(1)$, implying that the time that phase 2. lasts is very short $(O(\lambda^{-1}))$ compared to phase 1.

Phase 3. $t_2 < t < t_3$; spurt flow.

In this phase we assume $P > B(Q)$ (see Fig. 7.10), so $H(P - B(Q)) = 1$.

Since $(7.75)^3$ yields $R(t_2) = \alpha(1 + O(\lambda^{-1})) \approx \alpha$, for $\lambda^{-1} \to 0$, (7.68) renders $R(t) = \alpha$, for all $t \in (t_2, t_3)$.

With the new time scale $\tau = (t - t_2)$, (7.66)–(7.68) reduces to ($P = P(\tau)$, etc.)

$$\frac{dP}{d\tau} = Q_i - Q(\tau), \quad P(0) = Q_1,$$
$$Q(\tau) = (1 + 4\alpha)P(\tau). \tag{7.77}$$

The solution of this system reads

$$P(\tau) = \frac{Q(\tau)}{1 + 4\alpha} = \frac{Q_i}{1 + 4\alpha} + \left[Q_1 - \frac{Q_i}{1 + 4\alpha}\right]e^{-(1+4\alpha)\tau}. \tag{7.78}$$

Since phase 3. always runs along the line $P = Q/(1+4\alpha)$ in a $P-Q$-diagram, it must be so that if this line crosses the switch curve $B(Q)$ this happens in the point $P = B(Q_2) = Q_2/(1 + 4\alpha)$. According to (7.78), $Q(\tau) \to Q_i$, for $\tau \to \infty$. As $Q \to Q_i$, there are now two possibilities:

1. If $Q_i > Q_2$,

   then $P(\tau) \to Q_i/(1 + 4\alpha) > B(Q_i) = Q_2/(1 + 4\alpha)$
   In this case no transition takes place, and phase 3. tends to a final stationary spurt state, in which $(P(\tau), Q(\tau)) \to (Q_i/(1 + 4\alpha), Q_i))$, for $\tau \to \infty$.

2. If $Q_i < Q_2(< (1 + 4\alpha)Q_1$,

   then a transition to classical flow takes place when $Q(t)$ reaches $Q_2 > Q_i$, and then phase 4. starts at $t = t_3(< \infty)$. Here, t_3 is such that $Q(t_3) = Q_2$, yielding

$$t_3 = t_2 + \frac{1}{(1 + 4\alpha)} \ln\left(\frac{(1 + 4\alpha)Q_1 - Q_i}{Q_2 - Q_i}\right). \tag{7.79}$$

We assume case 2. to hold, and we proceed with phase 4. We shall see that in this case relaxation oscillations occur.

Phase 4. $t_3 < t < t_4$; transition from spurt to classical flow.

At $t = t_3$, $P(t)$ crosses the switch curve coming from above, so during this phase we assume $P < B(Q)$, and thus $H(P - B(Q)) = 0$

With the new time scale $\tau = \lambda(t - t_3)$, (7.66)–(7.68) reduces to ($R = R(\tau)$, etc.)

$$\frac{dR}{d\tau} + R(\tau) = 0, \quad R(0) = \alpha,$$
$$\frac{dP}{d\tau} = \frac{1}{\lambda}(Q_i - Q(\tau))(= O(\lambda^{-1})), \quad P(0) = Q_2/(1 + 4\alpha),$$
$$Q(\tau) = (1 + 4R(\tau))P(\tau). \tag{7.80}$$

The solution of this system reads

$$P(\tau) = \frac{Q_2}{(1 + 4\alpha)}(1 + O(\lambda^{-1})) \, ,$$

$$R(\tau) = \alpha e^{-\tau} \quad (\to 0) \, ,$$

$$Q(\tau) = \frac{Q_2}{(1 + 4\alpha)}(1 + 4\alpha e^{-\tau}) \, . \tag{7.81}$$

Analogous to phase 2., phase 4. ends at $t = t_4 = t_3(1 + O(\lambda^{-1}))$.
At $t = t_4$, phase 1. starts anew, but now not from $P(0) = Q(0) = 0$, but from

$$P = Q = Q_2/(1 + 4\alpha).$$

This brings us to:

Phase 5. $t_4 < t < t_5$; classical flow.

Analogous to Phase 1., the solution now reads

$$P(t) = Q(t) = Q_i + \left(\frac{Q_2}{1 + 4\alpha} - Q_i \right) e^{-(t - t_4)}. \tag{7.82}$$

This phase ends at $t = t_5$ when $Q(t_5) = Q_1$, yielding

$$t_5 = t_4 + \ln \left(\frac{Q_i - Q_2/(1 + 4\alpha)}{Q_i - Q_1} \right) \, , \tag{7.83}$$

after which a phase identical to phase 2. follows.

Thus, a loop is followed as depicted by the dashed line in Fig. 7.10. Since this dashed line is a closed loop, it represents a periodic phenomenon. Its behaviour is of relaxation type, because the phases 2. and 4. are extremely short. Therefore, we call this a relaxation oscillation. The period of one oscillation is $(t_3 - t_2) + (t_5 - t_4)$, or

$$T_{os} = \ln \frac{((1 + 4\alpha)Q_1 - Q_i)((1 + 4\alpha)Q_i - Q_2)}{(1 + 4\alpha)(Q_2 - Q_i)(Q_i - Q_1)} \, , \tag{7.84}$$

for $\lambda^{-1} \approx 0$. The behaviour of the pressure $P(t)$ and the volumetric flow rate $Q(t)$ during these relaxation oscillations is depicted in Fig. 7.11.

Results

Our model described by the system (7.66)–(7.68) is characterized by the parameter set $\{Q_1, Q_2, \alpha, \lambda\}$ plus the prescribed inlet flow rate Q_i . In the preceding section we have seen that dependent on the value of Q_i different types of capillary flow can occur. We distinguish three regimes for Q_i, knowing $Q_i < Q_1$, $Q_1 < Q_2 < Q_1$, and $Q_i > Q_2$. For the numerical results in this section, the following fixed values for the parameters are used:

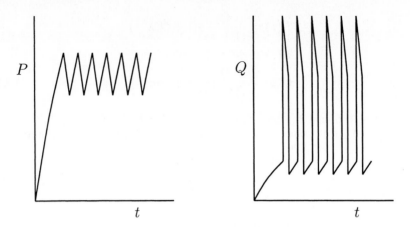

Fig. 7.11. Relaxation oscillations for the pressure $P(t)$ and the volumetric flow rate $Q(t)$ ($Q_1 = 1$, $Q_2 = 3$, $Q_i = 2$, $\alpha = 1$, and $\lambda = 1000$)

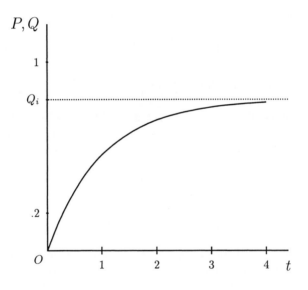

Fig. 7.12. The pressure $P(t)$ and the flow rate $Q(t)$ as function of time t for $Q_i = 0.8 < Q_1$ (classical flow)

$$Q_1 = 1 \, , Q_2 = 3 \, , \alpha = 1 \, , \lambda = 1000 \, .$$

We find:

- In the first regime, $Q_i < Q_1$, the flow is classical, just like a Poiseuille flow. Since there is no spurt zone, the velocity profile is smooth. The pressure and the flow rate tend monotone to their stationary values, according to (7.72). This behaviour is depicted in Fig. 7.12.
- In the second regime $Q_1 < Q_i < Q_2$, persistent relaxation oscillations occur. The flow periodically jumps from classical to spurt and vice versa, and large jumps in the pressure and, especially, the flow rate are found. The value of the flow rate is relatively high during spurt, and low during classical flow. Typical relaxation oscillations in $P(t)$ and $Q(t)$ in case $Q_i = 2$ are depicted in Fig. 7.11.
- In the third regime $Q_i > Q_2$, the flow again tends to a stationary state, but now to one in which spurt occurs. The spurt zone is fixed to $R = \alpha$, and pressure and flow reach the stationary values $\bar{P} = Q_i/(1+4\alpha)$, and $\bar{Q} = Q_i$, respectively, according to (7.78). In this regime we can further distinguish between $Q_i < (1 + 4\alpha)$ and $Q_i > (1 + 4\alpha)$. In case $Q_i < (1 + 4\alpha)$, an overshoot in both $P(t)$ and $Q(t)$ occurs, before they reach their final state. This overshoot at $t = t_1$ is depicted in Fig.7.13, for the case that $Q_i = 4$. If $Q_i > (1+4\alpha)$, no overshoot occurs; in this case the steady state is reached in a monotone way.

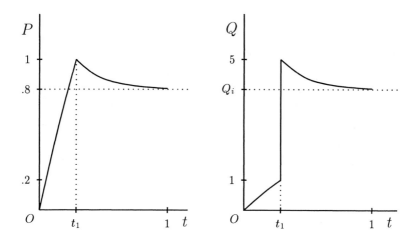

Fig. 7.13. The pressure $P(t)$ and the flow rate $Q(t)$ as function of time t for $Q_i = 4 > Q_2$ (spurt flow)

A finite range of prescribed inlet flow rates Q_i can be found for which relaxation oscillations can occur. This is illustrated by the simulation of a realistic loading process, depicted in Fig. 7.14, in which only for a restricted range of Q_i–values spurt oscillations show up. For Q_i–values below this range $(Q_i < Q_1)$ the flow tends to a stationary classical (Poiseuille) flow, whereas for values beyond this range $(Q_i > Q_2)$ the flow tends to a stationary spurt flow (having a fixed spurt layer).

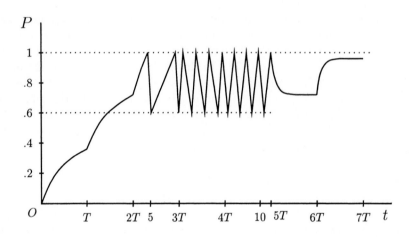

Fig. 7.14. The pressure $P(t)$ as function of time t for a loading process in which Q_i is step wise increased at time steps T, with $T = 2.1$, according to the sequence $\{0.4, 0.8, 1.2, 1.8, 2.4, 3.6, 4.8\}$

Improvement of the Model

A lot of experimental data on especially extrusion instabilities are known (in literature and from internal experiments at DOW). Comparison of our analytical results as depicted in Figs. 7.11 –7.14 with experimental results indicates a good *qualitative* agreement. However, for a *quantitative* agreement, some further modifications of the model presented here are needed. This is done by Den Doelder in his PhD-thesis ([2, Chapter 4]; see also [22]). These modifications concern:

– A nonlinear evolution equation, emanating from a nonlinear slip law. This nonlinear slip law relates the slip velocity to the *square* of the wall shear stress. Translated to our model, this would imply that relation (7.67) must become nonlinear in so far that $P(t)$ must be replaced by $P^2(t)$. A direct physical support for the exponent 2 does not exist yet; perhaps theories based on 3-dimensional models or on models for polymer-die wall slip-stick interactions can be developed in future to generate this parameter. At this moment, we can only say that the value 2 for the exponent of $P(t)$ in the slip law provides the best fit for the experiments performed by Den Doelder (see [2, Section 4.5])

- An improved constitutive model for the polymer fluid. The Newtonian model is replaced by a power law (shear thinning) model (generalised Newtonian ; no elasticity).
- Elastic effects in the die exit and entry region are incorporated in a pressure correction term (Bagley correction).
- The time dependence of the length l of the barrel ($l = l(t) = l(0) - V_0 t$) is taken into account.

The numerical results of this improved model are compared with experimental data from experiments performed at DOW. The calculated relaxation oscillations of the pressure are in very good *quantitative* correspondence with these data. The absolute pressure values, the amplitudes of the oscillations and the period of the oscillations compare very well with the experimental values (cf. [22]).

Conclusions

We state the following conclusions:

- The discrete models, either with no-slip and spurt zone or with a stick-slip law, explain relaxation oscillations in $P(t)$ and $Q(t)$.
- Comparison with experimental results indicates a good qualitative and quantitative agreement with experimentally observed relaxation oscillations in $P(t)$.

7.3 Injection Moulding

7.3.1 Introduction

The filling phase of an injection moulding process in a straight cavity, consisting of the narrow space between two adjacent parallel circular plates, inclusive the front motion at the free boundary of the flow, will be analysed. The cavity is *thin*, meaning that the distance between the plates is much smaller than their radius. The injection inlet is at the center of the plate and the resulting flow is rotationally symmetric. For some applications, the flow will be considered as being two-dimensional. Besides the flow fields (velocity, stresses) also the temperature and pressure fields must be calculated. Thermal effects are due to cooling at the walls and to viscous dissipation; pressure effects are due to compressibility and to interactions with thermal fields. The material coefficients of the melt (especially the viscosity) are strongly temperature and/or pressure dependent. The melt can be modelled as either a generalised Newtonian fluid or, if necessary, as a nonlinear viscoelastic fluid. In any case, the viscosity should be dependent on both shear rate (*shear thinning*) as well as temperature.

One specific application of injection moulding is in the production of compact discs (CD's). For an adequate description of this process, a thermoviscoelastic model is needed. Essential in this modelling is a simulation of the cooling of the hot polymer melt in the mould or cavity. Another point of utmost practical importance is the calculation of residual stresses and deformations in the final product.

Finally, we mention the so called flow front instability, which is observed in injection moulding as the wobbling of the flow front (up and down between the two plates). We are interested in an explanation of this instability by *analytical means*. For this, we need, amongst other things, an analytical solution for the temperature distribution in the melt, especially near to the upper and lower wall of the cavity. In order to be able to find analytical solutions, we have to make the model as simple as possible, though keeping the essential features. In our view, the most important feature is the coupling between the viscosity and the temperature of the fluid, as the latter is strongly influenced by the cooling at the walls of the cavity.

The flow front instability can be either due to thermal effects or to elasticity effects, or to combinations of both. We focus here purely on thermal effects. For this, we notice that an asymmetric flow front causes differences in temperature, and thus also in viscosity, due to different thermal contacts at upper and lower wall of the cavity. These differences in viscosities can induce an ongoing asymmetry of the flow front, which can result into instability. To explain our view further, let us assume there is initially an asymmetric flow front (for instance due to the asymmetric inlet of the mould, which is filled from above) and that at certain time t the contact point of the flow front with the upper wall is farther (in the r-direction) than the corresponding point at the lower wall (i.e. $R_f(+h, t) > R_f(-h, t)$). In that case, the fluid near the upper wall will cool more than the fluid below, and so the fluid will be stiffer (higher viscosity) at the upper wall than at the lower one. Consequently, the fluid will be retarded at the upper wall with respect to the fluid at the lower wall and, thus, the asymmetry will flip. This means that at a somewhat later time the flow front is farther at $z = -h$ than at $z = +h$. This cyclus will repeat itself and then the front is wobbling. We adhere to this phenomenon the *front instability* we are looking for.

7.3.2 Problem Description

The cylindrical cavity we consider here has height $2h$ and radius R, where $h \ll R$. The rotationally symmetric flow in the cavity is described in cylindrical coordinates (r, z). In the *filling phase* when the fluid is entering the cavity, the flow has a front which is an unknown free surface. The fluid occupies the flow region Ω behind the front (see Fig. 7.15). Let $r = R_f(z,t)$ denote the radial position of a point on the flow front on height z. Then the flow region is given by

$$\Omega(t) = \{(r,z) \mid R_n < R_i < r < R_f(z,t) < R, \ -h < z < h \} . \tag{7.85}$$

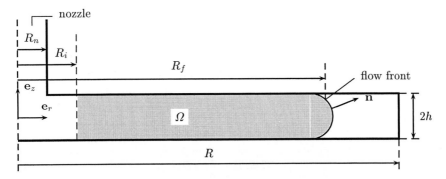

Fig. 7.15. The flow region G in the cavity.

To model the viscoelastic flow in the filling phase (see also [6]), we start with the formulation of the local balance equations, the constitutive equations and the boundary conditions.

Local Balance Equations. The fundamental unknowns in the injection process to be considered here are the velocity $\mathbf{v}(r,z,t) = v_r\mathbf{e}_r + v_z\mathbf{e}_z$, the pressure $p(r,z,t)$ and the temperature $T(r,z,t)$. They are governed by the balance laws of continuum mechanics, here written as

– *Equation of continuity*

$$\dot{\rho} + \rho\nabla \cdot \mathbf{v} = 0 , \tag{7.86}$$

where ρ is the density, and $\dot{\rho}$ is its material derivative.
– *Equation of motion*

$$\rho\dot{\mathbf{v}} = \nabla \cdot \mathcal{T} , \tag{7.87}$$

in the absence of body forces, where $\mathcal{T} = \mathcal{T}(r,z,t)$ is the total stress tensor.
– *Balance of energy*

$$\rho\dot{u} = \mathcal{T} : \mathcal{D} - \nabla \cdot \mathbf{q} , \tag{7.88}$$

where u is the internal energy density, $\mathcal{D}$ the rate-of-deformation tensor (see (2.35)), $\mathbf{q}$ the heat flux vector, while the heat source is assumed zero.

Constitutive Equations. The balance equations should be supplemented by constitutive equations for the stresses and the heat flux together with an explicit expression for the internal energy density and, finally, especially for compressible melts a constitutive equation for ρ as function of T and p:

– *Stresses.*

The total stress tensor is decomposed into a pressure part and an extra stress according to

$$\mathcal{T} = -p\mathcal{I} + \mathcal{S} . \tag{7.89}$$

The extra stress will be described by either a generalised Newtonian model or by a nonlinear viscoelastic model (e.g. Leonov model):
– For the generalised Newtonian model we have

$$\mathcal{S} = 2\eta \mathcal{D}^d , \tag{7.90}$$

where $\mathcal{D}^d$ is the deviatoric part of $\mathcal{D}$, i.e.

$$\mathcal{D}^d = \mathcal{D} - \frac{1}{3}tr(\mathcal{D})\mathcal{I} , \tag{7.91}$$

and where the viscosity η is a function of the shear rate $\dot{\gamma}$ defined by

$$\dot{\gamma} = \sqrt{2\mathcal{D}^d : \mathcal{D}^d} , \tag{7.92}$$

and the temperature ($\eta = \eta(\dot{\gamma}, T)$).
– For the Leonov model we have

$$\mathcal{S} = 2\eta_r \mathcal{D}^d + \sum_{k=1}^{K} \tau_k , \tag{7.93}$$

where η_r is the so called retardation viscosity and τ_k are the substresses, having K modes.
– *Heat flux.*
We use Fourier's law

$$\mathbf{q} = -\lambda \nabla T , \tag{7.94}$$

where the heat conduction coefficient λ still can depend on temperature and pressure.
– *Internal energy density.*
Using some relations from thermodynamics (a.o. *Gibbs law*), we can derive

$$\dot{u} = c_p \dot{T} - \frac{\alpha T}{\rho}\dot{p} + \frac{p}{\rho^2}\dot{\rho} , \tag{7.95}$$

where α is the coefficient of thermal expansion ($\rho\alpha = \partial\rho/\partial T$) and c_p is the heat capacity at constant pressure ($c_p = \partial u(T, p)/\partial T$).

– *Density*

Considering for a compressible melt $\rho = \rho(T, p)$, we find

$$\dot{\rho} = \rho \kappa \dot{p} - \rho \alpha \dot{T} , \tag{7.96}$$

where κ is the isothermal compressibility ($\rho \kappa = \partial \rho / \partial p$).

In case of an incompressible medium, we have $\rho = \rho_0$, and then the pressure $p = p(x, z, t)$ is an unknown field variable; as an extra equation, we then have the incompressibility condition div $\mathbf{v} = \nabla \cdot \mathbf{v} = 0$.

All the material coefficients appearing in the above relations are in general functions of temperature, pressure and shear rate.

Final Equations. Substituting the constitutive equations listed above into the local balance equations, we arrive at the following set of equations for the fundamental unknowns $p, \mathbf{v}$ and T:

$$\rho \kappa \dot{p} - \rho \alpha \dot{T} + \rho \nabla \cdot \mathbf{v} = 0 , \tag{7.97}$$

$$\rho \dot{\mathbf{v}} + \nabla p = \nabla \cdot \mathcal{S} , \tag{7.98}$$

$$\rho c_p \dot{T} - \alpha T \dot{p} = \mathcal{S} : \mathcal{D} + \nabla \cdot (\lambda \nabla T) . \tag{7.99}$$

Boundary Conditions. We proceed with the boundary conditions on $\partial \Omega$. The boundary $\partial \Omega$ of Ω consists of the following three segments:

1. *Inlet:* $r = R_i > R_n$; $-h < z < h$.
   Here, R_i is chosen a few times h larger than R_n (see Fig. 7.15), in order to get rid of inlet disturbances. At $r = R_i$ the flow is assumed fully developed and the pressure and the temperature are taken equal to their inlet values. This leads to the boundary conditions (p_i, $V(z)$ and T_i prescribed)

$$p(R_i, z, t) = p_i ,$$
$$\mathbf{v}(R_i, z, t) = V(z) \mathbf{e}_z ,$$
$$T(R_i, z, t) = T_i . \tag{7.100}$$

2. *Walls of the cavity:* $z = \pm h$; $R_i < r < R_f(\pm h, t)$.
   At these walls no-slip is assumed and the walls are cooled at a prescribed temperature T_w, which, in general, can still be a function of r and different at the upper and lower wall ($T_w^{\pm}(r)$ prescribed)

$$\mathbf{v}(r, \pm h, t) = \mathbf{0} ,$$
$$T(r, \pm h, t) = T_w^{\pm}(r) . \tag{7.101}$$

3. *Flow front*: $r = R_f(z,t)$; $-h < z < h$.

The flow front is a free boundary, which is assumed free of stress (the pressure in the environment is scaled to zero) and thermally insulated. Moreover, the front moves with a velocity V_f in r−direction. This leads to:

$$\mathcal{T}\mathbf{n} = \mathbf{0} \, ,$$
$$((\mathbf{v} - V_f \mathbf{e}_r), \mathbf{n}) = 0 \, ,$$
$$\frac{\partial T}{\partial n} = 0 \, , \qquad \text{at } r = R_f(z,t) \, . \qquad (7.102)$$

The equations and boundary conditions presented above form a complete and consistent system for the simulation of an injection moulding process in a cylindrical cavity between two parallel plates.

7.3.3 Zeroth Order Problem

In view of our aim to find by analytical means an explanation for the flow front instability, we need to simplify the general problem described in the preceding section as much as possible. In our zeroth order model we take a.o. all material coefficients, including the viscosity η, constant. In a first order perturbation we can then incorporate the temperature dependence of η in the model. However, here we only present the zeroth order model, in which we make the following simplifications:

- We assume the fluid to be incompressible ($\alpha = \kappa = \nabla \cdot \mathbf{v} = 0$) and Newtonian.
- We assume all material coefficients occurring in (7.97), (7.98) and (7.99) to be constant.
- We neglect the inertia term in the equation of motion with respect to the viscous term (quasi-static approximation).
- We neglect the viscous dissipation term in the energy equation.
- We consider the cavity to be a two-dimensional slit (this means that we have to replace $\mathbf{e}_r$ and r by $\mathbf{e}_x$ and x, respectively).
- We take the prescribed temperature at the upper and lower wall equal and uniform.
- We assume that the changes in temperature remain small (we are only aiming at the first order effect).

Based on the latter assumption, we split up our zeroth order problem in a basic problem in which the velocity field and the temperature are uncoupled.

This results in two separate basic problems:

1. *The velocity problem*, unknowns $\mathbf{v} = \mathbf{v}(x, z, t) = u\mathbf{e}_x + w\mathbf{e}_z$, and $p = p(x, z, t)$, (the temperature is here taken equal to the uniform reference temperature) governed by

$$\nabla \cdot \mathbf{v} = \frac{\partial u}{\partial x} + \frac{\partial w}{\partial z} = 0 \,,$$

$$-\frac{\partial p}{\partial x} + \eta \Delta u = 0 \,,$$

$$-\frac{\partial p}{\partial z} + \eta \Delta w = 0 \,, \tag{7.103}$$

for $(x, z) \in \Omega, \ t > 0$, together with the boundary conditions (here, $x = 0$ corresponds to $r = R_i$)

$$u(0, z, t) = V(z) = \frac{3}{2} V_f \left(1 - \left(\frac{z}{h} \right)^2 \right) \,,$$

$$u(x, \pm h, t) = 0 \,,$$

$$\mathcal{S} \mathbf{n} = p \, \mathbf{n} \,, \quad ((\mathbf{v} - V_f \mathbf{e}_x) \cdot \mathbf{n}) = 0 \,, \quad \text{at the front} \,. \tag{7.104}$$

From this problem not only the velocity and the pressure can be solved, but also a zeroth order version for the shape of the flow front, yielding a symmetric flow front.

2. *The temperature problem*, unknown $T(x, z, t)$ ($\mathbf{v}$ is known from 1.), with governing equation

$$\frac{\partial T}{\partial t} + (\mathbf{v} \cdot \nabla T) = \kappa \Delta T \,, \tag{7.105}$$

where $\kappa = \lambda / \rho c_p$: the thermal diffusion constant.
The boundary conditions are

$$T(0, z, t) = T_i \,,$$

$$T(x, \pm h, t) = T_w \,,$$

$$\frac{\partial T}{\partial n} = 0 \,, \quad \text{at the front} \,. \tag{7.106}$$

From this problem the basic temperature distribution can be solved.

NOTE
To obtain the perturbed problem, we must substitute the perturbed fields, i.e.

$$\mathbf{v} \to \mathbf{v} + \delta \mathbf{v} \,, \quad p \to p + \delta p \,, \quad T \to T + \delta T \,,$$

into the original equations (7.97)–(7.99) and linearise the thus obtained set with respect to the perturbations. Moreover, we must also introduce the perturbed flow front, assumed in an asymmetric shape, and linearise the free-boundary conditions at the front.

Zeroth Order Velocity Problem

As shown by Van Vroonhoven and Kuijpers, [23], the velocity problem formulated in (7.103) and (7.104) can be solved by use of complex function theory.

For this the velocity and the pressure are expressed in holomorphic functions of a complex variable $(z = x + iz)$. By a conformal mapping $z = m(\zeta)$, the flow region Ω is mapped onto the unit circle in the ζ-plane. The problem in the ζ-plane is then reformulated as a Hilbert problem, which is solved analytically. The, still unknown, mapping function $m(\zeta)$ is determined by the free-boundary conditions at the flow front. This function is then approximated by a polynomial of degree N,

$$m(\zeta) \approx m_N(\zeta) = \sum_{k=0}^{N} \mu_k \zeta^k . \tag{7.107}$$

It turns out that already for small values of N (i.e. $N = 2$ or 3) this polynomial renders a very good approximation of the mapping function in the vicinity of the free boundary (i.e. in the right half of the unit circle). With $m(\zeta)$, also the shape of the flow front is determined. From the expressions for the velocity, the flow lines can be calculated (see also [24]).

Zeroth Order Temperature Problem

In this section we present a solution of the temperature problem (7.105) and (7.106). This solution is quite different in different regions in Ω (e.g. far behind the flow front; near the walls; near the flow front). We first make the equations dimensionless by introducing the dimensionless variables $\hat{t}, \hat{x}, \hat{z}, \hat{\mathbf{v}}$ and $\hat{T}$ by

$$t = \frac{L}{V_f}\hat{t} , \quad x = L\hat{x} , \quad z = h\hat{z} , \quad \mathbf{v} = V_f\hat{\mathbf{v}} , \quad \text{and} \quad \hat{T} = \frac{T - T_w}{T_i - T_w} , \tag{7.108}$$

where L is a characteristic measure of length in the x-direction, to be specified further on. The specific choice for L depends on which region in Ω we are considering.

In these new variables (7.105) becomes

$$\frac{\partial \hat{T}}{\partial \hat{t}} + \hat{u}\frac{\partial \hat{T}}{\partial \hat{x}} + \hat{w}\frac{\partial \hat{T}}{\partial \hat{z}} = \hat{\kappa}\left[\frac{\partial^2 \hat{T}}{\partial \hat{z}^2} + \left(\frac{h}{L}\right)^2 \frac{\partial^2 \hat{T}}{\partial \hat{x}^2}\right] , \tag{7.109}$$

with

$$\hat{\kappa} = \frac{\kappa L}{V_f h^2} , \tag{7.110}$$

while the boundary conditions (7.106) become

$$\hat{T}(0, \hat{z}, \hat{t}) = 1 ,$$
$$\hat{T}(\hat{x}, \pm 1, \hat{t}) = 0 ,$$
$$\frac{\partial \hat{T}}{\partial n} = 0 , \quad \text{at the front} . \tag{7.111}$$

We consider first the:

Region Far Behind the Flow Front

$$(x, z) \in \Omega_0, \qquad \text{where } \Omega_0 = \{(x, z) \mid 0 < x < x_f - kh, \; -h < z < h \},$$

with kh a few times h, and $x_f = V_f t = R_f(h, t) - R_n$ (see Fig. 7.16).
Here, we choose $L \gg h$. Moreover, we assume the temperature to be stationary $(T = T(x, z))$ and the flow fully developed. The latter means that

$$\hat{u} = \frac{3}{2}(1 - \hat{z}^2) , \quad \hat{w} = 0 . \tag{7.112}$$

The factor $3/2$ in (7.112) is due to the normalisation, which is such that the dimensionless flow rate pertinent to $\hat{u}$ should be 2 (note that the dimensionful flow rate Q equals $2hV_f$).
Finally, we use the fact that $h/L \ll 1$, implying that the diffusion in x-direction can be neglected in the heat equation (7.109). All this leads to the thin-layer approximation for the heat diffusion equation, i.e. (omitting the hats; the factor $3/2$ is incorporated in κ, i.e $\kappa \to 2\hat{\kappa}/3$)

$$(1 - z^2)\frac{\partial T}{\partial x} = \kappa \frac{\partial^2 T}{\partial z^2} ,$$
$$T(0, z) = 1 ,$$
$$T(x, \pm 1) = 0 , \quad \text{at the front} . \tag{7.113}$$

In practice, the dimensionless diffusion constant κ $(\equiv \hat{\kappa})$ shows up to be rather small $(\kappa = O(10^{-2}))$. Therefore, we use asymptotics based on $\kappa \ll 1$. For the *inner solution* (i.e. z not too close to ± 1) we then get

$$\frac{\partial T}{\partial x} = 0 , \quad \text{and} \quad T(0, z) = 1 , \tag{7.114}$$

yielding

$$T(x, z) = 1 , \tag{7.115}$$

(in fact, of course, $T(x, z) = 1 + O(\kappa)$). Since this solution does not satisfy the boundary conditions at $z = \pm 1$, we expect that there is a boundary layer in the vicinity of $z = \pm 1$. Since the problem is symmetric with respect to z, we only have to consider the boundary layer near $z = 1$.
    For the *outer solution*, holding for z in a *boundary layer* at $z = 1$ (or $z = -1$), where the gradient $\partial T/\partial z$ becomes large, we replace z by

$$z = 1 - \varepsilon \zeta \quad \Longrightarrow \quad \zeta = \frac{1 - z}{\varepsilon} , \tag{7.116}$$

where the small parameter ε $(0 < \varepsilon \ll 1)$, representing the thickness of the boundary layer, needs still to be determined.
Substitution of (7.116) into (7.113) yields

$$\varepsilon\zeta(2 - \varepsilon\zeta)\frac{\partial T}{\partial x} = \frac{\kappa}{\varepsilon^2}\frac{\partial^2 T}{\partial \zeta^2} \,. \tag{7.117}$$

Since we want the left-hand side of (7.117) to be of the same order as the right-hand side, ε must satisfy the following equation

$$\varepsilon^3 = \kappa \qquad \Longrightarrow \qquad \varepsilon = \sqrt[3]{\kappa} \,. \tag{7.118}$$

With this choice for ε, (7.117) can be rewritten as follows

$$2\zeta\frac{\partial T}{\partial x} - \frac{\partial^2 T}{\partial \zeta^2} = \varepsilon\zeta^2\frac{\partial T}{\partial x} \approx 0. \tag{7.119}$$

The boundary conditions are given by

$$T(0, \zeta) = 1,$$
$$T(x, 0) = 0,$$
$$T(\zeta \to \infty) = 1. \tag{7.120}$$

We try as a solution of (7.119) and (7.120)

$$T_1(x, \zeta) = \tilde{T}(\xi) \,, \tag{7.121}$$

where $\xi = \zeta \, x^\alpha$ for some value of α. Substitution of (7.121) into (7.119) shows that $\alpha = -1/3$ must hold, and consequently the solution can be found as

$$T(x, \zeta) = T_1(x, \zeta) = C\int_0^{\frac{\zeta}{\sqrt[3]{x}}} e^{-\frac{2}{9}s^3}\,ds \,, \tag{7.122}$$

where C is given by

$$C = \left(\int_0^\infty e^{-\frac{2}{9}s^3}\,ds\right)^{-1} \,. \tag{7.123}$$

We can proceed in a completely analogous way to derive second and higher order (in ε) terms in this asymptotic solution. The second order term is given by

$$T_2(\zeta, \xi) = -\frac{C\zeta}{10}\left(\int_\xi^\infty e^{-\frac{2}{9}s^3}\,ds + \xi e^{-\frac{2}{9}\xi^3}\right) \,. \tag{7.124}$$

The total solution for the temperature in Ω_0 is now

$$T(x, \zeta) = T_1(x, \zeta) + \varepsilon T_2(x, \zeta) + O(\varepsilon^2) \,. \tag{7.125}$$

We compare the asymptotic solution of (7.125) with a numerical solution obtained by means of a finite difference method. In Table 7.1 the results are shown for the case that $\kappa = 0.01$. The results for $T(x, z)$ at $x = 1$ are shown (in dimensionful variables, $x = 1$ corresponds to $x = 6 \cdot 10^{-2}$m). In this table, T_1^{as}, T_2^{as}, ΔT_1^{as} and ΔT_2^{as} are defined by

$$T_1^{as}(x, z) = T_1 \left(x, \frac{1-z}{\varepsilon} \right),$$

$$T_2^{as}(x, z) = (T_1 + \varepsilon T_2) \left(x, \frac{1-z}{\varepsilon} \right),$$

$$\Delta T_1^{as} = \frac{T_1^{as} - T^{num}}{T^{num}},$$

$$\Delta T_2^{as} = \frac{T_2^{as} - T^{num}}{T^{num}}. \tag{7.126}$$

From Table 7.1 it follows that the asymptotic solution and the numerical

Table 7.1. Comparison between asymptotic and numerical results at $x = 1$

z value	Asympt. sol.		Num. sol.	Relative difference	
	T_1^{as}	T_2^{as}	T^{num}	ΔT_1^{as}	ΔT_2^{as}
0.0	1.000	1.000	1.000	$5.47 \cdot 10^{-7}$	$5.47 \cdot 10^{-7}$
0.1	1.000	1.000	1.000	$5.42 \cdot 10^{-7}$	$5.18 \cdot 10^{-7}$
0.2	1.000	1.000	1.000	$1.47 \cdot 10^{-5}$	$1.23 \cdot 10^{-5}$
0.3	1.000	1.000	1.000	$2.09 \cdot 10^{-4}$	$1.31 \cdot 10^{-4}$
0.4	0.999	0.998	0.997	$1.59 \cdot 10^{-3}$	$5.98 \cdot 10^{-4}$
0.5	0.990	0.985	0.984	$6.71 \cdot 10^{-3}$	$1.23 \cdot 10^{-3}$
0.6	0.946	0.932	0.930	$1.74 \cdot 10^{-2}$	$2.03 \cdot 10^{-3}$
0.7	0.824	0.803	0.799	$3.15 \cdot 10^{-2}$	$5.41 \cdot 10^{-3}$
0.8	0.603	0.585	0.578	$4.42 \cdot 10^{-2}$	$1.22 \cdot 10^{-2}$
0.9	0.313	0.303	0.297	$5.28 \cdot 10^{-2}$	$1.94 \cdot 10^{-2}$
1.0	0.000	0.000	0.000		

solution are in good agreement with each other, despite the fact that $\varepsilon = \sqrt[3]{\kappa} = 0.21$, which is not really very small.

Next, we consider the:
Flow front region

$$(x, z) \in \Omega_f, \qquad \text{where} \quad \Omega_f = \{(x, z) \mid x_f < x < x_{f0}, \ -h < z < h \},$$

with $x_{f0} = x_f + R_f(0, t) - R_f(h, t) = R_f(0, t) - R_n$ (see Fig. 7.16).
In this region, $x - x_f$ and z are of the same order of magnitude, implying

the choice $L = h$, and moreover also $(v_x - V_f)$ and v_z are of the same order. Therefore, we introduce new variables $\tilde{x}, \tilde{z}$ and $\tilde{t}$ as

$$\tilde{x} = \frac{x - x_f}{h} \ , \qquad \tilde{z} = \frac{z}{h} \ , \qquad \tilde{t} = \frac{V_f}{h} t \ , \tag{7.127}$$

write the velocity as (note that u and w are known from the preceding zeroth order velocity problem)

$$\mathbf{v} = V_f \left[\left(\frac{3}{2}(1 - \tilde{z}^2) + \tilde{u}(\tilde{x}, \tilde{z}) \right) \mathbf{e}_x + \tilde{w}(\tilde{x}, \tilde{z}) \mathbf{e}_z \right] \ , \tag{7.128}$$

and consider the temperature in the *moving* frame $(\tilde{x}, \tilde{z})$, i.e.

$$\frac{T(x, z, t) - T_w}{T_i - T_w} = \tilde{T}(\tilde{x}, \tilde{z}, \tilde{t}) \ . \tag{7.129}$$

With all this, the convective-diffusive heat equation (7.105) turns into

$$\frac{\partial \tilde{T}}{\partial \tilde{t}} + \left(\frac{1}{2}(1 - 3\tilde{z}^2) + \tilde{u} \right) \frac{\partial \tilde{T}}{\partial \tilde{x}} + \tilde{w} \frac{\partial \tilde{T}}{\partial \tilde{z}} = \tilde{\kappa} \Delta \tilde{T} \ , \tag{7.130}$$

where

$$\tilde{\kappa} = \frac{\kappa}{V_f h} = \frac{h}{R} \hat{\kappa} \ . \tag{7.131}$$

Since $\hat{\kappa}$ itself is already small, we thus have $\tilde{\kappa} \ll 1$, and therefore we may neglect the diffusive term in (7.130). Moreover, because

$$\tilde{\mathbf{v}} = \left(\frac{1}{2}(1 - 3\tilde{z}^2) + \tilde{u} \right) \mathbf{e}_x + \tilde{w} \mathbf{e}_z \ , \tag{7.132}$$

is precisely the velocity of a material particle *with respect to the flow front*, we may write

$$\frac{\partial \tilde{T}}{\partial \tilde{t}} + (\tilde{\mathbf{v}} \cdot \nabla \tilde{T}) = 0 \ , \tag{7.133}$$

expressing that the change in time of $\tilde{T}$ is purely due to convection. Consequently, the temperature of each material particle moving in the front region remains equal to the temperature with which it enters this region. Because (from the beginning of the filling of the cavity on) this temperature is always equal to T_i, this implies that the temperature in the whole front region will be uniform and equal to T_i. However, this solution then leads to two evident contradictions:

– the temperature at the tips of the flow region ($\tilde{z} = \pm 1$) should be T_w instead of T_i ($\Rightarrow$ $\tilde{T} = 0$, at $\tilde{z} = \pm 1$);

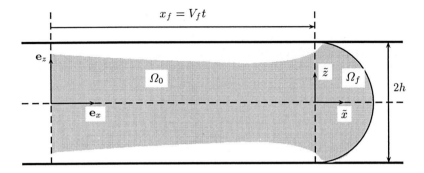

Fig. 7.16. Sketch of the temperature distribution in the cavity, behind and in the flow front region; in the gray part the temperature is T_i.

– the temperature in the boundary layer in Ω_0 directly behind the straight flow front boundary $x = x_f$ of Ω_f is smaller than T_i, causing the temperature across this boundary to become discontinuous.

From these discrepancies we conclude that there should also exist a boundary layer within Ω_f, extending behind Ω_f, and continuously merging into the boundary layer in Ω_0. In this boundary layer, the diffusion must again be taken into account. However, as $\tilde{\kappa} \ll \hat{\kappa}$, the boundary layer in Ω_f will be much $(O(h/R))$ narrower than the one in Ω_0. A sketch of the expected temperature distribution in the cavity at a certain position of the flow front is given in Fig. 7.16. This expectation is confirmed by the results of a forthcoming publication [25]. Knowing the zeroth order unperturbed solution, we hope to proceed with a study of the perturbed problem. The perturbation will be assumed due to an initial asymmetry in the flow front. We will examine how this asymmetry influences the flow. This influence will be most manifest in the boundary layers. Our main focus will be the evolution in time of the (asymmetric) shape of the flow front.

7.4 Closing Remarks

In this contribution, special attention is given to the mathematical modelling of industrial processes for polymers, and to the analytical and numerical evaluation of the models thus obtained. In Sect 7.2, we presented some results of two projects investigating the (in)stability of the extrusion process for polymer melts. The instability considered here is the so called spurt phenomenon. A 3-dimensional model describes spurt either as a sudden huge increase in

flow rate at a slight increase in pressure or as the onset of persistent oscillations in both the pressure and the flow rate. A discrete stick-slip model
describes spurt in terms of relaxation oscillations in the pressure and the flow
rate. The latter model is at one hand mathematically simple, but on the other
hand it yields a good correspondence, both qualitatively and quantitatively,
with existing experimental results.

In Sect 7.3, injection moulding between two parallel plates is considered.
A general model for the mechanical (flow, pressure, stresses) and thermal
features (temperature, heat flux) of this process is derived. Next, a very simplified zeroth order model of this general model is presented, and on the basis
of this the zeroth order velocity and temperature distribution are calculated.
The temperature distribution shows boundary layers near the walls of the
cavity. We hope that these results can be used for an explanation of the instability (wobbling) of the flow front. This hope is based on the expectation
that the wobbling could be driven by asymmetric temperature distributions.
Hence, the source for the wobbling of the front is sought in first instance in
thermal effects (and not in elasticity effects).

Besides this special aspect of injection moulding, we also look at a complete numerical simulation of the process. In this, we specifically aim at calculations of residual stresses and deformations, in connection with birefringence
(e.g. in compact discs).

References

1. Aarts, A.C.T. (1997) Analysis of the Flow Instabilities in the Extrusion of Polymeric Melts. Ph.D.Thesis, Eindhoven University of Technology
2. Doelder, C.F.J. den. (1999) Design and Implementation of Polymeric Melt Fracture Models. Ph.D.Thesis, Eindhoven University of Technology
3. Molenaar, J. Koopmans, R.J. (1994) Modeling Polymer Melt-flow Instabilities, Journal of Rheology, **38**, 99-109.
4. Greenberg, J.M., Demay, Y. (1994) A Simple Model of the Melt Fracture Instability, European Journal of Applied Mathematics **5**, 337-358
5. Kissi, N. el, Piau, J.M., Toussaint, F. (1997) Sharkskin and cracking of polymer melts extrudates, J. Non-Newtonian Fluid Mech. **68**, 271-290.
6. Kennedy, P. (1995) Flow analysis in injection molds. Hanser Publishers, Munich Vienna New York
7. Baaijens, F.P.T. (1991) Calculation of Residual Stresses in Injection Moulded Products, Rheologica Acta **30**, 284-299.
8. Tas, P.P., (1994) Film Blowing: from Polymer to Product, Ph.D.Thesis, Eindhoven University of Technology.
9. Ptitchkina, I.I., (1996) Blow Moulding Modelling, Final Report of Postgraduate Programme Mathematics for Industry, Eindhoven.
10. Binding, D.M., Blythe, A.R., Gunter, S., Mosquera, A.A., Townsend, P. (1996) Modelling Polymer Melt Flows in Wirecoating Processes, J. Non-Newtonian Fluid Mech. **64**, 191-206.

11. Mutlu, I., Townsend, P., Webster, M.F., (1998) Simulation of cable-coating viscoelastic flows with coupled and decoupled schemes, J. Non-Newtonian Fluid Mech. **74**, 1-23.
12. Fulchiron, R., Verney, V., Michel, A., (1995) Correlations between relaxation time and melt spinning behaviour of polypropylene. II: melt spinning simulation from relaxation time spectrum, Polymer Eng. Science, **35**, 518-527.
13. Fulchiron, R., Revenu, P., Kim, B.S., Carrot, C., Guillet, J., Extrudate swell and isothermal melt spinning analysis of linear low density polyethylene using the Wagner constitutive equation, J. Non-Newtonian Fluid Mech. **69**, 113-136.
14. Aarts, A.C.T., and Ven, A.A.F. van de. (1995) Transient Behaviour and Stability Points of the Poiseuille Flow of a KBKZ-fluid. J. Eng. Math. **29**, 371-392
15. Doelder, C.F.J.den, Koopmans, R.J., Molenaar, J., Ven, A.A.F. van de. (1998) Comparison of Wall Slip and Constitutive Instability Spurt Models. J. Non-Newtonian Fluid Mech. **75**, 25-41
16. Malkus,D.S., Nohel, J.A., Plohr, B.J. (1991) Analysis of new phenomena in shear flow of non-Newtonian fluids. SIAM Journal on Applied Mathematics **51**, 899-929
17. Aarts, A.C.T., Ven, A.A.F.van de. (1999) The Occurrence of Periodic Distortions in the Extrusion of Polymeric Melts. Continuum Mechanics and Thermodynamics, **11**, 113-139
18. Aarts, A.C.T., Ven, A.A.F. van de. (1996) Instabilities in the Extrusion of Polymers due to Spurt. In: Progress in Industrial Mathematics at ECMI94, Kaiserslautern; ed. H. Neunzert, Wiley-Teubner, Chichester, pp. 216-223.
19. Tanner, R.I. (1988) Engineering Rheology. Clarendon Press, Oxford
20. Ven, A.A.F. van de. (1999) Comparing Stick and Slip Models for Spurt in the Extrusion of Polymeric Melts. In: Progress in Industrial Mathematics at ECMI98, eds: L. Ackeryd, J. Bergh, P. Brenner, R. Petterson. Proceedings of 10th ECMI-Conference, ECMI 98, Göteborg (pp. 154-162). Teubner, Stuttgart
21. Ven, A.A.F. van de. (2000) Spurt in the Extrusion of Polymeric Melts; Discrete Models for Relaxation Oscillations. In: Complex Flows in Industrial Processes; ed. A. Fasano, Birkhäuser, Boston, pp. 125-145.
22. Doelder, C.F.J. den, Koopmans, R.J., Molenaar, J. (1998) Quantitative Modelling of HDPE Spurt Experiments Using Wall Slip and Generalised Newtonian Flow, J. Non-Newtonian Fluid Mech. **79**, 503-514.
23. Vroonhoven, J.C.W., Kuijpers, W.J.J. (1990) A Free-boundary Problem for Viscous Fluid Flow in Injection Moulding. J. Eng. Math. **24**, 151-165
24. Gramberg, H.J.J., Ven, A.A.F. van de. (forthcoming) Flow Patterns behind the Free Flow Front of a Fluid Injected between Two Parallel Plates.
25. Gramberg, H.J.J., Ven, A.A.F. van de. (forthcoming) Temperature Distribution of a Fluid Injected between Two Semi-infinite Plates.

Appendix

Color Plates

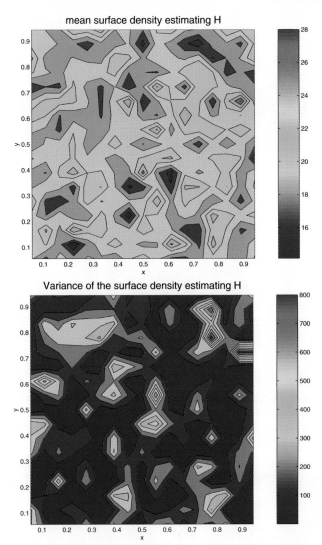

Plate 1. Mean surface density estimated locally over 46 simulations (top figure) and its variance (bottom figure). See also Fig. 5.8 Page 194

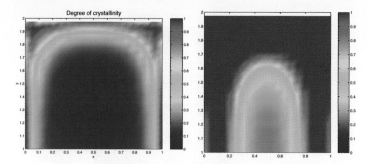

Plate 2. Degree of crystallinity in Example 1 after 5 and 15 minutes. See also Fig. 5.26 Page 222

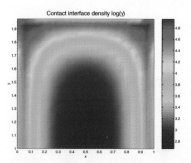

Plate 3. Plot of $\log(\gamma)$, where γ is the mean density of crystals interfaces in $\mathbf{R}^2$, in Example 1 after 5 minutes. See also Fig. 5.27 Page 222

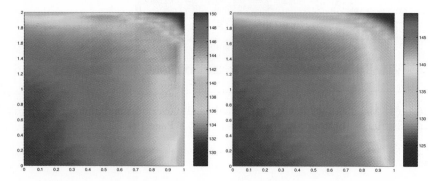

Plate 4. Temperature in Example 2. See also Fig. 5.28 Page 223

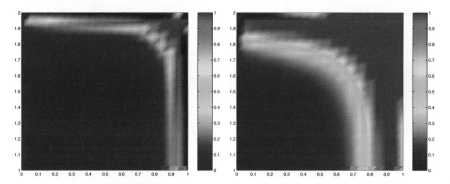

Plate 5. Degree of crystallinity in Example 2. See also Fig. 5.29 Page 223

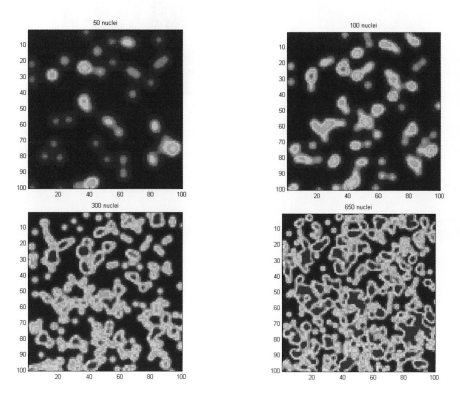

Plate 6. Particle System. Time evolution of the crystallization in the isothermal case. See also Fig. 6.1 Page 253

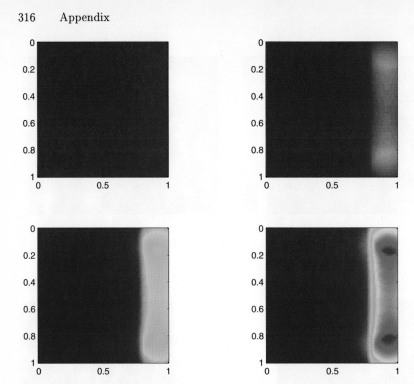

Plate 7. Particle System. Time evolution of the crystallization in the not isothermal case. See also Fig. 6.2 Page 254

Index